PROGRESS IN BRAIN RESEARCH

VOLUME 197

DOWN SYNDROME: FROM UNDERSTANDING THE NEUROBIOLOGY TO THERAPY

PROGRESS IN BRAIN RESEARCH

VOLUME 197

DOWN SYNDROME: FROM UNDERSTANDING THE NEUROBIOLOGY TO THERAPY

EDITED BY

MARA DIERSSEN

Genes and Disease Program, Center for Genomic Regulation (CRG), Barcelona Biomedical Research Park (PRBB), Barcelona, Spain
and
Centro de Investigación Biomédica en Red de Enfermedades Raras (CIBERER)

RAFAEL DE LA TORRE

Human Pharmacology and Clinical Neurosciences Research Group, Neurosciences Research Programme, IMIM-Institut de Recerca Hospital del Mar, and Universitat Pompeu Fabra (CEXS-UPF), Barcelona, Spain
and
Centro de Investigación Biomédica en Red Fisiopatología de la Obesidad y Nutrición (CIBEROBN), Santiago de Compostela, Spain

ELSEVIER

AMSTERDAM – BOSTON – HEIDELBERG – LONDON – NEW YORK – OXFORD
PARIS – SAN DIEGO – SAN FRANCISCO – SINGAPORE – SYDNEY – TOKYO

Elsevier
Radarweg 29, PO Box 211, 1000 AE Amsterdam, The Netherlands
Linacre House, Jordan Hill, Oxford OX2 8DP, UK
360 Park Avenue South, New York, NY 10010-1710

First edition 2012

Library of Congress Cataloging-in-Publication Data
A catalog record for this book is available from the Library of Congress

British Library Cataloguing in Publication Data
A catalogue record for this book is available from the British Library

ISBN: 978-0-444-54299-1
ISSN: 0079-6123

Printed and bound by CPI Group (UK) Ltd, Croydon, CR0 4YY

Transferred to Digital Print 2012

List of Contributors

S. Abekhoukh, Institute of Molecular and Cellular Pharmacology, CNRS-UMR6097, Université de Nice Sophia-Antipolis, Valbonne, France

M.M. Ahmed, Department of Pediatrics, University of Colorado Denver, Denver, CO, USA

X. Altafaj, Centro de Investigación Biomédica en Red de Enfermedades Raras (CIBERER), and Institute of Neuropathology, IDIBELL, L'Hospitalet de Llobregat, Barcelona, Spain

S.E. Antonarakis, Department of Genetic Medicine and Development, University of Geneva Medical School, University Hospitals of Geneva, and iGE3 Institute of Genetics and Genomics of Geneva, Geneva, Switzerland

B. Bardoni, Institute of Molecular and Cellular Pharmacology, CNRS-UMR6097, Université de Nice Sophia-Antipolis, Valbonne, France

P.V. Belichenko, Department of Neurosciences, and Down Syndrome Center for Research and Treatment, University of California, San Diego, La Jolla, CA, USA

V. Brault, Institut de Génétique et de Biologie Moléculaire et Cellulaire, Translational medicine and Neurogenetics program, IGBMC, CNRS, INSERM, Université de Strasbourg, UMR7104, UMR964, Illkirch, Strasbourg, France

D.G. Currier, Department of Physiology, The McKusick-Nathans Institute for Genetic Medicine, Johns Hopkins University School of Medicine, Baltimore, MA, USA

R. de la Torre, Neurosciences Research Programme, IMIM-Hospital del Mar Research Institute and CIBEROBN, Barcelona, Spain

M. Dierssen, Genes and Disease Programme, Centre de Regulació Genómica and CIBERER, Barcelona, Spain

A. Duchon, Institut de Génétique et de Biologie Moléculaire et Cellulaire, Translational medicine and Neurogenetics program, IGBMC, CNRS, INSERM, Université de Strasbourg, UMR7104, UMR964, Illkirch, Strasbourg, France

J.O. Edgin, University of Arizona, Down Syndrome Research Group, Tucson, AZ, USA

M. Esteller, Cancer Epigenetics and Biology Program (PEBC), Bellvitge Biomedical Research Institute, L'Hospitalet, Barcelona, Catalonia, Spain

A. Fernández, University of Arizona, Down Syndrome Research Group, Tucson, AZ, USA

C. Fillat, Institut d'Investigacions Biomèdiques August Pi i Sunyer (IDIBAPS), and Centro de Investigación Biomédica en Red de Enfermedades Raras (CIBERER), Barcelona, Spain

K.J. Gardiner, Department of Pediatrics; Computational Biosciences Program; Intellectual and Developmental Disabilities Research Center, and Human Medical Genetics and Neuroscience Programs, University of Colorado Denver, Denver, CO, USA

Y. Herault, Institut de Génétique et de Biologie Moléculaire et Cellulaire, Translational medicine and Neurogenetics program, IGBMC, CNRS, INSERM, Université de Strasbourg, UMR7104, UMR964; Transgenèse et Archivage Animaux Modèles, TAAM, CNRS, UPS44, Orléans, and Institut Clinique de la Souris, ICS, GIE CERBM, Illkirch, Strasbourg, France

D. Huertas, Cancer Epigenetics and Biology Program (PEBC), Bellvitge Biomedical Research Institute, L'Hospitalet, Barcelona, Catalonia, Spain

A.M. Kleschevnikov, Department of Neurosciences, and Down Syndrome Center for Research and Treatment, University of California, San Diego, La Jolla, CA, USA

T. Le, Department of Computer Science and Engineering, University of Colorado Denver, Denver, CO, USA

A. Letourneau, Department of Genetic Medicine and Development, University of Geneva Medical School, University Hospitals of Geneva, and iGE3 Institute of Genetics and Genomics of Geneva, Geneva, Switzerland

J. Levenga, Department of Physiology and Neuroscience, New York University, School of Medicine, New York, NY, USA

I.T. Lott, Department of Pediatrics and Neurology, School of Medicine, University of California Irvine (UCI), Orange, CA, USA

D. Maréchal, Institut de Génétique et de Biologie Moléculaire et Cellulaire, Translational medicine and Neurogenetics program, IGBMC, CNRS, INSERM, Université de Strasbourg, UMR7104, UMR964, Illkirch, Strasbourg, France

G.M. Mason, University of Arizona, Down Syndrome Research Group, Tucson, AZ, USA

M. Melko, Institute of Molecular and Cellular Pharmacology, CNRS-UMR6097, Université de Nice Sophia-Antipolis, Valbonne, France

L. Nadel, University of Arizona, Down Syndrome Research Group, Tucson, AZ, USA

R.C. Polk, Department of Physiology, The McKusick-Nathans Institute for Genetic Medicine, Johns Hopkins University School of Medicine, Baltimore, MA, USA

R.H. Reeves, Department of Physiology, The McKusick-Nathans Institute for Genetic Medicine, Johns Hopkins University School of Medicine, Baltimore, MA, USA

A. Salehi, VA Palo Alto Health Care System, Palo Alto, and Department of Psychiatry and Behavioral Sciences, Stanford Medical School, Stanford, CA, USA

J.V. Sanchez-Mut, Cancer Epigenetics and Biology Program (PEBC), Bellvitge Biomedical Research Institute, L'Hospitalet, Barcelona, Catalonia, Spain

G. Spanò, University of Arizona, Down Syndrome Research Group, Tucson, AZ, USA

X. Sturgeon, Department of Pediatrics, and Computational Biosciences Program, University of Colorado Denver, Denver, CO, USA

E. Velot, Institut de Génétique et de Biologie Moléculaire et Cellulaire, Translational medicine and Neurogenetics program, IGBMC, CNRS, INSERM, Université de Strasbourg, UMR7104, UMR964, Illkirch, Strasbourg, France

R. Willemsen, CBG-Department of Clinical Genetics, Erasmus MC, Rotterdam, The Netherlands

C. Wu, Department of Neurosciences, and Down Syndrome Center for Research and Treatment, University of California, San Diego, La Jolla, CA, USA

S. Zongaro, Institute of Molecular and Cellular Pharmacology, CNRS-UMR6097, Université de Nice Sophia-Antipolis, Valbonne, France

Preface

Down syndrome (DS) is the most common example of neurogenetic aneuploid disorder leading to mental retardation. In most cases, DS results from an extra copy of chromosome 21 (HSA21) producing deregulated gene expression in brain that gives rise to suboptimal intellectual functioning. Although several hypotheses have been put forward, it is unclear as to how candidate gene loci or genomic elements on HSA21 cause DS-associated features. Despite the prevalence of DS, relatively little research has been devoted to understanding its neurobiology or to developing therapeutics for cognitive deficits. This neglect has been due in part to the presumed global nature of the molecular and cellular abnormalities in DS that involves misexpression of hundreds of genes. Nevertheless, recent work in mouse models suggests that specific synaptic defects leading to mental retardation are amenable to pharmacological correction and may depend on a very few critical genes on HSA21. Besides, the understanding of the pathogenicity of the extragenomic material in trisomy 21 has been speed up in the past years due to the recent advances in genome sequencing, comparative genome analysis, functional genome exploration, and the use of model organisms. This has led to new evidence-based approaches to prevention or amelioration of DS effects on the brain and cognition and will impact other research areas on neurogenomics of cognition and behavior.

This volume focuses on DS as prototype of "genomic disorder." Exciting new findings demonstrating considerable plasticity of the human genome are helping to identify how deregulation/dysfunction of a myriad of different molecules in DS affect cellular mechanisms underlying neuronal development (neurite outgrowth and neuronal morphogenesis, shaping the three-dimensional architecture of neurons, synapse and network formation) and synapse rearrangement, allowing long-term memory formation and adaptation to the environment. Concepts such as metabolome or conectome are certainly important to be included as keys for the global understanding of DS and other mental retardation syndromes. The authors evaluate and interpret recent progress in the understanding of the role of genome-altering processes in the generation of mental retardation and neural phenotypes in DS and how does this concept reconcile with phenotypic variability associated with aneuploidy. We also review the new modeling strategies in mice and illustrate the potential that genetically engineered mouse models of DS can provide into the understanding of the molecular biology of human learning disorders and into the development of new therapies. Moreover, the increase in the life expectancy in DS individuals pose significant personal and societal burdens, since DS patients suffer early onset Alzheimer's disease, neurological symptoms.

This book brings together scientists of diverse backgrounds, integrating research directed at different levels of biological organization and highlighting translational research and the application of the existing scientific knowledge to develop improved DS treatments and cures. The authors are leaders in the field of DS clinical, basic neuroscience and neurogenomics, research, representing both basic and clinical aspects to promote translational aspects. They focus on the recent advances, challenges, and diagnosis/therapeutic implications of genomic structural and functional variation in brain systems related to cognition and disease-related dysfunction, focusing on both mouse and human studies.

Mara Dierssen
Rafael de la Torre

Contents

M. Dierssen and R. de la Torre (Eds.)
Progress in Brain Research, Vol. 197
ISSN: 0079-6123

CHAPTER 1

Therapeutic approaches in the improvement of cognitive performance in Down syndrome: past, present, and future

Rafael de la Torre[†,*] and Mara Dierssen[‡]

[†] *Neurosciences Research Programme, IMIM-Hospital del Mar Research Institute and CIBEROBN, Barcelona, Spain*
[‡] *Genes and Disease Programme, Centre de Regulació Genómica and CIBERER, Barcelona, Spain*

Abstract: Clinical trials with drugs aimed at treatment of Alzheimer disease to decelerate cognitive decline and translated mimetically to demented and young nondemented Down syndrome patients have been unable to demonstrate improvements in cognitive performance and functioning. Unfortunately, results from clinical trials do not support the use of NMDA antagonists like memantine and we should await at the development of safer GABA(A) antagonists to conclude about the efficacy of approaching Down syndrome therapeutics by modulating neurotransmission systems altered in this pathology. The use of folinic acid or antioxidants in DS patients is not supported by scientific evidence and do not provide improvement in cognitive performance to patients. Alternatively to the modulation of neurotransmission systems, future therapeutic approaches should focus at normalizing the expression levels or function of candidate molecules. Epigallocatechin gallate, a green tea polyphenol, that modulates DYRK1A functioning has already shown preliminarily that this approach may prove useful in therapeutics.

Keywords: therapeutics; folinic acid; donezepil; rivastagmine; galantamine; vitamins and minerals; acetyl-l-carnitine; epigallocatechin gallate; memantine; DYRK1A; acetylcholinesterase inhibitors; gabaergic GABA(A) antagonists; N-methyl-d-aspartate (NMDA) antagonists.

Introduction

Down syndrome (DS) is the most common genetic cause (trisomy of chromosome 21) of mental retardation appearing in about 9.65 of every 10,000 live births in the European countries (EUROCAT,

*Corresponding author.
Tel.: +34-933-160-484; Fax: +34-933-160-467
E-mail: rtorre@imim.es

DOI: 10.1016/B978-0-444-54299-1.00001-7

2010; Khoshnood et al., 2011) it is characterized by a range of physical and psychological conditions. The increased protein expression of genes on chromosome 21 leads to a cascade of effects in the development of fetal brain structure and subsequent structural and behavioral effects across the life span of individuals with DS (Lott and Dierssen, 2010). Three of the more affected areas are the cerebral cortex, the cerebellum, and the hippocampus that show reduced volume specially cerebellum. During the prenatal period, mouse and human studies have shown altered corticogenesis in trisomic mouse models (Haydar and Reeves, 2012) and an altered cortical lamination in fetal brains. (Golden and Hyman, 1994), smaller dendritic tree of pyramidal neurons (Benavides-Piccione et al., 2004), and a smaller and hypocellular hippocampal dentate gyrus. Postnatally, the DS brain exhibits degeneration of cortical pyramidal neurons (Schmidt-Sidor et al., 1990), profound dendrite and synapses abnormalities (Becker et al., 1986), and hypocellular hippocampus (Ferrer and Gullotta, 1990). The adult brain presents reduced numbers of granule cells, increased astrocytosis (see Lott and Dierssen for review) and neurodegeneration of the cholinergic and noradrenergic systems (see Lott and Dierssen, 2012).

DS subjects display several morphological features that can be mild and do not constitute a hallmark of the condition and a number of malformations and complications including cognitive impairment (intellectual disability, see also Chapter 7), heart defects, digestive malformations, congenzxital cataract, Hirschsprung disease, West syndrome, seizures, leukemia, sleep apnea, sensory deficiencies, autoimmune and endocrine pathologies, earlier aging, and an Alzheimer-type neuropathology. Despite medical interventions in DS have resulted in a significant increase of life expectancy of DS, cognitive and behavioral features still exert a hard impact on their functional ability throughout life. The neurocognitive profile of individuals with DS is characterized by a remarkable deficit in learning/memory and language abilities. The intelligence quotient (IQ) scale of patients with DS generally fall between 25 and 75 is less than average (average score in the population is set at 100).

The present chapter will focus at therapies oriented at improving cognitive performance of DS patients nondemented and with prevalent Alzheimer's disease (AD) symptoms (see Chapter 6). In Table 1, there is a summary of the main clinical trials undertaken in populations of DS patients.

Clinical trials in DS patients related with drugs used for the treatment of AD

In terms of cognitive symptoms (not brain pathology), the prevalence of AD in patients with DS is quite high. It is important to note that people with DS tend to present with AD at a much younger age than the normal population as well as having subtle differences in physiology (e.g., metabolism and heart rate) and may therefore have different requirements from the general population. Up to 10% of DS patients ages 30–39 also have diagnosable AD. Between ages 40 and 49 the rate is as high as 25% and in ages 50–59 the prevalence is 55%. AD rates as high as 75% may occur in patients between the ages of 60 and 69.3. For comparison, AD occurs in about 14% of people older than 65 in the general population.

Not surprisingly many clinical trials in DS patients have been oriented at testing drugs acting on altered neurotransmission systems in AD. Most of them are currently in use in therapeutics of AD, like acetylcholinesterase inhibitors (donepezil, rivastigmine, galantamine), gabaergic (GABAA) antagonists (e.g., pentylenetetrazol), and *N*-methyl-D-aspartate (NMDA) receptor antagonists (memantine) (Prasher, 2004). The evaluation of these treatments is not only justified on a basis of a higher prevalence of AD like neuropathology in DS subjects but also on the fact that AD and several neurodevelopmental diseases like DS share common molecular pathways and brain alterations. For these reasons, these drugs have been evaluated not only in DS demented patients but also in DS nondemented patients.

Table 1. Summary of main clinical trials performed in DS subjects

Drug	Dose	Design	Patients	Evaluations	Outcome	Reference
Donepezil	5mg (6 weeks) and 10mg up to 9 months	Open label	4 DS adults (2 demented)	Vineland adaptive behavior scales (VABS)	Some improvement first 3 months	Kishnani et al. (1999)
	5mg (4 weeks) and 10mg up to 6 months	Randomized, double blind placebo controlled	30 DS demented 16 donepezil and 14 placebo	Dementia scale for mentally retarded persons (DMR) Neuropsychiatry inventory (NPI) Adaptative behavior scale (ABS) Severe impairment battery	Inconclusive	Prasher et al. (2002)
	5mg (6 weeks) and 10mg up to 6 months	Open label	6 DS nondemented	Test of problem solving-TOPS Clinical evaluation of language fundamentals-revised-CELF-R	Some improvement in language performance	Heller et al. (2003)
	2.5–10mg up to 2 years	Open label	25 DS demented	Dementia scale for mentally retarded persons (DMR) Neuropsychiatry inventory (NPI) Adaptative behavior scale (ABS) severe impairment battery	Significantly less deterioration in global functioning and adaptive behavior in DS adults with DAD	Prasher et al. (2003)
	2.5mg (8 weeks) and 5mg up to 8 weeks	Open label	7 DS (8–13 years range)	Test of problem solving-TOPS Clinical evaluation of language fundamentals-revised-CELF-R	Some improvement in language performance	Heller et al. (2004)
	2.5mg (8 weeks) and 5mg up to 8 weeks final evaluation at 22 weeks	Open label	7 DS (8–13 years range)	Visual attention, memory for names Narrative memory subtests from the NEPSY Conners' parent rating scales-revised vineland adaptive behavior scales (VABS)	Improvement in memory (NEPSY memory for names and narrative memory) and sustained attention to tasks (Conners' parent rating scales)	Spiridigliozzi et al. (2007)
	5mg (6 weeks) and 10mg (6 weeks)	Randomized, double blind placebo controlled with open label extension	Nondemented patients (18–35 years) 123 double blind 87 open-label	Severe impairment battery (SIB) Rivermead behavioral memory Test for children (RBMT-C) Clinical evaluation of language fundamentals (CELF-3)	Efficacy in some but not all subjects	Kishnani et al. (2009)
	2.5–10mg (2.5mg/week escalation) (10weeks)	Randomized, double blind placebo controlled	129 DS patients (10–17years)	Vineland-II adaptive behavior scales (VABS-II) Parent/Caregiver rating form (PCRF)	No benefit demonstrated	Kishnani et al. (2010)

(*Continued*)

Table 1. Summary of main clinical trials performed in DS subjects (*Continued*)

Drug	Dose	Design	Patients	Evaluations	Outcome	Reference
Rivastigmine	1.5–12mg escalation in 8 weeks (24 weeks)	Open label comparison with a pervious placebo group (Prasher et al., 2002)	17 DS demented	Dementia scale for mentally retarded persons (DMR) Neuropsychiatry inventory (NPI) Adaptative behavior scale (ABS)	Inconclusive	Prasher et al. (2005)
	4.5mg	Open label 20 weeks follow-up 28 weeks	10 DS patients (mean age 15 years)	Vineland adaptive behavior scales (VABS) Clinical evaluation of language fundamentals (CELF-3) Leiter International performance scalerevised (Leiter-R) Memory subtests from the NEPSY	No improvement in cognitive and language performance	Heller et al. (2010)
Folinic acid vitamins and minerals	Selenium 10µg, zinc 5mg, vitamin A 0.9 mg, vitamin E100mg, and vitamin C 50mg, folinic acid 0.1mg, for 18 months	Randomized controlled trial with 2×2 factorial design, balanced with placedo: folinic acid versus antioxidants versus folinic acid+ antioxidants versus placebo	156 DS patients <7 months	Griffiths developmental quotient MacArthur communicative development inventory at 18 months after starting supplementation	No evidence to support the use of antioxidants or folinic acid supplements in DS children	Ellis et al. (2008)
Acetyl-L-carnitine	10mg/kg/day (1st month) 20mg/kg/day (2nd month) 30mg/kg/day (rest of study)	Randomized, double blind placebo controlled	40 DS patients (18–30 years)	Stanford–Binet Intelligence Scale Hiskey–Nebraska Visual Attention Span and Matching Familiar Figure tests Mazes and Coding subtests (WISC) Riddles subtest (KABC) Comprehension subtest (WISC) Vineland Adaptive Behavior Scales (VABS) Child Behavior Checklist	No improvement in cognition and functioning	Pueschel (2006)

Cholinergic neurotransmission

Neuropathological and neurochemical similarities between DS and AD disease are well established, among them, there is a loss of cholinergic input to cortical and limbic areas in both disorders. Cholinergic deficits have been linked to the cognitive (memory) and adaptive impairments in AD, and acetylcholinesterase inhibitors, such as donepezil hydrochloride, are proven symptomatic therapies improving cognitive performance and quality of life in individuals with dementia and AD. Pioneering studies already suggested that there was some potential in using this drug in DS patients (Kishnani et al., 1999).

Acetyl-L-carnitine

Acetyl-L-carnitine, derived from carnitine, has been reported to display properties which may be beneficial in AD. These include activity at cholinergic neurons, membrane stabilization, and mitochondrial antioxidant activity (Palacios et al., 2011). Preliminary studies suggested a beneficial effect of acetyl-L-carnitine on cognition and behavior in aging subjects. However, the evidence does not suggest that acetyl-L-carnitine is likely to prove an important therapeutic agent (Hudson and Tabet, 2003).

Acetyl-L-carnitine has been administered to DS patients and significant improvements of visual memory and attention were reported (De Falco et al., 1994). In a more recent double-blind clinical trial with 40 individuals with DS, acetyl-L-carnitine was administered during a 6-month period with psychological evaluations at baseline and at 3, 6, and 9 months. No improvement in cognitive functions was found (Pueschel, 2006).

Nicotine

The rationale for administering nicotine in the form of a transdermal patch are similarities between DS and AD, particularly cholinergic deficits (Fodale et al., 2006). The potential activity of a direct nicotinic stimulation in cognitive performance has been examined in a very limited number of DS patients. Results are conflicting and improvements in attention and information processing have been reported in some subjects (Bernert et al., 2001; Seidl et al., 2000).

Inhibitors of acetylcholinesterase

Pilot studies in DS and AD patients supported the view that the selective anticholinesterase inhibitor *donepezil* has some benefit (improvement in dementia scores) in mild to moderate AD patients. A number of side effects did occur including diarrhea, insomnia, fatigue, and nausea (Lott et al., 2002; Prasher et al., 2002).

Donepezil has been evaluated not only in DS and AD patients but also in nondemented DS adult patients and in DS children. Studies evaluated cognitive performance but were mainly focused in the treatment of language deficits (Heller et al., 2003, 2004; Johnson et al., 2003; Kishnani et al., 2009). With the exception of one area (language), no improvement was noted in any of the cognitive subtests, behavioral scores, or caregiver ratings. Increased irritability and/or assertiveness were noted in some DS children (Spiridigliozzi et al., 2007). In a recent report, donepezil versus placebo failed to demonstrate any benefit in cognitive performance in children and adolescents with DS (Kishnani et al., 2010). A Japanese study nevertheless reported improvements in mental functioning in females DS patients with severe cognitive impairment, raising the question on whether there are gender differences in the efficacy of the drug (Kondoh et al., 2011).

Rivastigmine, an inhibitor of acetylcholinesterase, has also been evaluated in DS patients (Heller et al., 2010; Prasher et al., 2005). Results are similar to those reported for donepezil, suggesting that the long-term use of rivastigmine does not improve cognitive and language performance.

Galantamine is a competitive and reversible cholinesterase inhibitor and an allosteric ligand at nicotinic acetylcholine receptors. It is used for the treatment of mild to moderate AD and various other memory impairments, in particular those of vascular origin. Although there are no ongoing clinical trials evaluating this drug in DS patients, it would not be surprising its administration in demented patients in the near future.

Glutamatergic neurotransmission

NMDA antagonists

Memantine is an uncompetitive NMDA receptor antagonist approved for the treatment of moderate to severe AD. In trisomic animal models (Ts65Dn), memantine has been able to normalize several phenotypic abnormalities, such as impaired cognition associated with DS and AD (Rueda et al., 2010). The rationale for evaluating memantine in humans is that this drug should reduce abnormal activation of glutamate neurotransmission, one of the factors contributing to the pathogenesis of AD. Its low affinity for NMDA receptors should not prevent learning and the formation of memory and in AD patients it has already shown improvements in cognitive function and to slow the decline of AD in the general population over time (Mohan et al., 2009). There is an ongoing clinical trial (NCT01112683) where memantine is administered to young adults with DS examining improvements in memory and other cognitive abilities. However the MEADOWS project demonstrates that people older than 40 years with DS despite promising indications, had no positive responses to treatment with memantine, suggesting that it is not an effective treatment in DS (Hanney et al., 2012).

GABAergic neurotransmission

GABAA receptors are the major inhibitory receptors in brain. They are involved in anxiety, depression, insomnia, learning and memory, and epilepsy. These processes are also disturbed in several neurodevelopmental disorders as fragile X patients and DS (D'Hulst et al., 2006). Alterations in excitatory–inhibitory balance occurring in AD are postulated to be a primary mechanism contributing to AD cognitive decline and to intellectual disabilities associated with DS and that GABAA antagonists may be useful therapeutic agents for this disorder (Rissman and Mobley, 2011).

Trisomic DS mouse models display an excessive inhibition in the dentate gyrus, a condition that could compromise synaptic plasticity and mnemonic processing. The chronic systemic treatment of these mice with GABAA antagonists (picrotoxin, bilobalide, and pentylenetetrazol) at non-epileptic doses causes a persistent postdrug recovery of cognition and long-term potentiation in the dentate gyrus (Fernandez et al., 2007). GABAA antagonist assayed until now in DS animal models cannot be easily translated to clinics, because their narrow therapeutic index. We must await for the development of new GABAA antagonists to test the efficacy of this therapeutic approach in DS patients.

Clinical trials in DS patients with miscellaneous drugs: Vitamins and mineral supplements, piracetam, and growth hormone

Supplementation with folate, antioxidants, vitamins, and minerals

The use of vitamin and mineral supplements is widespread in children with DS in Europe and the United States as a result of the marketing of commercial preparations that claim substantial benefits for children with DS. The rationale for this approach is based on the observation of an increased activity of two enzymes, copper/zinc superoxide dismutase and cystathionine β-synthase, both coded for chromosome 21. A misbalance between SOD-1 and glutathione perioxidase activity in DS has been linked to free radical generation. Cystathionine β-synthase catalyzes the condensation of homocysteine with serine to form cystathionine. Increased concentrations of this enzyme in DS lead to significantly reduced plasma concentrations of homocysteine, methionine, *S*-adenosylhomocysteine, and *S*-adenosylmethionine and thereby to a "folate trap" and a functional folate deficiency.

A systematic review of randomized controlled trials of dietary supplements reporting any assessment of cognitive function in subjects with DS concluded that the administration of these therapies was not justified (Salman, 2002). The poor design of clinical trials and the reduced number of participants were the main limitations to draw conclusions on the health benefits associated to these therapies in DS subjects. Recently, in an attempt to overcome these limitations, a randomized controlled trial with a 2×2 factorial design examining the effect of folinic acid and/or antioxidants was undertaken. A total of 156 DS infants aged less than 7 months participated in the study. Interventions consisted in a daily oral supplementation with antioxidants (selenium 10μg, zinc 5mg, vitamin A 0.9mg, vitamin E 100mg, and vitamin C 50mg), folinic acid (0.1mg), antioxidants and folinic acid combined, or placebo. The conclusion was that daily supplementation with antioxidants, folinic acid, or both did not alter psychomotor or language development in DS children (Ellis et al., 2008). Despite these negative results, recent studies in animal models still suggest that supplementation with antioxidants should be revisited (Shichiri et al., 2011).

DS is associated with short stature, obesity, and psychomotor delay. DS adults share some features with *growth hormone* (GH) deficient adult subjects. The GH/IGF-I axis is involved in the regulation of brain growth, development, and metabolism (Gasperi and Castellano, 2010). When GH is administered to DS during infancy and childhood, it normalizes growth velocity and improves fine motor skill performance (Annerén et al., 1999). DS adolescents treated during its childhood with GH display a greater head circumference (circumference standard deviation score) and better psychomotor performance. DS subjects may benefit from an early GH treatment (Myrelid et al., 2010).

Piracetam, a cyclic derivative of GABA, is a nootropic drug widely used for improving cognitive function in children with DS. Double blind clinical trials do not support an improvement on cognitive function in DS subjects; on the contrary, a number of side effects are derived from its consumption (Lobaugh et al., 2001).

Alternative therapeutic strategies in DS: Normalizing the expression levels or function of candidate molecules

DS has been considered a hopeless disease from the therapeutic point of view, as have been other developmental disorders. Previous sections in this chapter have already shown that most of the therapeutic approaches evaluated until now failed to provide evidence of improvement in the cognition of DS patients. However, there is increasing scientific evidence demonstrating that the DS brain holds potential for therapeutic intervention. Different lines of research have demonstrated that treatments aimed at normalizing the expression levels or function of candidate molecules may be beneficial to some DS phenotypes, and may lead, in conjunction with specific interventions, to long-lasting cognitive effects. In this book, there is a chapter dedicated to the Sonic hedgehog signaling as a therapeutic target in DS (see Chapter 11). In this chapter, we will discuss the therapeutic potential of normalizing the function of one gene located in the DCR-1 region (DS Critical Region for the pathogenesis of SD) that plays a pivotal role in neurogenesis: Dyrk1A.

Dyrk1A *overexpression: Intellectual disability and AD*

The genetic basis of DS morphogenetic and cognitive phenotype has been mostly associated with the overexpression of a specific region within the HAS 21 (Delabar et al., 1993; Yamamoto et al., 2011) containing 13 genes (Ronan et al., 2007). *Dyrk1A* (dual specificity tyrosine-phosphorylation-regulated kinase 1A gene) is one of the affected genes, which is involved in brain morphogenesis and in synaptic plasticity (Branchi et al., 2004). DYRK1A is upregulated ~1.5-fold in DS brains (Dowjat et al., 2007) as well as in DS mice models (Amano et al., 2004). Its overexpression has been repeatedly associated with molecular alterations (pFKHR, cyclin B1, pCreb, BDNF) and a variety of cognitive deficits (Martinez de Lagran et al.,

2004).Therefore, DYRK1A satisfies the criteria for contributing to the DS phenotype; that is, (1) normally expressed in the brain, (2) expressed at an increased level in DS, and (3) evidence of participation in relevant neurobiological processes (Belichenko et al., 2007).

Recent studies indicate that *DYRK1A* may play a significant role in developmental brain defects and in early onset neurodegeneration, neuronal loss, and dementia in DS. *Dyrk1A* overexpression in DS brains may contribute to early onset neurofibrillary degeneration directly through specific hyperphosphorylation of tau and indirectly through phosphorylation of alternative splicing factors, leading to an imbalance between 3R-tau and 4R-tau (Azorsa et al., 2010). The several-fold increases in the number of DYRK1A-positive and 3R-tau-positive neurofibrillary tangles in DS support this hypothesis (Wegiel et al., 2011a,b). Increased *dyrk1A* expression has also been reported in human AD brains (Ferrer et al., 2005). The enhanced phosphorylation of amyloid precursor protein by overexpressed *dyrk1a* facilitates amyloidogenic amyloid precursor protein cleavage elevating Aβ40 and 42 levels and leading to brain β-amyloidosis. Presenilin 1 (PS1) a key component of the γ-secretase complex in the generation of β-amyloid (Aβ), an important trigger protein in the pathogenesis of AD is involved in this process. PS1 is phosphorylated by DYRK1A at Thr (354) and this phosphorylation increases γ-secretase activity (Ryu et al., 2010).

Normalization of Dyrk1A *function as a therapeutic approach to improve cognitive performance and decelerate AD like progression: Epigallocatechin gallate as therapeutic tool*

Normalization of *Dyrk1A* expression has been proven to rescue the morphological and behavioral phenotypes in transgenic mice (Ortiz-Abalia et al., 2008); further, it has been shown that *Dyrk1A* is necessary to produce the altered phenotype in DS (see Dierssen et al., 2009 for review; Olson et al., 2007). Therefore, we have now a relevant potential to develop evidence-based therapies. Gene function correction by using Dyrk1A inhibitors could also be a therapeutic approach to reverse some phenotypes related to cognitive and motor dysfunctions seen in subjects suffering DS.

Dyrk1A kinase activity is inhibited by the green tea natural compound epigallocatechin gallate (EGCG) as well as the psychoactive compound harmine. EGCG offers a safety profile very interesting for the translation of observations made in mouse models to humans. EGCG functions as a noncompetitive inhibitor against ATP (Bain et al., 2003). Polyphenols from green tea have already been used to rescue the neuromorphological phenotypes observed in transgenic *Dyrk1A* animal models (YACtgDyrk1A; Lepagnol-Bestel et al., 2009) including brain weight, total brain volume, and thalamus–hypothalamus volume. In addition, BDNF levels were corrected to normal values and the impairment of long-term memory assessed by a novel object recognition paradigm was also rescued. Treating gestating mothers and litters up to adult age successfully produced significant improvement of the DS phenotype (morphogenesis and learning impairment; Guedj et al., 2009). Therefore, inhibiting DYRK1A activity in DS may serve to counteract the phenotypic effects of its overexpression and is a potential method of treatment of developmental defects and the prevention of age-associated neurodegeneration, including Alzheimer-type pathology.

Additional therapeutic contribution of EGCG over AD pathogenesis

A neuropathological hallmark of AD are the amyloid plaques, which are extracellular protein deposits found in the brain of AD patients. The major constituent of the plaques is the ≈4kDa amyloid β peptide (Aβ), which are considered the culprit of the disease. Aβ is a 38–42 amino acid long peptide, which is hydrophobic and prone to aggregation. Small Aβ oligomers are neurotoxic and trigger a pathological cascade of events, leading to the clinical onset of the disease and ultimately death. As the generation of Aβ is at the beginning of the cascade, the mechanisms of Aβ generation and Aβ prevention are a target therapeutic approach. Formation of Aβ requires APP to be cleaved by the two proteases β- and γ-secretase. A third protease, α-secretase, cleaves APP within the Aβ domain and is assumed to preclude Aβ generation. Recently, it has been shown that α-secretase further contributes to the non-amyloidogenic processing of APP by generating an inhibitor of γ-secretase (Tian et al., 2010). An increase in α-secretase cleavage is considered as a therapeutic approach for AD (Lichtenthaler, 2011). The constitutively cleaving α-secretase activity is selectively mediated by ADAM10 (Kuhn et al., 2010). This enzyme is activated *in vivo* by SIRT1 (Donmez et al., 2010).

Among genes that are overexpressed in DS (APP, DSCAM, MNB/DYRK1A, and RCAN1), there is APP that codifies the amyloid precursor protein. In a transgenic mouse model of AD, cleavage of APP by the α-secretase ADAM10 prevented amyloid plaque formation and alleviated cognitive deficits. Further, ADAM10 overexpression increased the cortical synaptogenesis. These results suggested that the upregulation of ADAM10 in the brain has beneficial effects on AD pathology (Prinzen et al., 2009).

EGCG promotes the non-amyloidogenic (α-secretase cleavage) APP processing through ADAM10 activation (Obregon et al., 2006). EGCG-mediated enhancement of non-amyloidogenic processing of APP through ADAM10 maturation is mediated via an estrogen receptor-α (ERα)/phosphoinositide 3-kinase/Ak-transforming dependent mechanism, independent of furin-mediated ADAM10 activation (Fernandez et al., 2010).

The contribution of EGCG over the non-amyloidogenic of APP is of relevance since the pathogenesis of AD in DS subjects may be not the same that in the general population. While studies in cerebrospinal fluid and plasma show alterations in cerebral amyloidosis, not always other alterations are shown in other key biomarkers like Tau protein, suggesting that other aging processes are involved in the AD like development in DS subjects (Englund et al., 2007; Petersen and Trojanowski, 2009; Schupf et al., 2010. In any case, the activity of EGCG over APP in DS subjects may be of relevance in preventing the progression to AD.

Contribution of EGCG in improving cognitive performance in DS: Preliminary results from a pilot study

The availability of inhibitors of DYRK1A activity of dietary origin, like EGCG from green tea, allows the translation of studies in animal models that already showed the efficacy of this drug to clinical studies.

In a pilot study, the safety and clinical efficacy of EGCG for cognitive functioning in DS by means of a 3 months follow-up trial under continued EGCG treatment plus a 3 months follow-up once drug administration was discontinued, was evaluated. Study design was double blind with randomized assignment of treatments and balanced with placebo. A total of 31 DS individuals younger than 29 years were randomly assigned to two treatment conditions: EGCG (7–9 mg/kg) and placebo. Treatments lasted 3 months, and during this time period subjects were evaluated in terms of cognitive condition and biochemical parameters at baseline, 1 month and 3 months. Two follow-up sessions provided information about residual effects on biomarkers and cognitive performance. This is the summary of findings of this pilot study

(i) Active treatment is well tolerated; enzyme hepatic biomarkers of safety (SGOT/GPT) do not suggest alterations clinically relevant.
(ii) Although globally all participants in the study were healthy in terms of their lipid profile, the EGCG treatment reduced concentrations of cholesterol, LDL cholesterol and oxidized-

LDL. Changes in these biomarkers can be considered globally as health benefits for subjects treated with EGCG.

(iii) Homocysteine a secondary biomarker of DYRK1A activity shows expected alterations in the group of subjects treated with the active treatment (EGCG). The inhibition of DYRK1A should result in an increase of homocysteine plasma concentrations (sustained while subjects treated with EGCG). After the discontinuation of treatment (control performed in the 4th–5th month), homocysteine comes back to basal concentrations. Concentrations of homocysteine reached are in the normal range and do not imply placing subjects at risk of cardiovascular diseases.

(iv) Concerning cognitive performance, individuals treated with EGCG have a better performance for immediate and working memory (PRM, SSP). Concerning executive function, subjects treated perform more errors (psychomotor speed, MOT). A short-term stable effect was found neither on those cognitive variables which had revealed any significant or marginal effect related to EGCG active treatment nor on other cognitive measures after discontinuing the treatment (3 months later).

(v) Subjective positive impressions of mild cognitive enhancement and/or improved behavioral control over a wide array of cognitive/behavioral skills were mostly reported by parents of EGCG treated individuals against placebo.

Limitations of past clinical studies and future perspectives

Many clinical trials undertaken until now intended at evaluating the therapeutic efficacy of drugs in DS patients, suffer from several limitations, among them,

1. *Poor design of clinical trial.* To perform double blind randomized balanced with placebo clinical trials, the only design that can provide scientific evidence of treatment efficacy, it is necessary to undertake multicenter clinical trials at the international level, using common exploration protocols. Nevertheless, many parents/caregivers are reluctant to participate in clinical trials if the patient may be treated with placebo as a result of the study design, rendering a complex task to perform trials following this design.
2. *Reduced number of participants.* Many studies evaluate drugs with a very limited number of subjects following an open-label design. The heterogeneity of subjects in cognitive terms, gender, and age further limit conclusions can be drawn for these studies. Some authors postulate that results should be evaluated on an individual basis rather than on a group treatment basis. Some DS patients are responders to treatment while others not, consistent with phenotypic variability of DS (Kishnani et al., 2009). In the future, the number of subjects participating in studies has to be increased substantially, an objective only attainable in the context of International Multicenter Clinical Trials as stated earlier.
3. *Lack of common instruments for neuropsychological evaluation of subjects.* As an example, there is not a single test or set of tests that definitively diagnoses AD, and many tests that may be used in persons of average or higher intelligence are notoriously unreliable in persons with DS. The major focus of these tests are to identify changes from the patient's specific baseline rather than comparing performance to some average level of functioning. This issue is addressed in one of the chapters of the present work (see Chapter 7). There are some proposals (Edgin et al., 2010) and ongoing initiatives that address this issue.
4. *Dependence on IQ of individuals that determine prognosis in a given cognitive area.* As

an example, IQ plays a substantial role in determining the prognosis of language learning in children with DS. An IQ level of greater than 50 is critical to ensuring substantial growth when maturational factors permit it (Fowler, 1988).

5. In the same way that in the therapy of AD, researchers and pharmaceutical companies find relevant the availability of validated biomarkers for disease diagnosis and progression (Hampel et al., 2010). Clinical studies in DS subjects need to define biomarkers of drug efficacy and safety. Beware that as an example, biomarkers of AD cannot be mimetically translated to demented DS patients. Preliminary results already suggest that new biomarkers and/or criteria of evaluation of current ones are needed (Jones et al., 2009; Prasher et al., 2010).

Concluding remarks

Past clinical trials with drugs aimed at treatment of AD to decelerate cognitive decline and translated mimetically to demented DS patients have been unable to demonstrate improvements in cognitive performance and functioning. The same is applicable to studies performed in young non-demented DS patients when looking at particular cognitive areas like language skills. Globally, it cannot be discarded that some of these drugs produce some health benefits in selected patients taking into account the heterogeneity of DS patients. Also, it has to be reminded that clinical trials in AD patients were performed in large populations and this is not the case for SD patients. We should await for results from clinical trials with NMDA antagonists like memantine and at the development of safer GABAA antagonists to conclude about the efficacy of approaching SD therapeutics by modulating neurotransmission systems altered in this pathology.

The earlier the pharmacotherapy is initiated (e.g., pediatric population), the more relevant improvements would be seen in neurogenesis and cognitive behavior (Bianchi et al., 2010).

The use of folinic acid or antioxidants in SD patients is not supported by scientific evidence and does not provide any improvement in cognitive performance to patients.

Future therapeutic approaches should focus at normalizing the expression levels or function of candidate molecules. EGCG that modulates DYRK1A functioning has already shown that this approach may prove useful in therapeutics.

Acknowledgments

This work was supported by Fondation Jérôme Lejeune and by Fondo de Investigaciones Sanitarias-ISCIII PI11/00744.

References

Amano, K., Sago, H., Uchikawa, C., Suzuki, T., Kotliarova, S. E., Nukina, N., et al. (2004). Dosage-dependent over-expression of genes in the trisomic region of Ts1Cje mouse model for Down syndrome. *Human Molecular Genetics*, *13*, 1333–1340.

Annerén, G., Tuvemo, T., Carlsson-Skwirut, C., Lönnerholm, T., Bang, P., Sara, V. R., et al. (1999). Growth hormone treatment in young children with Down's syndrome: Effects on growth and psychomotor development. *Archives of Disease in Childhood*, *80*, 334–338.

Azorsa, D. O., Robeson, R. H., Frost, D., Meec hoovet, B., Brautigam, G. R., Dickey, C., et al. (2010). High-content siRNA screening of the kinome identifies kinases involved in Alzheimer's disease-related tau hyperphosphorylation. *BMC Genomics*, *11*, 25.

Bain, J., McLauchlan, H., Elliott, M., & Cohen, P. (2003). The specificities of protein kinase inhibitors: An update. *The Biochemical Journal*, *371*, 199–204.

Becker, L. E., Armstrong, D. L., & Chan, F. (1986). Dendritic atrophy in children with Down's syndrome. *Annals of Neurology*, *20*, 520–526.

Belichenko, P. V., Kleschevnikov, A. M., Salehi, A., Epstein, C. J., & Mobley, W. C. (2007). Synaptic and cognitive abnormalities in mouse models of Down syndrome: Exploring genotype–phenotype relationships. *The Journal of Comparative Neurology*, *504*, 329–345.

Benavides-Piccione, R., Ballesteros-Yáñez, I., de Lagrán, M. M., Elston, G., Estivill, X., Fillat, C., et al. (2004). On dendrites in

Down syndrome and DS murine models: A spiny way to learn. *Progress in Neurobiology*, *74*, 111–126.

Bernert, G., Sustrova, M., Sovcikova, E., Seidl, R., & Lubec, G. (2001). Effects of a single transdermal nicotine dose on cognitive performance in adults with Down syndrome. *Journal of Neural Transmission Supplementum*, *61*, 237–245.

Bianchi, P., Ciani, E., Guidi, S., Trazzi, S., Felice, D., Grossi, G., et al. (2010). Early pharmacotherapy restores neurogenesis and cognitive performance in the Ts65Dn mouse model for Down syndrome. *Journal of Neuroscience*, *30*, 8769–8779.

Branchi, I., Bichler, Z., Minghetti, L., Delabar, J. M., Malchiodi-Albedi, F., Gonzalez, M. C., et al. (2004). Transgenic mouse in vivo library of human Down syndrome critical region 1: Association between DYRK1A overexpression, brain development abnormalities, and cell cycle protein alteration. *Journal of Neuropathology and Experimental Neurology*, *63*, 429–440.

De Falco, F. A., D'Angelo, E., Grimaldi, G., Scafuro, F., Sachez, F., & Caruso, G. (1994). Effect of the chronic treatment with L-acetylcarnitine in Down's syndrome. *Clinica Terapeutica*, *144*, 123–127.

Delabar, J. M., Theophile, D., Rahmani, Z., Chettouh, Z., Blouin, J. L., Prieur, M., et al. (1993). Molecular mapping of twenty-four features of Down syndrome on chromosome 21. *European Journal of Human Genetics*, *1*, 114–124.

D'Hulst, C., De Geest, N., Reeve, S. P., Van Dam, D., De Deyn, P. P., Hassan, B. A., et al. (2006). Decreased expression of the GABAA receptor in fragile X syndrome. *Brain Research*, *1121*, 238–245.

Donmez, G., Wang, D., Cohen, D. E., & Guarente, L. (2010). SIRT1 suppresses beta-amyloid production by activating the alpha-secretase gene ADAM10. *Cell*, *142*, 320–332.

Dowjat, W. K., Adayev, T., Kuchna, I., Nowicki, K., Palminiello, S., Hwang, Y. W., et al. (2007). Trisomy-driven overexpression of DYRK1A kinase in the brain of subjects with Down syndrome. *Neuroscience Letters*, *413*, 77–81.

Edgin, J. O., Mason, G. M., Allman, M. J., Capone, G. T., Deleon, I., Maslen, C., et al. (2010). Development and validation of the Arizona Cognitive Test Battery for Down syndrome. *Journal of Neurodevelopmental Disorders*, *2*, 149–164.

Ellis, J. M., Tan, H. K., Gilbert, R. E., Muller, D. P., Henley, W., Moy, R., et al. (2008). Supplementation with antioxidants and folinic acid for children with Down's syndrome: Randomised controlled trial. *British Medical Journal*, *336*, 594–597.

Englund, H., Annerén, G., Gustafsson, J., Wester, U., Wiltfang, J., Lannfelt, L., et al. (2007). Increase in beta-amyloid levels in cerebrospinal fluid of children with Down syndrome. *Dementia and Geriatric Cognitive Disorders*, *24*, 369–374.

European Surveillance of Congenital Anomalies (EUROCAT) 2010. http://www.eurocat-network.eu/.

Fernandez, F., Morishita, W., Zuniga, E., Nguyen, J., Blank, M., Malenka, R. C., et al. (2007). Pharmacotherapy for cognitive impairment in a mouse model of Down syndrome. *Nature Neuroscience*, *10*, 411–413.

Fernandez, J. W., Rezai-Zadeh, K., Obregon, D., & Tan, J. (2010). EGCG functions through estrogen receptor-mediated activation of ADAM10 in the promotion of non-amyloidogenic processing of APP. *FEBS Letters*, *584*, 4259–4267.

Ferrer, I., Barrachina, M., Puig, B., Martínez de Lagrán, M., Martí, E., Avila, J., et al. (2005). Constitutive Dyrk1A is abnormally expressed in Alzheimer disease, Down syndrome, Pick disease, and related transgenic models. *Neurobiology of Disease*, *20*, 392–400.

Ferrer, I., & Gullotta, F. (1990). Down's syndrome and Alzheimer's disease: Dendritic spine counts in the hippocampus. *Acta Neuropathologica*, *79*, 680–685.

Fodale, V., Mafrica, F., Caminiti, V., & Grasso, G. (2006). The cholinergic system in Down's syndrome. *Journal of Intellectual Disabilities*, *10*, 261–274.

Fowler, A. E. (1988). Determinants or rate of language growth in children with DS. In L. Nadel (Ed.), *The psychobiology of Down syndrome* (pp. 217–245). Cambridge, MA: MIT Press.

Gasperi, M., & Castellano, A. E. (2010). Growth hormone/insulin-like growth factor I axis in neurodegenerative diseases. *Journal of Endocrinological Investigation*, *33*, 587–591.

Golden, J. A., & Hyman, B. T. (1994). Development of the superior temporal neocortex is anomalous in trisomy 21. *Journal of Neuropathology and Experimental Neurology*, *53*, 513–520.

Guedj, F., Sébrié, C., Rivals, I., Ledru, A., Paly, E., Bizot, J. C., et al. (2009). Green tea polyphenols rescue of brain defects induced by overexpression of DYRK1A. *PLoS One*, *4*, e4606.

Hampel, H., Frank, R., Broich, K., Teipel, S. J., Katz, R. G., Hardy, J., et al. (2010). Biomarkers for Alzheimer's disease: Academic, industry and regulatory perspectives. *Nature Reviews. Drug Discovery*, *9*, 560–574.

Hanney, M., Prasher, V., Williams, N., Jones, E. L., Aarsland, D., Corbett, A., et al. (2012). Memantine for dementia in adults older than 40 years with Down's syndrome (MEADOWS): a randomised, double-blind, placebo-controlled trial. *Lancet*, *379*(9815), 528–536.

Haydar, T. F., & Reeves, R. H. (2012). Trisomy 21 and early brain development. *Trends Neuroscience*, *35*(2), 81–91.

Heller, J. H., Spiridigliozzi, G. A., Crissman, B. G., McKillop, J. A., Yamamoto, H., & Kishnani, P. S. (2010). Safety and efficacy of rivastigmine in adolescents with Down syndrome: Long-term follow-up. *Journal of Child and Adolescent Psychopharmacology*, *20*, 517–520.

Heller, J. H., Spiridigliozzi, G. A., Doraiswamy, P. M., Sullivan, J. A., Crissman, B. G., & Kishnani, P. S. (2004).

Donepezil effects on language in children with Down syndrome: Results of the first 22-week pilot clinical trial. *American Journal of Medical Genetics. Part A, 130A*, 325–326.

Heller, J. H., Spiridigliozzi, G. A., Sullivan, J. A., Doraiswamy, P. M., Krishnan, R. R., & Kishnani, P. S. (2003). Donepezil for the treatment of language deficits in adults with Down syndrome: A preliminary 24-week open trial. *American Journal of Medical Genetics. Part A, 116A*, 111–116.

Hudson, S., & Tabet, N. (2003). Acetyl-L-carnitine for dementia. *Cochrane Database of Systematic Reviews, 2*, CD003158.

Johnson, N., Fahey, C., Chicoine, B., Chong, G., & Gitelman, D. (2003). Effects of donepezil on cognitive functioning in Down syndrome. *American Journal of Mental Retardation, 108*, 367–372.

Jones, E. L., Hanney, M., Francis, P. T., & Ballard, C. G. (2009). Amyloid beta concentrations in older people with Down syndrome and dementia. *Neuroscience Letters, 451*, 162–164.

Khoshnood, B., Greenlees, R., Loane, M., & Dolk, H. EUROCAT Project Management Committee. EUROCAT Working Group. (2011). Paper 2: EUROCAT public health indicators for congenital anomalies in Europe. *Birth Defects Research. Part A Clinical and Molecular Teratology, 91* (Suppl. 1), S16–S22.

Kishnani, P. S., Heller, J. H., Spiridigliozzi, G. A., Lott, I., Escobar, L., Richardson, S., et al. (2010). Donepezil for treatment of cognitive dysfunction in children with Down syndrome aged 10-17. *American Journal of Medical Genetics. Part A, 152A*, 3028–3035.

Kishnani, P. S., Sommer, B. R., Handen, B. L., Seltzer, B., Capone, G. T., Spiridigliozzi, G. A., et al. (2009). The efficacy, safety, and tolerability of donepezil for the treatment of young adults with Down syndrome. *American Journal of Medical Genetics. Part A, 149A*, 1641–1654.

Kishnani, P. S., Sullivan, J. A., Walter, B. K., Spiridigliozzi, G. A., Doraiswamy, P. M., & Krishnan, K. R. (1999). Cholinergic therapy for Down's syndrome. *Lancet, 353*, 1064–1065.

Kondoh, T., Kanno, A., Itoh, H., Nakashima, M., Honda, R., Kojima, M., et al. (2011). Donepezil significantly improves abilities in daily lives of female Down syndrome patients with severe cognitive impairment: A 24-week randomized, double-blind, placebo-controlled trial. *International Journal of Psychiatry in Medicine, 41*, 71–89.

Kuhn, P. H., Wang, H., Dislich, B., Colombo, A., Zeitschel, U., Ellwart, J. W., et al. (2010). ADAM10 is the physiologically relevant, constitutive alpha-secretase of the amyloid precursor protein in primary neurons. *EMBO Journal, 29*, 3020–3032.

Lepagnol-Bestel, A. M., Zvara, A., Maussion, G., Quignon, F., Ngimbous, B., Ramoz, N., et al. (2009). DYRK1A interacts with the REST/NRSF-SWI/SNF chromatin remodelling complex to deregulate gene clusters involved in the neuronal phenotypic traits of Down syndrome. *Human Molecular Genetics, 18*, 1405–1414.

Lichtenthaler, S. F. (2011). Alpha-secretase in Alzheimer's disease: Molecular identity, regulation and therapeutic potential. *Journal of Neurochemistry, 116*, 10–21.

Lobaugh, N. J., Karaskov, V., Rombough, V., Rovet, J., Bryson, S., Greenbaum, R., et al. (2001). Piracetam therapy does not enhance cognitive functioning in children with down syndrome. *Archives of Pediatrics & Adolescent Medicine, 155*, 442–448.

Lott, I. T., & Dierssen, M. (2010). Cognitive deficits and associated neurological complications in individuals with Down's syndrome. *Lancet Neurology, 9*, 623–633.

Lott, I. T., Osann, K., Doran, E., & Nelson, L. (2002). Down syndrome and Alzheimer disease: Response to donepezil. *Archives of Neurology, 59*, 1133–1136.

Martinez de Lagran, M., Altafaj, X., Gallego, X., Martí, E., Estivill, X., Sahún, I., et al. (2004). Motor phenotypic alterations in TgDyrk1a transgenic mice implicate DYRK1A in Down syndrome motor dysfunction. *Neurobiology of Disease, 15*, 132–142.

Mohan, M., Bennett, C., & Carpenter, P. K. (2009). Memantine for dementia in people with Down syndrome. *Cochrane Database of Systematic Reviews, 1*, CD007657.

Myrelid, A., Bergman, S., Elfvik Strömberg, M., Jonsson, B., Nyberg, F., Gustafsson, J., et al. (2010). Late effects of early growth hormone treatment in Down syndrome. *Acta Paediatrica, 99*, 763–769.

Obregon, D. F., Rezai-Zadeh, K., Bai, Y., Sun, N., Hou, H., Ehrhart, J., et al. (2006). ADAM10 activation is required for green tea (-)-epigallocatechin-3-gallate-induced alpha-secretase cleavage of amyloid precursor protein. *Journal of Biological Chemistry, 281*, 16419–16427.

Olson, L. E., Roper, R. J., Sengstaken, C. L., Peterson, E. A., Aquino, V., Galdzicki, Z., et al. (2007). Trisomy for the Down syndrome 'critical region' is necessary but not sufficient for brain phenotypes of trisomic mice. *Human Molecular Genetics, 16*, 774–782.

Ortiz-Abalia, J., Sahún, I., Altafaj, X., Andreu, N., Estivill, X., Dierssen, M., et al. (2008). Targeting Dyrk1A with AAVshRNA attenuates motor alterations in TgDyrk1A, a mouse model of Down syndrome. *American Journal of Human Genetics, 83*, 479–488.

Palacios, H. H., Yendluri, B. B., Parvathaneni, K., Shadlinski, V. B., Obrenovich, M. E., Leszek, J., et al. (2011). Mitochondrion-specific antioxidants as drug treatments for Alzheimer disease. *CNS & Neurological Disorders Drug Targets, 10*, 149–162.

Petersen, R. C., & Trojanowski, J. Q. (2009). Use of Alzheimer disease biomarkers: Potentially yes for clinical trials but not yet for clinical practice. *Journal of the American Medical Association, 302*, 436–437.

Prasher, V. P. (2004). Review of donepezil, rivastigmine, galantamine and memantine for the treatment of dementia in Alzheimer's disease in adults with Down syndrome: Implications for the intellectual disability population. *International Journal of Geriatric Psychiatry*, *19*, 509–515.

Prasher, V. P., Fung, N., & Adams, C. (2005). Rivastigmine in the treatment of dementia in Alzheimer's disease in adults with Down syndrome. *International Journal of Geriatric Psychiatry*, *20*, 496–497.

Prasher, V. P., Adams, C., & Holder, R. (2003). Down Syndrome Research Group. Long term safety and efficacy of donepezil in the treatment of dementia in Alzheimer's disease in adults with Down syndrome: open label study. *International Journal of Geriatric Psychiatry*, *18*, 549–551.

Prasher, V. P., Huxley, A., & Haque, M. S. (2002). Down syndrome Ageing Study Group. A 24-week, double-blind, placebo-controlled trial of donepezil in patients with Down syndrome and Alzheimer's disease–pilot study. *International Journal of Geriatric Psychiatry*, *17*, 270–278.

Prasher, V. P., Sajith, S. G., Mehta, P., Zigman, W. B., & Schupf, N. (2010). Plasma beta-amyloid and duration of Alzheimer's disease in adults with Down syndrome. *International Journal of Geriatric Psychiatry*, *25*, 202–207.

Prinzen, C., Trümbach, D., Wurst, W., Endres, K., Postina, R., & Fahrenholz, F. (2009). Differential gene expression in ADAM10 and mutant ADAM10 transgenic mice. *BMC Genomics*, *10*, 66.

Pueschel, S. M. (2006). The effect of acetyl-L-carnitine administration on persons with Down syndrome. *Research in Developmental Disabilities*, *27*, 599–604.

Rissman, R. A., & Mobley, W. C. (2011). Implications for treatment: GABAA receptors in aging, Down syndrome and Alzheimer's disease. *Journal of Neurochemistry*, *117*, 613–622.

Ronan, A., Fagan, K., Christie, L., Conroy, J., Nowak, N. J., & Turner, G. (2007). Familial 4.3 Mb duplication of 21q22 sheds new light on the Down syndrome critical region. *Journal of Medical Genetics*, *44*, 448–451.

Rueda, N., Llorens-Martín, M., Flórez, J., Valdizán, E., Banerjee, P., Trejo, J. L., et al. (2010). Memantine normalizes several phenotypic features in the Ts65Dn mouse model of Down syndrome. *Journal of Alzheimer's Disease*, *21*, 277–290.

Ryu, Y. S., Park, S. Y., Jung, M. S., Yoon, S. H., Kwen, M. Y., Lee, S. Y., et al. (2010). Dyrk1A-mediated phosphorylation of Presenilin 1: A functional link between Down syndrome and Alzheimer's disease. *Journal of Neurochemistry*, *115*, 574–584.

Salman, M. (2002). Systematic review of the effect of therapeutic dietary supplements and drugs on cognitive function in subjects with Down syndrome. *European Journal of Paediatric Neurology*, *6*, 213–219.

Schmidt-Sidor, B., Wisniewski, K. E., Shepard, T. H., & Sersen, E. A. (1990). Brain growth in Down syndrome subjects 15 to 22 weeks of gestational age and birth to 60 months. *Clinical Neuropathology*, *9*, 181–190.

Schupf, N., Zigman, W. B., Tang, M. X., Pang, D., Mayeux, R., Mehta, P., et al. (2010). Change in plasma Aβ peptides and onset of dementia in adults with Down syndrome. *Neurology*, *75*, 1639–1644.

Seidl, R., Tiefenthaler, M., Hauser, E., & Lubec, G. (2000). Effects of transdermal nicotine on cognitive performance in Down's syndrome. *Lancet*, *356*, 1409–1410.

Shichiri, M., Yoshida, Y., Ishida, N., Hagihara, Y., Iwahashi, H., Tamai, H., et al. (2011). α-Tocopherol suppresses lipid peroxidation and behavioral and cognitive impairments in the Ts65Dn mouse model of Down syndrome. *Free Radical Biology & Medicine*, *50*, 1801–1811.

Spiridigliozzi, G. A., Heller, J. H., Crissman, B. G., Sullivan-Saarela, J. A., Eells, R., Dawson, D., et al. (2007). Preliminary study of the safety and efficacy of donepezil hydrochloride in children with Down syndrome: A clinical report series. *American Journal of Medical Genetics. Part A*, *143A*, 1408–1413.

Tian, Y., Crump, C. J., & Li, Y. M. (2010). Dual role of alpha-secretase cleavage in the regulation of gamma-secretase activity for amyloid production. *Journal of Biological Chemistry*, *285*(42), 32549–32556.

Wegiel, J., Gong, C. X., & Hwang, Y. W. (2011). The role of DYRK1A in neurodegenerative diseases. *FEBS Journal*, *278*, 236–245.

Wegiel, J., Kaczmarski, W., Barua, M., Kuchna, I., Nowicki, K., Wang, K. C., et al. (2011). Link between DYRK1A overexpression and several-fold enhancement of neurofibrillary degeneration with 3-repeat tau protein in Down syndrome. *Journal of Neuropathology and Experimental Neurology*, *70*, 36–50.

Yamamoto, T., Shimojima, K., Nishizawa, T., Matsuo, M., Ito, M., & Imai, K. (2011). Clinical manifestations of the deletion of Down syndrome critical region including DYRK1A and KCNJ6. *American Journal of Medical Genetics. Part A*, *155A*, 113–119.

M. Dierssen and R. de la Torre (Eds.)
Progress in Brain Research, Vol. 197
ISSN: 0079-6123

CHAPTER 2

Genomic determinants in the phenotypic variability of Down syndrome

Audrey Letourneau and Stylianos E. Antonarakis*

Department of Genetic Medicine and Development, University of Geneva Medical School, University Hospitals of Geneva, and iGE3 Institute of Genetics and Genomics of Geneva, Geneva, Switzerland

Abstract: Down syndrome caused by trisomy 21 is a collection of phenotypes with variable expressivity and penetrance. The significant advances in exploring the human genome now provide the tools to better understand the contribution of trisomy 21 in the different manifestations of Down syndrome, and the functional links between the genome variability and the phenotypic variability.

Keywords: Down syndrome; trisomy 21; human chromosome 21; phenotypic variability; genomic variability; gene expression.

Introduction

The phenotypes associated with Down syndrome (DS) are complex and highly variable among the patients. DS patients are characterized by some phenotypes found in every affected individual, while other phenotypes are highly variable in incidence and severity from patient to patient. The significant advances in exploring the human genome now provide the tools to better understand the functional links between the extra copy of chromosome 21 and the phenotypic variability of DS. All DS individuals are characterized by cognitive impairment and intellectual disability (Nadel, 2003). Trisomy 21 is considered as the most frequent genetic cause of intellectual disability (Pulsifer, 1996). DS newborns present characteristic dysmorphic features and a muscle hypotonia. A delayed cognitive impairment and a language production deficiency appear early in the infancy (Chapman and Hesketh, 2001; Chapman et al., 1998; Silverman, 2007). In the fourth decade, this impairment is always associated with early onset of the Alzheimer's disease neuropathology (Zigman and Lott, 2007; Zigman et al., 2008). Interestingly, the severity of the neurological disorders varies from mild to severe among the DS population. Other phenotypic characteristics are found only in a fraction of DS patients.

*Corresponding author.
Tel.: +41 22 379 5707; Fax: +41 22 379 5706
E-mail: stylianos.antonarakis@unige.ch

DOI: 10.1016/B978-0-444-54299-1.00002-9

For example, congenital heart defects occur in ~40% of the DS newborns. Affected individuals present an increased risk of developing Hirschsprung disease (40×) or acute megakaryocytic leukemia (200–400×) compared to the general population. Other phenotypes such as short stature, brachycephaly, or dysplastic ears are found only in some DS patients (Antonarakis et al., 2004; Roizen and Patterson, 2003).

Human chromosome 21

The complete characterization and functional analysis of human chromosome 21 (HSA21) are crucial to understand the molecular mechanisms responsible for the phenotypic variability.

Gene content

HSA21 is the smallest human chromosome. It is ~48Mb long and represents ~1.5% of the human genome. The ~33Mb long arm (21q) was almost completely sequenced in 2000 (99.7% coverage) (Hattori et al., 2000). This initial analysis of 21q revealed 225 genes and 59 pseudogenes. Of these, 43% were novel and 41% had no functional attribution. The GENCODE project (part of the ENCODE project) aims to annotate all the gene features in the human genome (http://www.gencodegenes.org) (Harrow et al., 2006). It combines data from the HAVANA manual annotation and the Ensembl automatic annotation. In 2011, GENCODE (release 8) identifies 696 genes on HSA21, including 235 protein-coding genes and 142 pseudogenes. The majority of those genes and pseudogenes are already known; only ~3% are novel. GENCODE also includes 207 processed transcripts that do not contain an open-reading frame. The gene content of HSA21 is proportional to its size and average compared to the other chromosomes (~14.5 genes per megabase) (Fig. 1). However, the gene content is not uniform along the chromosome. Some regions are gene poor, whereas other regions, mostly corresponding to the Giemsa light chromosomal bands, are gene rich. A major research goal in trisomy 21 (TS21) is to identify the genes that are dosage sensitive and thus likely to be involved in some of the phenotypes of the syndrome.

Gene ontology (GO) analysis (UniProtKB-GOA, http://www.ebi.ac.uk/GOA/) of the proteins encoded by HSA21 reveals that they are involved in 636 different biological processes, have 304 different molecular functions, and are present in 163 different cellular components (each gene being often associated to several GO annotations). The most frequent cellular localizations are the nucleus and the cytoplasm (112 and 99 proteins, respectively), and the most represented molecular function is protein binding (100 HSA21 proteins associated to this function). In the biological processes category, 31 proteins are involved in regulation of transcription, 26 in multicellular organismal development, and 23 in signal transduction (Fig. 2).

Interestingly, HSA21 shows a significant enrichment for genes located into cytoskeletal structures compared to the rest of the genome. Indeed, the following categories are over-represented in HSA21: intermediate filament (p-value$=3.17\times10^{-39}$), intermediate filament cytoskeleton ($p=3.17\times10^{-39}$), cytoskeletal part ($p=2.08\times10^{-34}$), cytoskeleton (4.64×10^{-16}), and keratin filament ($p=6.39\times10^{-13}$) (GOstat analysis, http://gostat.wehi.edu.au/). Cytoskeletal proteins are known to play a role in the development of neurological disorders, in particular Alzheimer's neuropathology (Bamburg and Bloom, 2009; Terry, 1998). In 2007, Lockstone et al. analyzed the gene expression profiles of adult DS brains and observed an upregulation of genes involved in cytoskeletal function (Lockstone et al., 2007). This upregulation was specific to adult DS brains and not found in fetal samples, suggesting a role in the late-onset pathologies like Alzheimer's disease.

Transcription factors encoded on HSA21 are important elements to explain the DS phenotypes because they are good candidates to influence the expression of other genes in the genome. In

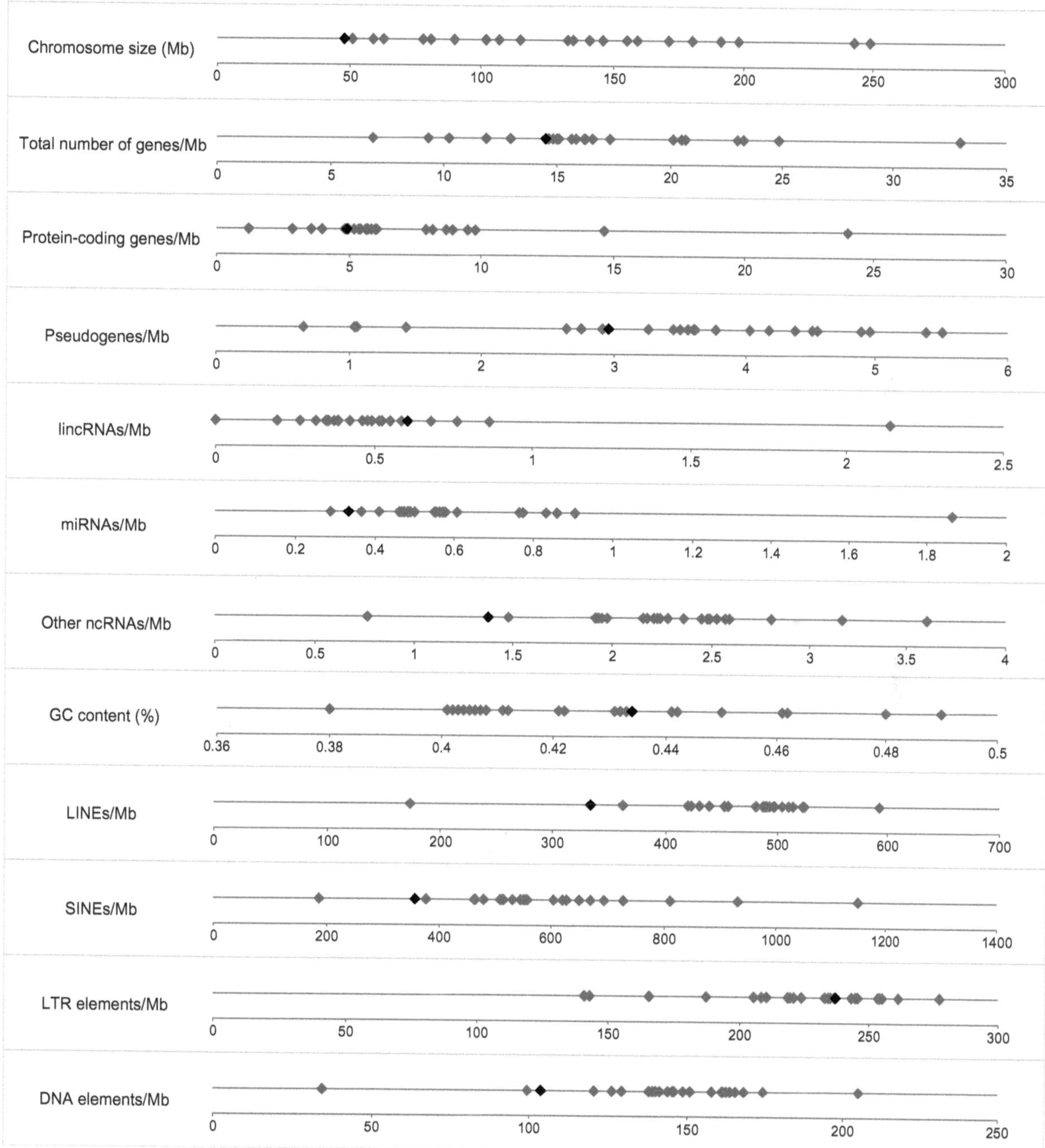

Fig. 1. Features of HSA21 relative to other chromosomes in the genome: chromosome size in megabases (Mb), gene density (number of genes per Mb), GC content, interspersed elements density (number of elements per Mb). Each dot represents one chromosome; the orange dot shows position of HSA21. (For interpretation of the references to color in this figure legend, the reader is referred to the Web version of this chapter.)

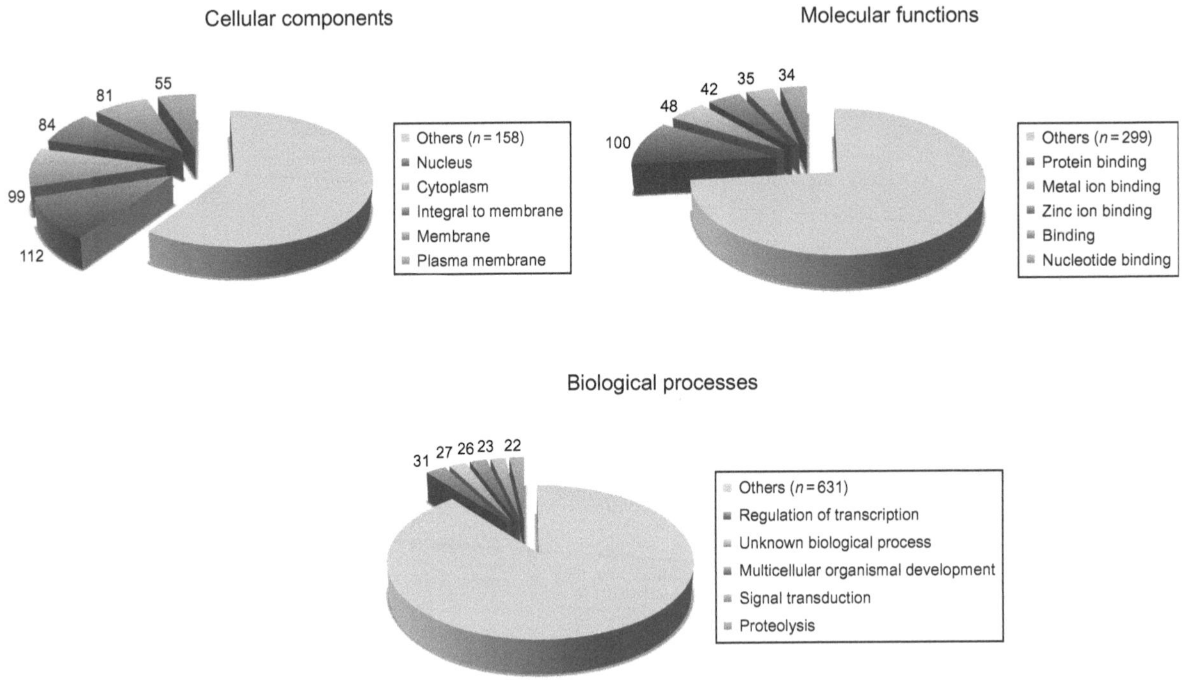

Fig. 2. Gene ontology analysis of HSA21 proteins. Each pie chart gives the five most represented GO categories on HSA21. Number of genes in each category is indicated in black. (For color version of this figure, the reader is referred to the Web version of this chapter.)

trisomic cells, their overexpression is likely to impact positively or negatively the transcription of their target genes. HSA21 contains 18 transcription factors and coregulators of transcription (Gardiner, 2006). Some of them have been individually studied and found to have a role in DS phenotypes. For instance, RUNX1 (RUNT domain transcription factor) is likely to contribute to the increased risk of leukemia in DS patients (De Vita et al., 2010; Osato and Ito, 2005). BACH1 (BTB and CNC homology 1 transcription factor) and ERG (Ets family of transcription factors) may play a role in the development of Alzheimer's disease associated to DS (Ferrando-Miguel et al., 2003; Shim et al., 2003).

Noncoding RNAs (ncRNAs) are transcripts that do not code for proteins but play important regulatory roles in the cells. According to GENCODE v8, HSA21 contains 111 of those ncRNAs, including 16 microRNAs (miRNAs), 29 long intergenic noncoding RNAs (lincRNAs), 21 small nucleolar RNAs (snoRNAs), and 21 small nuclear RNAs (snRNAs). A growing number of studies have demonstrated the importance of ncRNAs in the development of diseases (Faghihi et al., 2008; Gupta et al., 2010; Kocerha et al., 2009). It is possible that HSA21 ncRNAs participate to the development of some DS features. The five most studied HSA21 miRNAs (Hsa-miR-99a, Hsa-let-7c, Hsa-miR-125b-2, Hsa-miR-155, and Hsa-miR-802) are known to be overexpressed in TS21 tissues and may contribute to some DS phenotypes (Kuhn et al., 2008; Sethupathy et al., 2007). For instance,

Hsa-miR-155 is known to downregulate the *AGTR1* gene (Angiotensin receptor 1) in an allele-specific manner by binding the 1166A allele but not the hypertension-associated 1166C allele. The lower blood pressure observed in DS patients (Draheim et al., 2002; Morrison et al., 1996) could be in part explained by the overexpression of Hsa-miR-155 and the allele-specific underexpression of *AGTR1* (Sethupathy et al., 2007). In 2011, Costa et al. analyzed the nonribosomal transcriptome of human trisomic and euploid endothelial progenitor cells by RNA-Sequencing (Costa et al., 2011). They observed a significant differential expression between DS and normal cells for 15 miRNAs and 88 snoRNAs including 5 on HSA21. In particular, *SNORA80* showed a strong differential expression in trisomic cells and the highest expression on HSA21. Interestingly, among the five HSA21 snoRNAs differentially expressed, three were found downregulated in DS samples, despite the presence of three copies of HSA21. In addition, they observed a significant differential expression for 45 lincRNAs. In general, little is known about the role of ncRNAs in the development of the DS phenotypes and their study would probably help to further understand the molecular mechanisms involved in TS21.

GC content

HSA21 GC (guanine and cytosine) content is almost identical to other chromosomes in the genome (43%, Fig. 1). The GC content usually correlates with the gene density and GC-rich regions on HSA21 are associated to gene-rich domains of the chromosome (Fig. 3). In 2002, Takai et al. analyzed the CpG islands in HSA21 (Takai and Jones, 2002). A CpG island is defined as a 200-bp region of DNA with a GC content higher than 50% and an observed CpG versus expected CpG ratio greater or equal to 0.6 (Gardiner-Garden and Frommer, 1987). They identified 5′039 CpG islands on HSA21, mostly corresponding to *Alu* elements. In 2011, analysis of HSA21 sequence using the same criteria reveals 1′008 CpG islands (cpgplot, http://emboss.bioinformatics.nl/cgi-bin/emboss/cpgplot).

Conserved noncoding regions

Conserved noncoding elements (CNCs) are sequences that display a high evolutionary conservation. Their function is still poorly understood, but the conservation among species suggests

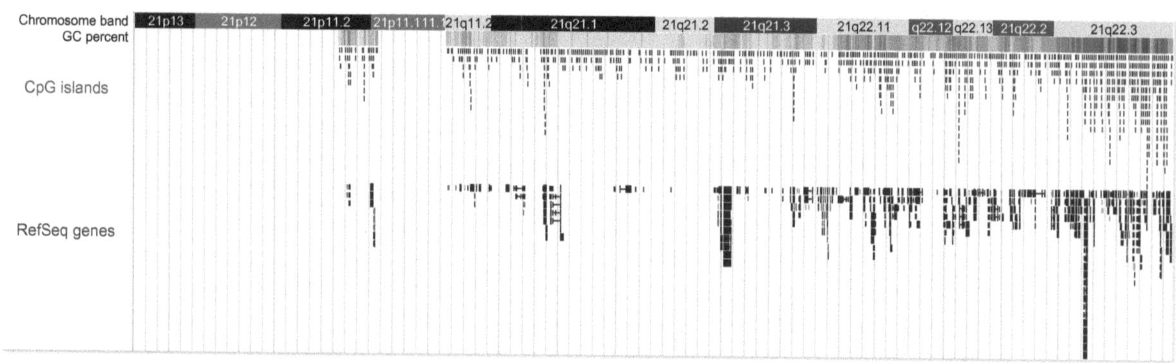

Fig. 3. GC content of HSA21. UCSC genome browser tracks (http://genome.ucsc.edu/) show the Giemsa chromosome banding, the percentage of GC, the CpG islands (cpgplot analysis, in green) and the RefSeq genes (in blue). (For interpretation of the references to color in this figure legend, the reader is referred to the Web version of this chapter.)

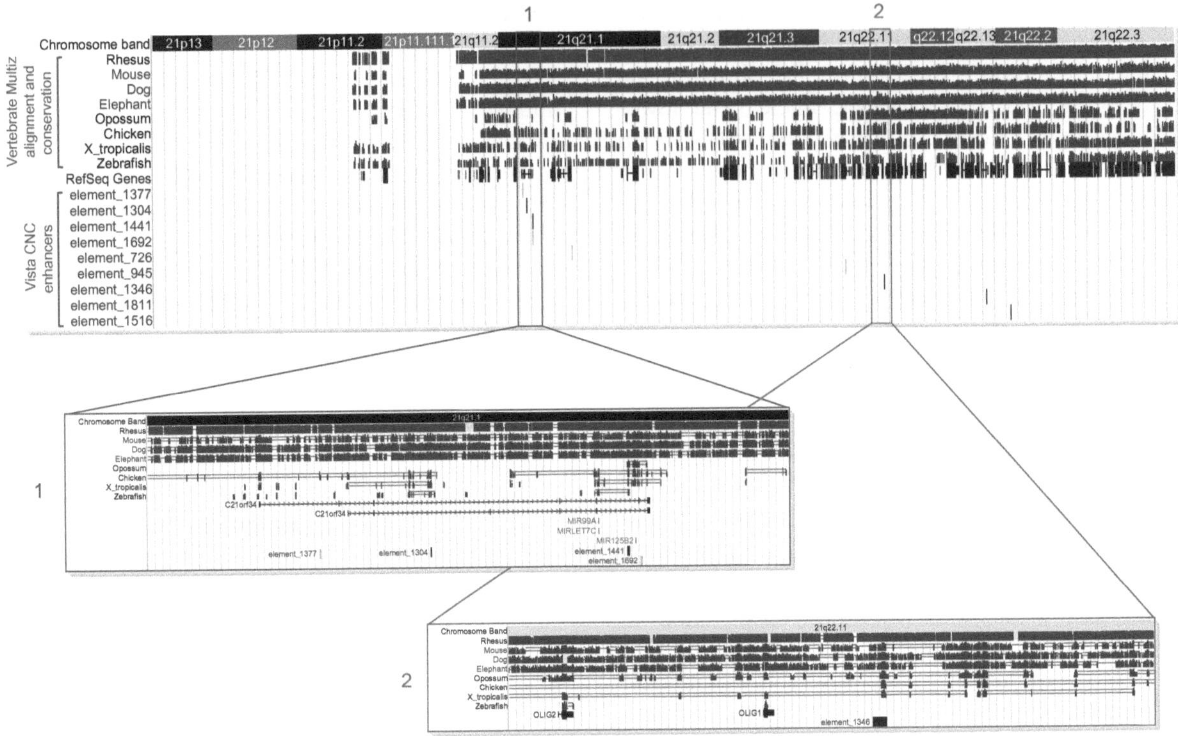

Fig. 4. CNC enhancers on HSA21. UCSC genome browser tracks show the Giemsa chromosome banding, the multiple alignments and conservation of vertebrate species (in green), the RefSeq genes (in blue), and the nine CNC enhancers identified by the Vista Enhancer Browser. Elements tested positive for an enhancer activity are displayed in black. Insert 1 shows the position of two positive CNCs (elements 1304 and 1441) located within introns of *C21orf34* gene. Insert 2 shows the location of one intergenic positive CNC (element 1346) close to *OLIG* genes. (For interpretation of the references to color in this figure legend, the reader is referred to the Web version of this chapter.)

that they bring a functional contribution to the genome. In 2002, Dermitzakis et al. identified 2′262 CNCs on HSA21 (Dermitzakis et al., 2002). In 2006, Pennacchio et al. tested the *in vivo* enhancer activity of the human CNCs, using a transgenic mouse enhancer assay (Pennacchio et al., 2006). The VISTA Enhancer Browser (UCSC build hg19) identifies five CNCs with an enhancer activity on HSA21 (Fig. 4). Interestingly, two of those CNC enhancers are located in introns of the *C21orf34* gene, close to three miRNAs (Hsa-miR-99a, Hsa-let-7c, Hsa-miR-125b-2). Another enhancer CNC is located in an intergenic region close to the oligodendrocyte genes *OLIG1* and *OLIG2*, two transcription factors of HSA21 involved in the development of the central nervous system (Ligon et al., 2006). In 2011, Robyr et al. used circular chromosome conformation capture (4C) to identify physical interactions between 10 CNCs of HSA21 and other loci in the genome (Robyr et al., 2011). They show that those CNCs likely interact with other CNCs in the genome as well as protein-coding genes, in particular *OLIG1* and *OLIG2*. Importantly, one CNC (presented in insert 2 of Fig. 4) was able to drive the expression of a reporter gene in the mouse embryo. The pattern of expression was similar to that of *OLIG* genes, suggesting a role in their regulation.

Other HSA21 features

The sequencing of HSA21 allowed the identification of four duplicated regions within the chromosome. Those regions are 10–200kb long and the number of repeats varies from 2 to 17 (Hattori et al., 2000). The same study estimated that interspersed repeats represent 38.01% of the HSA21. Those repeated elements are divided into two classes according to the intermediate they use for the transposition: the DNA elements (DNA intermediate) and the retroelements (RNA intermediate). This second category is split into two groups: the LTR (long terminal repeat) elements and the non-LTR elements encompassing SINEs (short interspersed nuclear elements) and LINEs (long interspersed nuclear elements). The sequencing of HSA21 showed that the majority of interspersed repeats on this chromosome are SINEs (15′748, i.e., 15.51%) and LINEs (12′723, i.e., 10.84%). LTR and DNA elements represent 9.21% (9′598) and 2.40% (3′950), respectively. In 2011, RepeatMasker (hg19 assembly) based on RepBase library estimates that HSA21 counts 17′109 SINEs, 16′016 LINEs, 11′380 LTR elements, and 4′983 DNA elements. All together, those four types of elements represent 33% of the HSA21. The density of LTR elements is average compared to other chromosomes. However, HSA21 contains less SINEs, LINEs, and DNA elements (Fig. 1).

The ENCODE project has identified by ChIP-Seq (Chromatin immunoprecipitation followed by high-throughput sequencing), the binding sites of several transcription factors in different cell types. For instance, in GM12878 cells, HSA21 counts 545 binding sites for p300, 473 for BCL1, and 50 for GABP.

In the same cell line, ENCODE (University of Washington) identified 3′199 hot spots for DNAseI hypersensitivity on HSA21. The Broad Institute (for ENCODE) reports peaks of signal enrichment for different histone modifications on HSA21. For example, they identified 872 regions enriched for H3K4me3, 612 for H3K27me3, and 594 for H3K9ac.

Comparison with other species

HSA21 is homologous to three different regions of the mouse genome, in three different chromosomes (MMU16, MMU17, and MMU10). Those homologous regions have been used to create different DS mouse models that reproduce some of the characteristic DS features.

The genome sequencing of *Drosophila melanogaster* in 2000 (Adams et al., 2000) identified 23 HSA21 genes with high similarity to *Drosophila* proteins (Gardiner and Davisson, 2000). Among them, SIM2 (single-minded 2) was already known to be a key regulator of the midline cells development in the central nervous system. Transgenic mice and expression studies in human confirmed the role of SIM2 in the development of key brain structures and its contribution to some DS features (Chrast et al., 2000; Rachidi et al., 2005).

Sturgeon and Gardiner (2011) compared the transcript catalog of HSA21 with orthologous regions of chimpanzee and mouse genomes. They found that 162 protein-coding genes are highly conserved between human and mouse. Ts65Dn mouse model is trisomic for 88 of those conserved genes. They also identified 151 genes (over 161 HSA21 RefSeq genes tested) completely conserved in Ptr21q (chimpanzee HSA21).

Genomic variability

SNPs (single nucleotide polymorphisms) or CNVs (copy number variants) are common genetic variations within the genome. Some of those variants likely contribute to disease susceptibility. Stranger et al. have shown that SNPs and CNVs are responsible for 83.6% and 17.7% of the gene expression variation between individuals, respectively (Stranger et al., 2007). The existence of such variation on HSA21 may be particularly important to explain the variability of DS phenotypes.

The Database of Genomic Variants (DGV v10, http://projects.tcag.ca/variation/) reports 820

CNVs on HSA21, in addition to 486 indels (insertions/deletions) and 25 inversions. Those CNVs cover 66% of HSA21. In 2010, de Smith et al. provided a list of CNVs on HSA21 that may affect the phenotypic variability in TS21 (de Smith et al., 2010). For instance, CNVs located within the *APP* gene may be associated to Alzheimer's disease (McKernan et al., 2009; Wang et al., 2008), and CNVs located within the *RUNX1* gene may be related to acute megakaryoblastic leukemia (Gusev et al., 2009) (Fig. 5, insert 1). In a 2007 review, Beckmann et al. discuss the contribution of CNVs to trisomy phenotypes and suggest that they might act as potential modifiers of the phenotypic variability of trisomies (Beckmann et al., 2007).

According to dbSNP (build 132), HSA21 contains 498′264 polymorphic sites including 412′289 SNPs and 108 microsatellites. Among the single nucleotide changes, 72 are nonsense (i.e., generate a stop codon) and 1′970 are missense (change an amino acid codon). Those single variations may contribute to the susceptibility of DS patients to particular diseases. In 2011, Patel et al. showed that some variants located within *APOE*, *SORL1*, *RUNX1*, *BACE1*, and *ALDH18A1* are associated with Alzheimer's disease dementia in DS individuals (Patel et al., 2011). Stranger et al. (2007) identified several SNPs (defined as eQTLs, expression quantitative trait loci) on HSA21 associated with the expression of their neighboring genes. They found for instance that rs1041456 is associated to the expression of *CSTB* (Fig. 5, insert 2).

Gene expression

Gene expression studies

It is likely that the majority of the DS phenotypes are related to alteration of gene expression. Over the years, several studies comparing gene expression in trisomic versus control cells or tissues have been reported (Ait Yahya-Graison et al., 2007; Altug-Teber et al., 2007; Conti et al., 2007; Costa et al., 2011; Lockstone et al., 2007; Prandini et al., 2007; Sommer et al., 2008; Sultan et al., 2007). Different methods including SAGE (Serial Analysis of Gene Expression), microarrays, qPCR, or more recently, RNA-Sequencing has been used to assess the gene expression changes in human or mouse. In the absence of any dosage compensation mechanism, HSA21 genes should theoretically be expressed 1.5 times more in TS21 tissues than in normal tissues. However, it has been shown that the change in copy number does not always correlate with the level of gene expression (Kahlem et al., 2004; Lyle et al., 2004). Most of the DS transcriptome studies reported an average upregulation close to the expected 1.5-fold. Nevertheless, only a subset of genes on HSA21 seems affected by the triplication (9–62% depending on the study). Moreover, some genes appear to be more overexpressed than others. Those observations suggest that some mechanisms compensate the gene dosage effect on HSA21.

The same studies have also shown changes of gene expression outside of HSA21, suggesting direct or indirect functional interactions between HSA21 elements (for instance, transcription factors) and other chromosomes (see also Chapters 3 and 5).

Because gene expression is known to be tissue specific, TS21 will affect differently the transcriptome of different tissues, modifying the list of overexpressed genes depending on the tissue studied. Several groups have reported the case of HSA21 genes differentially affected by the genomic imbalance in different tissues (Li et al., 2006; Mao et al., 2005; O'Leary et al., 2004; Rachidi et al., 2009). A deeper investigation of this tissue specificity in trisomic cells will be essential to understand the DS features associated with a particular organ (e.g., the congenital heart defects).

Whole transcriptome sequencing (RNA-Seq) now provides the opportunity to investigate in more details the perturbations of gene expression in TS21 cells and tissues. In the near future, RNA-Seq experiments will contribute to a better understanding of the molecular changes occurring in DS.

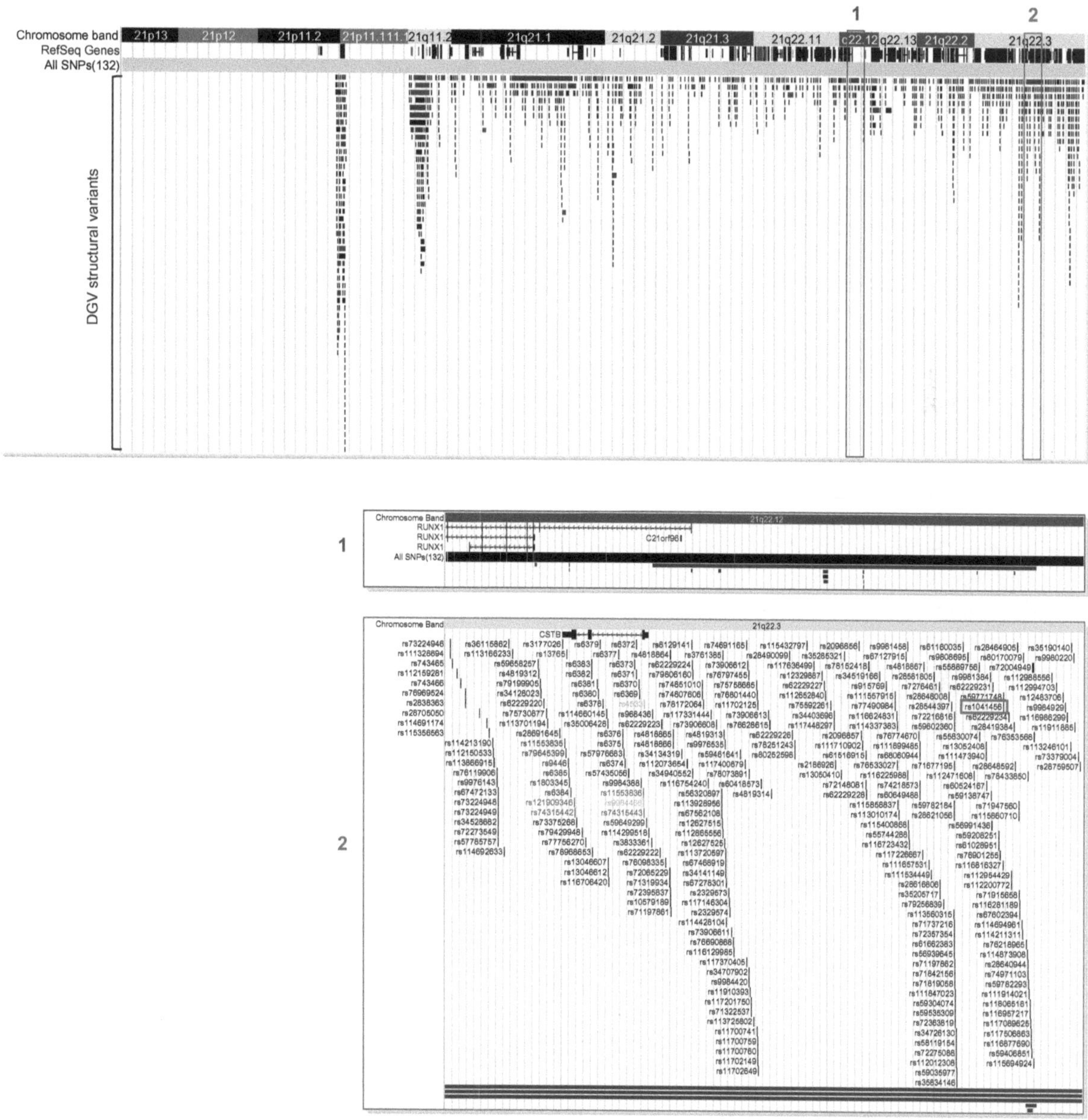

Fig. 5. Single nucleotide variants and structural variants on HSA21. UCSC genome browser tracks show the Giemsa chromosome banding, the RefSeq genes (in blue), the single nucleotide polymorphisms (SNPs) reported by dbSNP build 132 and the structural variants annotated by the Database of Genomic Variants (DGV v10). Deletions are represented in red, insertions in blue and inversions in purple. Insert 1 shows a reported deletion covering the first exon of the *RUNX1* gene. Insert 2 shows a reported deletion covering the *CSTB* gene as well as the position of the rs1041456 SNPs associated to the expression of *CSTB* in Stranger et al.'s (2007) study. (For interpretation of the references to color in this figure legend, the reader is referred to the Web version of this chapter.)

Gene expression variation

Prandini et al. (2007) used qPCR to study the gene expression variation in fibroblasts and lymphoblastoid cell lines from DS individuals and matched controls. They could identify two categories of HSA21 genes according to their natural gene expression variation. Some genes are almost equally expressed from individual to individual, whereas other genes show important gene expression variations within the DS population. This natural gene expression variation may contribute to the phenotypic variability of DS patients. Importantly, Prandini and colleagues also distinguished three groups of genes on HSA21 depending on the degree of overlap of gene expression between the DS and the normal populations. Genes presenting a minimal expression overlap between DS and euploid samples are good candidates to explain the constant DS features. On the contrary, genes with an extensive overlap between the two distributions are considered as dosage insensitive and are probably not contributing to the development of the DS phenotypes. The intermediate category would participate to the variable DS features. In related experiments, Sultan et al. (2007) established the same prioritization of candidate genes for DS phenotypes by studying the gene expression variation in Ts65Dn mouse model. They also discriminate the genes involved in the constant features of DS and the genes participating to the variable phenotypes.

Allele-specific expression

Allele-specific expression (ASE) is defined as the preferential expression of one allele versus the other one. It has been shown that ASE may be implicated in disease development (Feng et al., 2006; van Bilsen et al., 2008). Antonarakis et al. (2004) hypothesized that the effects of some HSA21 genes might be allele specific, meaning that the apparition of some DS features would depend on particular combinations of alleles and on the level of expression associated to each allele. Further investigations are needed to explore this hypothesis.

Protein levels

Most of DS studies on gene expression have focused on detection of RNA levels. However, changes in transcripts expression do not necessarily correlate with changes in protein levels. If the DS phenotypes result from the dosage imbalance of HSA21 genes, changes should also occur at the protein level. Unfortunately, this detection is challenging and no large-scale studies have been reported yet. In a series of articles published in 2003, Cheon et al. evaluated the expression levels of proteins encoded on HSA21 in fetal cerebral cortex of DS and control individuals, using Western blot analyses (Cheon et al., 2003a,b,c,d; Ferrando-Miguel et al., 2004). On 31 proteins analyzed in fetal brains, only four were found differentially expressed in DS tissues. In 2008, the same group analyzed protein expression levels of BACE1, BACE2, and APP in fetal and adult DS brains versus controls, using the same technique (Cheon et al., 2008). Only APP was found significantly increased in adult DS brains. These data suggest that the sensitivity of Western blotting is limited and additional methods should be used to determine the protein differences in the different tissues of DS.

Molecular mechanisms

Several hypotheses, not mutually exclusive, have been proposed to explain the genotype–phenotype correlation in DS: In the "gene dosage effect" hypothesis, the phenotypes directly result from the dosage imbalance of individual genes (Korenberg et al., 1994; Pritchard and Kola, 1999; Rachidi and Lopes, 2007). In other words, a subset of dosage-sensitive genes on HSA21 would be responsible for specific DS features. The DS critical region (DSCR) described in the 1990s matches with this hypothesis (Delabar et al., 1993; Korenberg et al., 1990). This region was defined by the analysis of genotype–phenotype correlation in individuals presenting partial TS21. The majority of the DS features could be attributed to a subset of critical genes located within the so-called DSCR, suggesting that the DS phenotypes may mainly result from the dosage imbalance of only few genes on HSA21. Recent studies by Lyle et al. and Korbel et al. also used segmental trisomies of various regions of HSA21 to identify portions of chromosome involved in the development of DS phenotypes (Korbel et al., 2009; Lyle et al., 2009). They created high-resolution genetic maps of DS features by detecting genomic regions that may contribute to the presence of specific DS phenotypes. In order to test the DSCR hypothesis in mice, Olson et al. (2004) created a specific mouse strain trisomic for the mouse chromosome segment orthologous to the DSCR. The genes in this region were not sufficient to produce the expected DS phenotypes. Few years later, the same group crossed Ts65Dn mice with mice deleted for the critical region, returning the gene dosage to normal for this region (Olson et al., 2007). Those mice showed an attenuated phenotype compared to the Ts65Dn mice. Those experiments demonstrate that the critical region is partially necessary to cause the phenotypes but not sufficient to explain all of them. It is likely that the dosage imbalance of some individual genes on HSA21 directly contributes to some phenotypes, but additional more global mechanisms are needed to fully explain the complexity of the DS manifestations. One explanation could be the additive effect of multiple HSA21 genes affected (even moderately) by the dosage imbalance.

The "Amplified developmental instability" hypothesis proposes a more global mechanism in which the dosage imbalance of hundreds of HSA21 genes together leads to a nonspecific disturbance of the cellular homeostasis (Pritchard and Kola, 1999; Rachidi and Lopes, 2007). The extra chromosome material itself could also contribute to the phenotypes by disturbing the nuclear organization and/or the chromosome territories. Some unpublished data on monozygotic twins discordant for TS21 show that the differential expression between the normal and the trisomic twins may be organized in domains along the chromosomes. This study suggests that some DS phenotypes could be explained by a modification of the chromatin architecture within the nucleus.

Interestingly, some studies reported the case of monozygotic twins affected by Down syndrome but presenting discordant phenotypes (Beattie et al., 1993; Grynberg et al., 2007; Telfer et al., 1972). Those cases suggest the role of epigenetics in the phenotypic variability of DS. For instance, it was shown that DNA methylation, known to play a role in the control of gene expression, was altered in TS21 samples (Kerkel et al., 2010). The analysis of a case of DS monozygotic twins discordant for a congenital heart defect (unpublished data) will help to further investigate this hypothesis (see also Chapter 5).

Conclusion

The sequencing of HSA21 and more generally the exploration of the entire human genome help to better understand the phenotypic variability associated with DS. The emerging technologies such as deep sequencing now provide the ability to test new hypotheses and to further investigate the genotype–phenotype association in TS21 patients.

Acknowledgments

We thank D. Robyr, M. Guipponi, K. Popadin, S. Nikolaev, J.L. Blouin, M. Nelis, R. Sailani, A. Nicoulaz, F. Santoni, and P. Makrythanasis for their contribution.

References

Adams, M. D., et al. (2000). The genome sequence of Drosophila melanogaster. *Science*, *287*(5461), 2185–2195.

Ait Yahya-Graison, E., et al. (2007). Classification of human chromosome 21 gene-expression variations in Down syndrome: Impact on disease phenotypes. *The American Journal of Human Genetics*, *81*(3), 475–491.

Altug-Teber, O., et al. (2007). Specific transcriptional changes in human fetuses with autosomal trisomies. *Cytogenetic and Genome Research*, *119*(3–4), 171–184.

Antonarakis, S. E., et al. (2004). Chromosome 21 and Down syndrome: From genomics to pathophysiology. *Nature Reviews. Genetics*, *5*(10), 725–738.

Bamburg, J. R., & Bloom, G. S. (2009). Cytoskeletal pathologies of Alzheimer disease. *Cell Motility and the Cytoskeleton*, *66*(8), 635–649.

Beattie, R. B., Manson, I. W., & Whittle, M. J. (1993). Identical twins with trisomy 21 discordant for exomphalos. *Prenatal Diagnosis*, *13*(11), 1067–1070.

Beckmann, J. S., Estivill, X., & Antonarakis, S. E. (2007). Copy number variants and genetic traits: Closer to the resolution of phenotypic to genotypic variability. *Nature Reviews. Genetics*, *8*(8), 639–646.

Chapman, R. S., & Hesketh, L. J. (2001). Language, cognition, and short-term memory in individuals with Down syndrome. *Down's Syndrome, Research and Practice: The Journal of the Sarah Duffen Centre/University of Portsmouth*, *7*(1), 1–7.

Chapman, R. S., et al. (1998). Language skills of children and adolescents with Down syndrome: II. Production deficits. *Journal of Speech, Language, and Hearing Research*, *41*(4), 861–873.

Cheon, M. S., et al. (2003a). Protein levels of genes encoded on chromosome 21 in fetal Down syndrome brain: Challenging the gene dosage effect hypothesis (Part II). *Amino Acids*, *24*(1–2), 119–125.

Cheon, M. S., et al. (2003b). Protein levels of genes encoded on chromosome 21 in fetal Down syndrome brain: Challenging the gene dosage effect hypothesis (Part III). *Amino Acids*, *24*(1–2), 127–134.

Cheon, M. S., et al. (2003c). Protein levels of genes encoded on chromosome 21 in fetal Down syndrome brain: Challenging the gene dosage effect hypothesis (Part I). *Amino Acids*, *24*(1–2), 111–117.

Cheon, M. S., et al. (2003d). Protein levels of genes encoded on chromosome 21 in fetal Down syndrome brain: Challenging the gene dosage effect hypothesis (Part IV). *Amino Acids*, *25*(1), 41–47.

Cheon, M. S., et al. (2008). Protein expression of BACE1, BACE2 and APP in Down syndrome brains. *Amino Acids*, *35*(2), 339–343.

Chrast, R., et al. (2000). Mice trisomic for a bacterial artificial chromosome with the single-minded 2 gene (Sim2) show phenotypes similar to some of those present in the partial trisomy 16 mouse models of Down syndrome. *Human Molecular Genetics*, *9*(12), 1853–1864.

Conti, A., et al. (2007). Altered expression of mitochondrial and extracellular matrix genes in the heart of human fetuses with chromosome 21 trisomy. *BMC Genomics*, *8*, 268.

Costa, V., et al. (2011). Massive-scale RNA-Seq analysis of non ribosomal transcriptome in human trisomy 21. *PLoS One*, *6*(4), e18493.

de Smith, A., Trewick, A., & Blakemore, A. (2010). Implications of copy number variation in people with chromosomal abnormalities: Potential for greater variation in copy number state may contribute to variability of phenotype. *The HUGO Journal*, *4*(1), 1–9.

De Vita, S., et al. (2010). Trisomic dose of several chromosome 21 genes perturbs haematopoietic stem and progenitor cell differentiation in Down's syndrome. *Oncogene*, *29*(46), 6102–6114.

Delabar, J. M., et al. (1993). Molecular mapping of twenty-four features of Down syndrome on chromosome 21. *European Journal of Human Genetics*, *1*(2), 114–124.

Dermitzakis, E. T., et al. (2002). Numerous potentially functional but non-genic conserved sequences on human chromosome 21. *Nature*, *420*(6915), 578–582.

Draheim, C. C., McCubbin, J. A., & Williams, D. P. (2002). Differences in cardiovascular disease risk between nondiabetic adults with mental retardation with and without Down syndrome. *American Journal of Mental Retardation*, *107*(3), 201–211.

Faghihi, M. A., et al. (2008). Expression of a noncoding RNA is elevated in Alzheimer's disease and drives rapid feed-forward regulation of beta-secretase. *Nature Medicine*, *14*(7), 723–730.

Feng, X., et al. (2006). Allele-specific silencing of Alzheimer's disease genes: The amyloid precursor protein genes with Swedish or London mutations. *Gene*, *371*(1), 68–74.

Ferrando-Miguel, R., Cheon, M. S., & Lubec, G. (2004). Protein levels of genes encoded on chromosome 21 in fetal Down syndrome brain (Part V): Overexpression of phosphatidyl-inositol-glycan class P protein (DSCR5). *Amino Acids*, *26*(3), 255–261.

Ferrando-Miguel, R., et al. (2003). Overexpression of transcription factor BACH1 in fetal Down syndrome brain. *Journal of Neural Transmission. Supplementum*, *67*, 193–205.

Gardiner, K. (2006). Transcriptional dysregulation in Down syndrome: Predictions for altered protein complex stoichiometries and post-translational modifications, and consequences for learning/behavior genes ELK, CREB, and the estrogen and glucocorticoid receptors. *Behavior Genetics*, *36*(3), 439–453.

Gardiner, K., & Davisson, M. (2000). The sequence of human chromosome 21 and implications for research into Down syndrome. *Genome Biology*, *1*(2) REVIEWS0002.

Gardiner-Garden, M., & Frommer, M. (1987). CpG islands in vertebrate genomes. *Journal of Molecular Biology*, *196*(2), 261–282.

Grynberg, M., et al. (2007). Prenatal diagnosis in monozygotic twins with Down syndrome who had different phenotypes. *Prenatal Diagnosis*, *27*(6), 552–554.

Gupta, R. A., et al. (2010). Long non-coding RNA HOTAIR reprograms chromatin state to promote cancer metastasis. *Nature*, *464*(7291), 1071–1076.

Gusev, A., et al. (2009). Whole population, genome-wide mapping of hidden relatedness. *Genome Research*, *19*(2), 318–326.

Harrow, J., et al. (2006). GENCODE: Producing a reference annotation for ENCODE. *Genome Biology*, *7*(Suppl. 1), S41–S49.

Hattori, M., et al. (2000). The DNA sequence of human chromosome 21. *Nature*, *405*(6784), 311–319.

Kahlem, P., et al. (2004). Transcript level alterations reflect gene dosage effects across multiple tissues in a mouse model of Down syndrome. *Genome Research*, *14*(7), 1258–1267.

Kerkel, K., et al. (2010). Altered DNA methylation in leukocytes with trisomy 21. *PLoS Genetics*, *6*(11), e1001212.

Kocerha, J., Kauppinen, S., & Wahlestedt, C. (2009). microRNAs in CNS disorders. *Neuromolecular Medicine*, *11*(3), 162–172.

Korbel, J. O., et al. (2009). The genetic architecture of Down syndrome phenotypes revealed by high-resolution analysis of human segmental trisomies. *Proceedings of the National Academy of Sciences of the United States of America*, *106* (29), 12031–12036.

Korenberg, J. R., et al. (1990). Molecular definition of a region of chromosome 21 that causes features of the Down syndrome phenotype. *The American Journal of Human Genetics*, *47*(2), 236–246.

Korenberg, J. R., et al. (1994). Down syndrome phenotypes: The consequences of chromosomal imbalance. *Proceedings of the National Academy of Sciences of the United States of America*, *91*(11), 4997–5001.

Kuhn, D. E., et al. (2008). Human chromosome 21-derived miRNAs are overexpressed in Down syndrome brains and hearts. *Biochemical and Biophysical Research Communications*, *370*(3), 473–477.

Li, C. M., et al. (2006). Cell type-specific over-expression of chromosome 21 genes in fibroblasts and fetal hearts with trisomy 21. *BMC Medical Genetics*, *7*, 24.

Ligon, K. L., et al. (2006). Olig gene function in CNS development and disease. *Glia*, *54*(1), 1–10.

Lockstone, H. E., et al. (2007). Gene expression profiling in the adult Down syndrome brain. *Genomics*, *90*(6), 647–660.

Lyle, R., et al. (2004). Gene expression from the aneuploid chromosome in a trisomy mouse model of Down syndrome. *Genome Research*, *14*(7), 1268–1274.

Lyle, R., et al. (2009). Genotype-phenotype correlations in Down syndrome identified by array CGH in 30 cases of partial trisomy and partial monosomy chromosome 21. *European Journal of Human Genetics*, *17*(4), 454–466.

Mao, R., et al. (2005). Primary and secondary transcriptional effects in the developing human Down syndrome brain and heart. *Genome Biology*, *6*(13), R107.

McKernan, K. J., et al. (2009). Sequence and structural variation in a human genome uncovered by short-read, massively parallel ligation sequencing using two-base encoding. *Genome Research*, *19*(9), 1527–1541.

Morrison, R. A., et al. (1996). Low blood pressure in Down's syndrome, a link with Alzheimer's disease? *Hypertension*, *28*(4), 569–575.

Nadel, L. (2003). Down's syndrome: A genetic disorder in biobehavioral perspective. *Genes, Brain, and Behavior*, *2*(3), 156–166.

O'Leary, D. A., et al. (2004). Tissue-specific overexpression of the HSA21 gene GABPalpha: Implications for DS. *Biochimica et Biophysica Acta*, *1739*(1), 81–87.

Olson, L. E., et al. (2004). A chromosome 21 critical region does not cause specific Down syndrome phenotypes. *Science*, *306*(5696), 687–690.

Olson, L. E., et al. (2007). Trisomy for the Down syndrome 'critical region' is necessary but not sufficient for brain phenotypes of trisomic mice. *Human Molecular Genetics*, *16*(7), 774–782.

Osato, M., & Ito, Y. (2005). Increased dosage of the RUNX1/AML1 gene: A third mode of RUNX leukemia? *Critical Reviews in Eukaryotic Gene Expression*, *15*(3), 217–228.

Patel, A., et al. (2011). Association of variants within APOE, SORL1, RUNX1, BACE1 and ALDH18A1 with dementia in Alzheimer's disease in subjects with Down syndrome. *Neuroscience Letters*, *487*(2), 144–148.

Pennacchio, L. A., et al. (2006). In vivo enhancer analysis of human conserved non-coding sequences. *Nature*, *444* (7118), 499–502.

Prandini, P., et al. (2007). Natural gene-expression variation in Down syndrome modulates the outcome of gene-dosage imbalance. *The American Journal of Human Genetics*, *81* (2), 252–263.

Pritchard, M. A., & Kola, I. (1999). The "gene dosage effect" hypothesis versus the "amplified developmental instability" hypothesis in Down syndrome. *Journal of Neural Transmission. Supplementum*, *57*, 293–303.

Pulsifer, M. B. (1996). The neuropsychology of mental retardation. *Journal of the International Neuropsychological Society*, *2*(2), 159–176.

Rachidi, M., & Lopes, C. (2007). Mental retardation in Down syndrome: From gene dosage imbalance to molecular and cellular mechanisms. *Neuroscience Research*, *59*(4), 349–369.

Rachidi, M., et al. (2005). Spatial and temporal localization during embryonic and fetal human development of the transcription factor SIM2 in brain regions altered in Down syndrome. *International Journal of Developmental Neuroscience*, *23*(5), 475–484.

Rachidi, M., et al. (2009). A quantitative assessment of gene expression (QAGE) reveals differential overexpression of DOPEY2, a candidate gene for mental retardation, in Down syndrome brain regions. *International Journal of Developmental Neuroscience*, *27*(4), 393–398.

Robyr, D., et al. (2011). Chromosome conformation capture uncovers potential genome-wide interactions between human conserved non-coding sequences. *PLoS One*, *6*(3), e17634.

Roizen, N. J., & Patterson, D. (2003). Down's syndrome. *Lancet*, *361*(9365), 1281–1289.

Sethupathy, P., et al. (2007). Human microRNA-155 on chromosome 21 differentially interacts with its polymorphic target in the AGTR1 3′ untranslated region: A mechanism for functional single-nucleotide polymorphisms related to phenotypes. *The American Journal of Human Genetics*, *81* (2), 405–413.

Shim, K. S., Ferrando-Miguel, R., & Lubec, G. (2003). Aberrant protein expression of transcription factors BACH1 and ERG, both encoded on chromosome 21, in brains of patients with Down syndrome and Alzheimer's disease. *Journal of Neural Transmission. Supplementum*, *67*, 39–49.

Silverman, W. (2007). Down syndrome: Cognitive phenotype. *Mental Retardation and Developmental Disabilities Research Reviews*, *13*(3), 228–236.

Sommer, C. A., et al. (2008). Identification of dysregulated genes in lymphocytes from children with Down syndrome. *Genome*, *51*(1), 19–29.

Stranger, B. E., et al. (2007). Relative impact of nucleotide and copy number variation on gene expression phenotypes. *Science*, *315*(5813), 848–853.

Sturgeon, X., & Gardiner, K. J. (2011). Transcript catalogs of human chromosome 21 and orthologous chimpanzee and mouse regions. *Mammalian Genome*, *22*(5–6), 261–271.

Sultan, M., et al. (2007). Gene expression variation in Down's syndrome mice allows prioritization of candidate genes. *Genome Biology*, *8*(5), R91.

Takai, D., & Jones, P. A. (2002). Comprehensive analysis of CpG islands in human chromosomes 21 and 22. *Proceedings of the National Academy of Sciences of the United States of America*, *99*(6), 3740–3745.

Telfer, M. A., Baker, D., & Bergman, M. (1972). Twins, probably monozygotic, displaying Down's syndrome, physical and functional mirror-imaging, and discordance for congenital heart disease. *American Journal of Mental Deficiency*, *76* (4), 391–396.

Terry, R. D. (1998). The cytoskeleton in Alzheimer disease. *Journal of Neural Transmission. Supplementum*, *53*, 141–145.

van Bilsen, P. H., et al. (2008). Identification and allele-specific silencing of the mutant huntingtin allele in Huntington's disease patient-derived fibroblasts. *Human Gene Therapy*, *19* (7), 710–719.

Wang, J., et al. (2008). The diploid genome sequence of an Asian individual. *Nature*, *456*(7218), 60–65.

Zigman, W. B., & Lott, I. T. (2007). Alzheimer's disease in Down syndrome: Neurobiology and risk. *Mental Retardation and Developmental Disabilities Research Reviews*, *13*(3), 237–246.

Zigman, W. B., et al. (2008). Alzheimer's disease in adults with Down syndrome. *International Review of Research in Mental Retardation*, *36*, 103–145.

M. Dierssen and R. de la Torre (Eds.)
Progress in Brain Research, Vol. 197
ISSN: 0079-6123

CHAPTER 3

Intellectual disabilities, neuronal posttranscriptional RNA metabolism, and RNA-binding proteins: three actors for a complex scenario

Barbara Bardoni*, Sabiha Abekhoukh, Samantha Zongaro and Mireille Melko*

Institute of Molecular and Cellular Pharmacology, CNRS-UMR6097, Université de Nice Sophia-Antipolis, Valbonne, France

Abstract: Intellectual disability (ID) is the most frequent cause of serious handicap in children and young adults and interests 2–3% of worldwide population, representing a serious problem from the medical, social, and economic points of view. The causes are very heterogeneous. Genes involved in ID have various functions altering different pathways important in neuronal function. Regulation of mRNA metabolism is particularly important in neurons for synaptic structure and function.

Here, we review ID due to alteration of mRNA metabolism. Functional absence of some RNA-binding proteins—namely, FMRP, FMR2P, PQBP1, UFP3B, VCX-A—causes different forms of ID. These proteins are involved in different steps of RNA metabolism and, even if a detailed analysis of their RNA targets has been performed so far only for FMRP, it appears clear that they modulate some aspects (translation, stability, transport, and sublocalization) of a subset of RNAs coding for proteins, whose function must be relevant for neurons.

Two other proteins, DYRK1A and CDKL5, involved in Down syndrome and Rett syndrome, respectively, have been shown to have an impact on splicing efficiency of specific mRNAs. Both proteins are kinases and their effect is indirect. Interestingly, both are localized in nuclear speckles, the nuclear domains where splicing factors are assembled, stocked, and recycled and influence their biogenesis and/or their organization.

Keywords: intellectual disability; Fragile X; Down syndrome; FRAXE mental retardation; RNA-binding proteins; RNA metabolism; neurons.

*Corresponding authors.
Tel.: 0033 (4) 93957766/62; Fax: 0033 (0) 9395 7708
E-mail: bardoni@ipmc.cnrs.fr; melko@ipmc.cnrs.fr

DOI: 10.1016/B978-0-444-54299-1.00003-0

Introduction

Gene expression is a multistep process that involves transcription, mRNA maturation and splicing, translation and degradation of mRNAs and proteins. Interestingly, a recent study reported that the cellular abundance of proteins is mainly regulated at the level of translation, thus showing the importance of the role of messenger ribonucleoprotein particles (mRNPs) in gene expression (Schwanhausser et al., 2011). RNPs containing RNAs and proteins are essential complexes in all living cells, and they are involved in most steps of mRNA synthesis, transport, activity, and degradation, which contribute to development and cell functions. Regulation of mRNA metabolism is particularly important for highly specialized cells like neurons. Indeed, during development, trafficking of mRNAs to both axonal and dendritic growth cones regulates neuronal growth, thus playing a key role in neuron physiology. After synapse formation, mRNAs continue to be transported to dendrites and axons where they are locally translated at the synapse. The consolidation of memory and several distinct forms of synaptic plasticity considered to be substrates of memory require mRNA translation and may be associated with changes in spine morphology (Martin and Ephrussi, 2009; Swanger and Bassell, 2011). mRNA transport along dendrites and axons is performed by RNA granules, a particular class of RNPs that contain also ribosomes and processing bodies (PB) proteins. These granules are transported along the dendrites and axons by "sliding" on microtubules until synapses, where repressed mRNA translation is locally reactivated under specific stimuli (e.g., stimulation of the mGluRI pathway; Aschrafi et al., 2005; Cougot et al., 2008; Krichevsky and Kosik, 2001). Decreased capacity to transport mRNA and control local translation into distal processes could alter the ability to respond to injury or local stimuli within axons and dendrites. In all these functions, RNA-binding proteins are the principal actors and, often, they are involved in different steps of mRNA metabolism. Mutations inactivating the function of these proteins have critical consequences for the cell and in particular for neurons, when these mutations destabilize the process that is at the molecular basis of synaptic plasticity and underlies learning and memory processes (Swanger and Bassell, 2011). For this reason, it is not surprising to discover that several RNA-binding proteins when mutated cause intellectual disability (ID), as well as other neurological disorders (Lukong et al., 2008; Ule, 2008).

To study ID means also to study brain functioning and development. Here we will review a class of ID caused by the absence of proteins that directly modulate different aspects of posttranscriptional mRNA metabolism and thus the expression and/or sublocalization of neuronal proteins, whose function is important for synapse formation and function (Gecz et al., 2009; Swanger and Bassell, 2011). Indeed, in some cases abnormal morphology of dendritic spines or abnormal synaptic plasticity is associated with the functional absence of proteins causing ID (Ramakers, 2002). It is clear that a major effort is needed in order to define all the mRNAs that are targeted by these RNA-binding proteins and, in particular, those that could be common targets of two or more RNA-binding proteins involved in ID and that could be involved in common pathways with a key role in learning and memory. On the other hand, other proteins involved in RNA metabolism have a critical role in synaptic development and function, even if a link with a human disease was never identified (Long et al., 2010; Swanger and Bassell, 2011). Thus, it is not surprising that two recent studies, describing 67 new genes causing ID, have reported that the functions of proteins encoded by these genes are mostly involved in transcription and translation (Kalscheuer et al., 2001; Najmabadi et al., 2011).

On the basis of the present knowledge, we can identify two classes of proteins causing ID and having a direct (RNA-binding proteins) or an indirect impact on mRNA metabolism.

Direct role on RNA metabolism. The RNA-binding proteins

FMRP—Fragile X mental retardation protein

Fragile X syndrome is the most common form of inherited ID (1:4000 newborn males and 1:7000 newborn females) and is caused by the silencing of the *FMR1* gene due to a triplet expansion in its 5′ region. However, sporadic cases that are caused by point mutations in *FMR1* have been described (MIM 309550). The syndrome is characterized by a mild to moderate form of intellectual disability, hyperactivity, attention deficit, language dysfunction, and autistic behavior. Around 20% of patients are also affected by epilepsy. Fragile X patients display a characteristic elongated facies with a broad forehead and large prominent ears. Furthermore, some physical abnormalities, such as flat feet, hyperextensible joints, and macroorchidism in postpubertal males, are in general present. All these hallmarks are less evident in young patients while they become more evident in adult patients, where also a decline of intellectual functioning is often present (Hagerman et al., 2010). The main phenotype observed in the brain of fragile X patients is due to the presence of dendritic spines that are longer, thinner, and denser that normal which seems to be the consequence of maturation, stabilization, and pruning problems (Swanger and Bassell, 2011). Recently, neuroanatomical studies have shown that some brain abnormalities observed in FXS patients are directly correlated with the lower level of fragile X mental retardation protein (FMRP)—the protein encoded by *FMR1*-and more pronounced cognitive deficits and aberrant behaviors (Hoeft et al., 2010). FMRP displays a high neuronal expression and is endowed with three domains (two KH and one RGG box) that can mediate its interaction with RNA and its ability to bind to RNA. FMRP binds with high-affinity to three different RNA structures (the G-quadruplex, the kissing complex, and SoSLIP; Bechara et al., 2009; Darnell et al., 2005; Schaeffer et al., 2001) and the poly-U stretch (Dolzhanskaya et al., 2003). Furthermore, the metabolism (translation, sublocalization) of hundreds of mRNAs has been shown to be altered in the absence of FMRP (Bardoni et al., 2006; Melko and Bardoni, 2010). FMRP is a component of mRNPs that are associated to polyribosomes in neuronal soma and at the synapse and its implication in translation as a repressor (when bound to G-quadruplex or to kissing complex) or as activator (when bound to SoSLIP) was described (Fig. 1). The repressor activity of FMRP seems to be carried out during translation elongation or by moving mRNA targets from polyribosomes and accumulating them in mRNP not associated to polyribosomes (Bardoni et al., 2006; Melko and Bardoni, 2010). Indeed, the expression of a subset of mRNAs has recently been published as modulated by FMRP by polysomal stalling (Darnell et al., 2011). While several mRNA targets of FMRP have been shown to contain a G-quadruplex and SoSLIP was identified in *SOD1* mRNA, no natural mRNA harboring a kissing complex has been identified so far (Melko and Bardoni, 2010; Subramanian et al., 2011; Fig. 1). To date, the double role of FMRP as repressor of translation and as translational enhancer was demonstrated (Bechara et al., 2009; Darnell et al., 2005; Lu et al., 2004; Schaeffer et al., 2001). The SoSLIP structure mediates the interaction between the *SOD1* mRNA and the RGG box of FMRP, and it is *per se* an enhancer of translation, whose action is also depending on factor/s other than FMRP and also works as IRES (internal ribosome binding site). FMRP enhances SoSLIP activity in cap-dependent translation but does not appear to affect cap-independent translation (Bechara et al., 2009). FMRP was reported to act as a translational activator also for the mRNA coding the protein Kv4.2 (potassium voltage-gated channel, Shal-related subfamily member; Gross et al., 2011), and the expression level of the two proteins SAPAP2 (SAP90/PSD95-associated protein 2) and NR1 (NMDA receptor subunit1) is increased in the hippocampus of *Fmr1* KO mice, while SAPAP1 and (again)

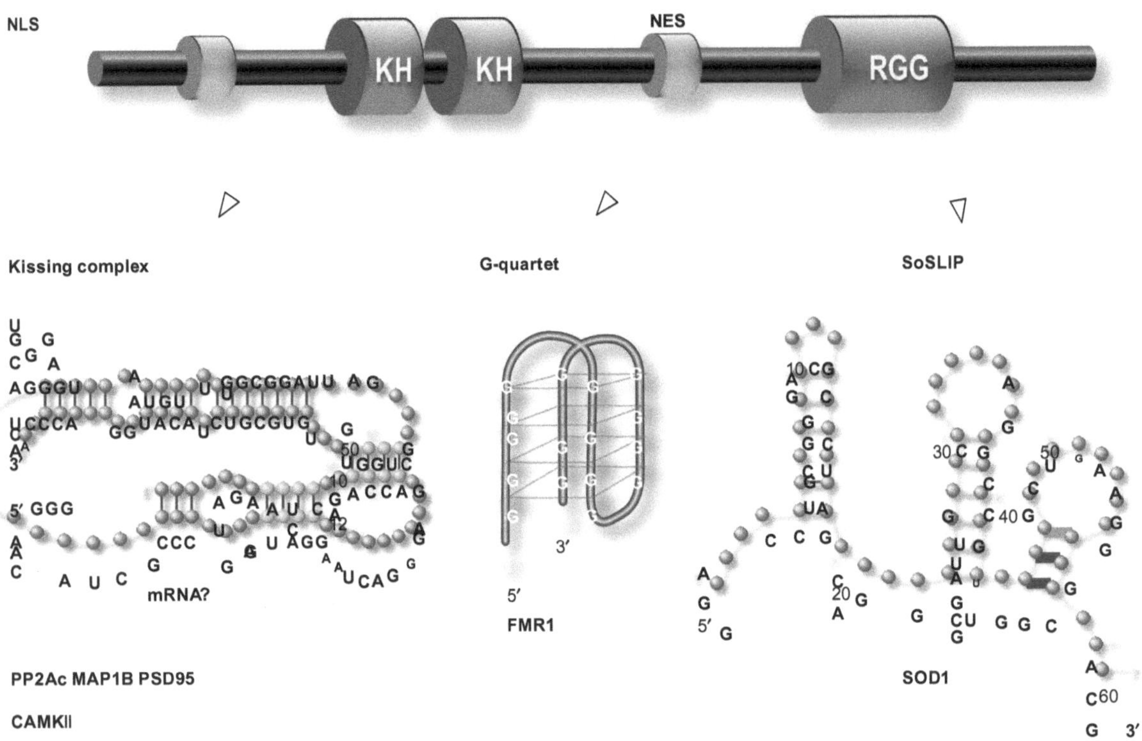

Fig. 1. Architectural organization of the fragile X mental retardation protein (FMRP). The principal domains are indicated: nuclear localization signal (NLS), nuclear export signal (NES), KH1, KH2 (binding to kissing complex), RGG box (binding to G-quadruplex and SoSLIP). RNAs harboring the RNA structures specifically recognized by FMRP are listed (Bechara et al., 2009; Darnell et al., 2005; Melko and Bardoni, 2010; Schaeffer et al., 2001; Subramanian et al., 2011). (For color version of this figure, the reader is referred to the Web version of this chapter.)

NR1 are increased in the cortex of the same animals, and the mRNAs coding for these proteins are associated with FMRP (Schütt et al., 2009). It is interesting to underline that the mRNAs positively modulated by the absence of FMRP encode for some proteins that are relevant for the pathophysiology of FXS. Indeed, Kv4.2 is the major potassium channel regulating hippocampal neuronal excitability (Gross et al., 2011), SAPAP2 is an adapter protein of postsynaptic density having connection with scaffolding and cytoskeletal components of the synapse and NR1 is a subunit of NMDA receptor (Schütt et al., 2009). Sod1 is an enzyme responsible for the suppression of free superoxide radicals in the cell. Thus, it is important to remember that brain of *Fmr1*-null mice have been reported to be more sensitive to oxidative stress (Davidovic et al., 2011; el Bekay et al., 2007) and that by treating these mice with some antioxidant drugs, it is possible to improve their cognitive phenotype (de Diego-Otero et al., 2009; Romero-Zerbo et al., 2009).

These findings collectively suggest that the mechanism of action of FMRP in translational control needs further studying and that the paradigm of FMRP being only a translational repressor must be revisited (Bechara et al., 2009). Interestingly, FMRP was also reported to bind to noncoding RNAs, miR-125b and miR-132 (Edbauer et al., 2010). More recently, it was reported that miR-125a expression level at synapses and its association with AGO2 are reduced in *Fmr1* KO mice. Inhibition of miR-125a increased PSD-95 levels (e.g., mRNA of PSD-95 is known to be a target of FMRP) in dendrites and altered dendritic spine morphology. To explain this result, it was shown that FMRP phosphorylation promotes the formation of an AGO2-miR-125a inhibitory complex on PSD-95 mRNA. On the other hand, mGluR signaling determines FMRP dephosphorylation and the release of AGO2 from the mRNA (Muddashetty et al., 2011). In conclusion, the role of FMRP in translation and in particular its interaction with miRNAs should be better studied, and it will probably be important to evaluate it in the context of different mRNPs where this protein is present. Finally, the interaction between FMRP and BC1 seems resolved after a long controversy (Zhong et al., 2010).

FMRP was also described as a shuttling protein involved in the transport of mRNAs between the nucleus and the cytoplasm since it is endowed with a Nuclear Export Signal (NES) and a (non classical) Nuclear Localization Signal (NLS) (Bardoni et al., 2006). Although the ability of FMRP to enter the nucleus was demonstrated, the effect of this shuttling action is not evident for any RNA target, suggesting that in the nucleus FMRP may be assembled in some mRNP complexes, whose specific function will be developed in the cytoplasm. FMRP enters also in the composition and formation of stress granules (SG; Didiot et al., 2009; Mazroui et al., 2003), the cytoplasmic domains that are observed in cells under stress conditions when the mRNA is protected and cap-dependent translation is blocked. FMRP is also located in PB, the cytoplasmic regions were mRNA can be degraded—being the site where decapping and mRNA decay occur and where mRNAs submitted to the RNA-induced silencing complex (RISC) action also accumulate or are stocked. In agreement with this latter localization, FMRP was found to interact with RISC, a key regulator of translation. However, the work by Didiot et al. indicated that FMRP and RISC are associated to distinct pools of mRNAs and, for instance, FMRP, unlike the RISC machinery, associates with the pool of mRNAs that eventually enters SG upon cellular stress. These findings lead to the conclusion that FMRP seems more involved in the storage process during stress conditions than in RNA degradation (Didiot et al., 2009). Furthermore, FMRP is also a component of RNA granules where it mediates the binding between mRNA and molecular motors, such as kinesins, promoting transport upon specific stimuli (Davidovic et al., 2007). This mechanism seems to influence also the synaptic abundance of a subset of mRNAs. Indeed, brains obtained from *Fmr1*-null mice have been reported to show a reduced pool of kinesin-associated mRNAs (Dictenberg et al., 2008; Fig. 2).

FXRP—Fragile X related proteins: FXR1P and FXR2P

All data collected so far suggest that FMRP is a multifaceted protein involved in the transport of mRNAs among different RNP complexes having different roles (e.g., polyribosomes and PB/SG/RNA granules). This consideration suggests that if FMRP is involved in the intracellular transport of mRNA, it should be a component of different mRNPs, and its function should be dependent on its interacting mRNA and protein partners. In this context, it is important to mention that a long list of proteins directly interacting with FMRP is known. Most of them are RNA-binding proteins, like FMRP, while a few of them are involved in cytoskeleton remodeling and molecular motors. Even if a huge effort has been performed to date

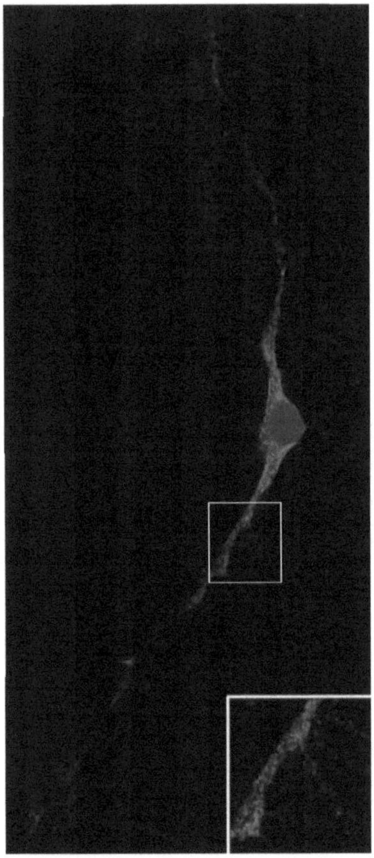

Fig. 2. Neuronal distribution of FMRP. (a) Immunofluorescence experiment using a polyclonal anti-FMRP on rat cortical neurons (red). Nuclei were counterstained by Dapi (blue). The white box indicates the area that is enlarged in (b). (For interpretation of the references to color in this figure legend, the reader is referred to the Web version of this chapter.)

to identify the interactors of FMRP, the functional significance of such interactions is in general missing or poorly understood (Bardoni et al., 2006). This is the case, for instance, of the two paralogs of FMRP, FXR1P (fragile X related 1 protein) and FXR2P, that are also its interactors. Both proteins share a high level of homology with FMRP and have been predicted to compensate, at least in part, for its absence in fragile X patients and they could be the cause of other form of ID when mutated. A few studies have been performed to unravel the molecular role of *Fxr2*. Its inactivation in mouse results in a phenotype very similar to that observed in *Fmr1* KO mice, suggesting that this gene, when mutated, could impact learning and memory. Indeed, a genome-wide expression analysis was performed in the brain of *Fxr2* KO mice showing that a relevant proportion of deregulated mRNAs in the absence of Fxr2 are important for neuronal function (Cavallaro et al., 2008). Furthermore, even if a solid study showing a compensatory action of other FXR proteins when FMRP is absent is still missing, some common functions have been described for the two proteins. For instance, *FMR1* and *FXR2* are both required for the regulation of behavioral circadian rhythms (Zhang et al., 2008). In particular, *FXR2* expression level could contribute to traits of the fragile X phenotype that are linked to the biological clock, such as sleep problems. The role of FXR1P has been more studied in muscle due to its high expression level in this tissue. Indeed, FXR1P plays an important role in normal muscle development and has been implicated in facioscapulohumeral muscular dystrophy (FSHD; Davidovic et al., 2008). Its absence also causes cardiac abnormalities in both mice and zebrafish (Mientjes et al., 2004; Van't Padje et al., 2009). Like FMRP, FXR1P muscular isoforms are able to bind to G-quadruplex RNA, and some cardiac specific mRNA targets have been described (Whitman et al., 2011). In this case, FXR1P acts as a translational repressor while, in other cell lines, it was described as activator of translation (Vasudevan and Steitz, 2007).

A neuronal phenotype for FXR1P has not been described. Concerning the function of this protein in neurons, it is interesting to underline that FXR1P isoforms most expressed in brain (namely, Isoa and Isod) reduce FMRP's affinity for G-quadruplex-containing RNAs. In this context, mechanisms modulating the homo- and heterodimerization of FXR proteins should be elucidated, since they are important in the regulation of the ability of FMRP to bind to RNA and also play a role, in a more general way, in the composition of FMRP-containing complexes.

FMR2P—FRAXE mental retardation 2 protein

Fragile X E (FRAXE) ID is a form of mild to moderate ID associated with a fragile site on chromosome Xq28 which is the cause of non-syndromic X-linked mental retardation in 1/50,000 newborn males. The disorder is due to the silencing of the *FMR2* gene as a consequence of a CCG expansion located upstream of this gene (MIM 300806). The disease is associated with learning difficulties, communication deficits, attention problems, hyperactivity, and autistic behavior (Knight et al., 1993). Interestingly, in a study performed on 26 patients affected by obsessive–compulsive disorder, a chronic psychiatric disease characterized by recurrent obsessions, compulsions, or both, one male patient was shown to carry a FRAXE full mutation (more than 200 repeats). Another member of the family was found to have a FRAXE full mutation (630–1200 copies of the CCG repeat), with speech impairment as clinical phenotype; two other members of the family with normal phenotypes had no FRAXE expansion. These findings seem to enlarge the spectrum of clinical phenotypes associated with this dynamic mutation (Wang et al., 2003). It was shown that the FRAXE full mutation is also associated to the inactivation of a noncoding RNA (ncRNA), called *FMR3* that is transcribed on the opposite strand to the *FMR2* gene (Gecz, 2000). Even if the role of *FMR3* in the cell is not known and the impact of its absence in FRAXE patients is not understood, this ncRNA should contribute to the definition of the phenotype of FRAXE patients.

FMR2 encodes a large protein of 1311-amino acids expressed in fetal and adult brain, in fibroblasts, and in placenta (Gecz et al., 1996; Gu et al., 1996). An alternative transcript from the FRAXE-associated CpG island exists (OX19) which encodes a truncated isoform of the *FMR2* gene with an alternative 3′ end. Both isoforms show a similar pattern of expression, with OX19 being expressed at a much lower level (Gecz et al., 1997b).

FMR2-null mice have been generated and display impairment in conditioned fear and hot plate analgesia tests, as well as enhanced long-term potentiation (LTP, for a definition see Chapter 8). The authors suggested a role for *FMR2* in regulating synaptic plasticity and that its absence in humans and mice can alter neuronal function and memory formation (Gu et al., 2002). *FMR2* and its paralogs (see below) have a common ancestor in *Drosophila melanogaster*, the *lilliputian* gene (*lilli*). Mutations in *lilli* were identified in a screen for a dominant suppressor of the rough eye phenotype caused by constitutive activation of Raf during eye development (Dickson et al., 1996). *lilli* has been shown to play a partially redundant function downstream of Raf in cell fate specification in the developing eye. Complete loss of *lilli* function leads to a reduction in cells and organ size (Wittwer et al., 2001). Moreover, *lilli* directs early developmental events, such as segmentation, cellularization, and gastrulation, through transcriptional activation of a specific group of zygotic regulators. These genes (*Sxl*, *ftz*, *sry-α*, and *hkb*) are normally activated at very early stages since they are essential for cellularization and embryonic patterning, and in all four cases this activation is reduced, if not abrogated, in the absence of maternally provided *lilli* (Tang et al., 2001). *lilli* plays also a critical role in Runt-dependent activation of the pair-rule gene *ftz* and in the activation of *Sex-lethal*, an early target of Runt in the sex determination pathway (Vanderzwan-Butler et al., 2007). To our knowledge, it has never been assessed whether the mammalian homologs of *lilli*'s targets are also target of FMR2P or another member of AFF family.

Subcellular localization of FMR2P was first studied by using green fluorescent protein as a reporter and showed that the protein is localized in the nucleus with a possible nucleolar sub-localization. FMR2P was shown to have a transcription transactivation potential (Gecz et al., 1997a). Recently, FMR2P was described to colocalize with the splicing factor SC35 in nuclear speckles, the nuclear regions where splicing factors are concentrated, assembled, and modified. As reported for splicing factors, blocking splicing or transcription led to the accumulation of FMR2P in enlarged rounded speckles and, in particular, it was also localized in the nucleolus when splicing was blocked (Bensaid et al., 2009) Interestingly, the presence of a nucleolar localization signal (NoLS) in the sequence of FMR2P as well as in that of several of its paralogs was shown (Melko et al., 2011). FMR2P is able to specifically bind to G-quadruplex-forming RNA structure with high affinity. By cloning a G-quadruplex sequence in a minigene, its role of exonic splicing enhancer (ESE) was confirmed. FMR2P reduced the action of G-quadruplex as ESE. Interestingly, in *FMR1* mRNA (whose absence causes the fragile X syndrome), a G-quadruplex RNA forming structure is located in proximity of an alternative splicing acceptor site in exon 15 and is involved in the alternative splicing of exon 14. This exon encodes the NES sequences of FMRP, a protein that is mainly localized in the cytoplasm, even if some nuclear isoforms are expressed at very low level (Sittler et al., 1996). FMR2P modulates the inclusion of exon 14 in *FMR1* mRNA and thus affects the amount of cytoplasmic FMRP isoforms (Bensaid et al., 2009). Interestingly, the two functional properties we described (subnuclear localization and RNA-binding activity) are retained by the C-terminal half of the protein. The interaction between FMRP and the G-quadruplex RNA is mediated by the RGG box domain, while no RNA-binding domain was defined in the sequence of FMR2P (Bensaid et al., 2009). Analyzing the FMR2P sequence by the software http://bindr2.gdcb.iastate.edu/RNABindR/ (Terribilini et al., 2006) one amino acid stretch (between residues 794 and 818) is predicted to have RNA-binding properties.

The C-terminal region of FMR2P displays the higher identity with the other members of the AFF (AF4/FMR2) family to which FMR2P belongs. The family includes other three members: AFF1 (also named ALL1-fused gene from chromosome 4, AF4), AFF3 (also named lymphoid nuclear protein related to AF4, LAF4), and AFF4 (also named ALL1-fused gene from 5q31, AF5q31) or major CDK9 elongation factor-associated protein, MCEF). It was shown that two of these proteins (AFF3 and AFF4), similarly to FMR2P, specifically bind to G-quadruplex-forming RNA structure and modulate the splicing efficiency of a minigene reporter with a G-quadruplex structure in its alternatively spliced exon. Furthermore, AFF3, AFF4, and FMR2P are localized in nuclear speckles and their overexpression affects the dynamics and/or biogenesis of these structures. Thus, it was proposed that the functional redundancy of AFF proteins can explain the mild nature of the cognitive impairment observed in FRAXE patients (Melko et al., 2011). This functional redundancy could be relied on coexpression studies during mouse brain development. Indeed, at embryonic day E13.5, when neurogenesis peaks in the developing mouse brain, the expression level of *Aff2*, *Aff3*, and *Aff4* genes is high (GEO at NCBI). In this context, the AFF proteins could have a role in ID; however, mutations in these genes have not been definitively linked to diseases characterized by ID (Melko et al., 2011). Furthermore, an animal model exists for AFF4, but learning and memory characterization of this mutant mouse was not performed (Urano et al., 2005).

PQBP1—POLYQ-binding protein 1

Defects in PQBP1 are the cause of Renpenning syndrome 1 (RENS1) (MIM 309500), also known as Sutherland-Haan X-linked mental retardation syndrome (SHS) or Golabi-Ito-Hall syndrome. The clinical features are intellectual disability, microcephaly, short stature, and small testes. The craniofacies tends to be narrow and tall with upslanting palpebral fissures, abnormal nasal configuration, cupped ears, and short philtrum. The nose may appear long or bulbous, with overhanging columella. Less consistent manifestations include ocular colobomas, cardiac malformations, cleft palate, and anal anomalies (Lubs et al., 2006; Stevenson et al., 2005). Several animal models have been generated confirming the role of this protein in cognitive development. Indeed, the knock down of PQBP1 in mouse causes a mild phenotype characterized by an abnormal anxiety-related behavior in the light/dark exploration and open-field tests. Interestingly, treatment of adult mice with 4-phenylbutyric acid, an HDAC (histone deacetylase) inhibitor, rescued the abnormal phenotypes in adult *Pqbp1*-KD mice (Ito et al., 2009). In *Drosophila*, the repression of *dPqbp1* (the fly homolog of *PQBP1*) results in the normal morphology of mushroom bodies and antennal lobes and in a significant impairment of learning acquisition, while memory retention was normal. Remarkably, the expression of NR1 was reduced in projection neurons and molecules upregulating NR1 expression (HDAC inhibitors, suberoylanilide hydroxamic acid, and phenylbutyrate) could partially rescue the abnormal learning in these flies (Tamura et al., 2010). In *Caenorhabditis elegans*, mutation of the homolog of *PQBP1* (*pqbp-1-1*) causes an abnormal lipid metabolism, which makes a link with the lean body in patients affected by ID and carrying mutated *PQBP1*. However, these researchers did not study the impact of such defect in neurons (Takahashi et al., 2009).

PQBP1 is a protein abundantly expressed in the central nervous system that was initially identified as a protein interacting with the polyglutamine tract (polyQ), the glutamine expansion that when present in a protein causes some forms of neurodegenerative disorders (e.g., Huntington disease, spinocerebellar ataxia type 1) (Okazawa et al., 2002). It is interesting to mention that transgenic mice overexpressing PQBP1 are affected by neurodegeneration (Okuda et al., 2003). This phenotype is consistent with the ability of this protein to mediate toxicity of the polyglutamine disease protein in cultured cells. PQBP1 has been described as a transcription factor having also a role in splicing regulation (Okazawa et al., 2002). Indeed, the protein is endowed with a WW domain, which is among the smallest protein modules that mediate protein–protein interactions by recognizing proline-rich sequence motifs in the cognate ligands (Bork and Sudol, 1994; Chen and Sudol, 1995). Recently, certain WW domains were shown to form homo- or heterodimers (reviewed in Südol and Harvey 2010).

The WW domain of PQBP1 mediates the interaction with RNA polymerase II (RNA PolII; Okazawa et al., 2002) and the splicing factor SIPP1 (splicing factor interacting with Ser/Thr phosphatase-1 (PP1) protein; Llorian et al., 2005; Fig. 3) being localized in nuclear speckles (Kunde et al., 2011), while its C-terminal domain (CTD) interacts with the splicing factor U5-15kD (Waragai et al., 2000). Interestingly, mutations causing ID and affecting the WW domain reduce the interaction with SIPP1 and the pre-mRNA splicing efficiency of the protein (Tapia et al., 2010; Fig. 3). These data suggest a direct role of PQBP1 in splicing modulation through the interaction with two splicing factors or, alternatively, this protein might have a mechanism of action similar to that described for SWI/SNF, whose subunit Brm modulates alternative splicing by directly interacting with RNA PolII and decreasing RNA PolII elongation rate (Allemand et al., 2008). Furthermore, most of the mutations described in the polyglutamine binding protein-1 (*PQBP1*) gene cause a truncation of its CTD and result in ID in human patients (Kalscheuer et al., 2003). The *Drosophila* homolog of PQBP1 lacks a CTD that interacts with a splicing factor whereas it conserves the WW domain. For this reason, transgenic flies expressing human PQBP1 were generated to examine whether *Drosophila* PQBP1 is involved in memory. Surprisingly, human PQBP1 impairs long-term memory in the transgenic flies, suggesting that dPQBP1 utilizes a distinct system from that of human PQBP1 (Yoshimura et al., 2006).

Some evidence also suggests that PQBP1 could have a cytoplasmic role. Indeed, it can move to

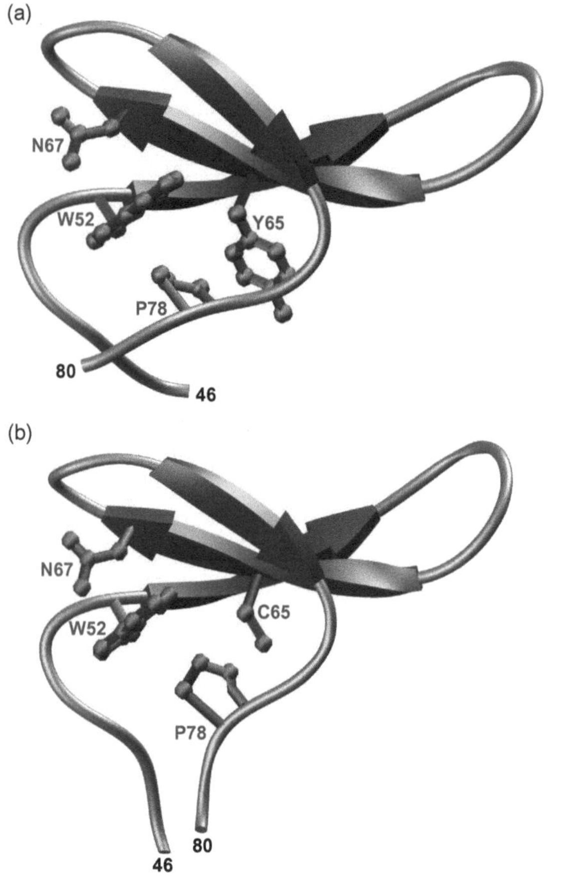

Fig. 3. Three-dimensional atomic models of wild-type (a) and Y65C-mutant (b) WW domains of PQBP1. The triple-stranded, antiparallel β-sheet fold of the WW domains, is shown in blue and the intervening loops in gray. The side chain atoms of residues W52, Y65/C65, N67, and P78 that constitute the hydrophobic core of the WW domains are colored red. Note that the hydrophobic core is located beneath the shallow binding pocket of the WW domain. The numerals indicate the boundary of the WW domains within the amino acid sequence of human PQBP1. The models were built using the MODELLER software based on homology modeling (Marti-Renom et al., 2000). In each case, the crystal structure of the WW domain of FE65 was used as a template (PDB# 2IDH). The authors decided to use FE65 structure as a template, instead of Pin1 template, used before (Tapia et al., 2010), because the binding profile of FE65 resembles that of PQBP1 (Kato et al., 2004; Südol M., unpublished observations; Zhang et al., 2000). The modeled structures were rendered using RIBBONS (Carson, 1991). This figure is a courtesy of Drs. Marius Südol and Amjad Farooq. (For color version of this figure, the reader is referred to the Web version of this chapter.)

the cytoplasm when it makes a complex with its interactor SIPP1. While in the cytoplasm the role of this complex is not studied, in the nucleus SIPP1 and PQBP1 form inclusion bodies similar to those detected in polyglutamine diseases (Llorian et al., 2005). Indeed, recently PQBP1 has been shown to interact with cytoplasmic RNA-binding proteins (KSRP, SFPQ/PSF, DDX1 and Caprin-1) and colocalizes with them in RNA granules and in polyribosomes. PQBP1 interacts also with two subunits of the intracellular transport-related dynactin complex, p150 (Glued) and p27. Collectively, these findings suggest a role of this protein in RNA translation and transport. Furthermore, PQBP1 is also involved in generation of SG, a function also described for FMRP. Interestingly, the two proteins interact and are partially colocalized in RNA granules (Kunde et al., 2011).

It will be interesting to study the mRNA targets of PQBP1. Indeed, a switch in alternative splicing and other aspects of RNA metabolism encoding neuron-specific proteins or synaptic-specific isoforms is known to impact neuronal differentiation/function.

UPF3B—Regulator of nonsense transcript homolog B

Mutations in the UPF3B protein have been found in patients affected by nonsyndromic and syndromic intellectual disability. Interestingly, the two forms of syndromic ID were in a first diagnosis defined as Lujan–Fryns syndrome (LFS) (OMIM 309520) and as Opitz-Kaveggia syndrome—also known as FG syndrome (OMIM 305450). The LFS syndrome is characterized by mild to moderate intellectual disability, typical facial dysmorphism (prominent forehead, maxillary hypoplasia, small mandible, long nose with short and deep philtrum, thin upper lip and low-set normally shaped ears, hypernasal voice, marfanoid-like features, and hypotonia). Patients affected by FG syndrome display intellectual disability, macrocephaly, imperforate anus, hypotonia, and facial anomalies (high broad forehead with frontal cowlick, hypertelorism and a prominent lower lip). In both cases, the physical features of patients are commonly observed in other individuals with ID (Tarpey et al., 2007). Furthermore, more recent studies showed that mutations in UPF3B are associated not only with intellectual disability but also with autism, attention deficit, hyperactivity, and schizophrenia (Addington et al., 2011; Laumonnier et al., 2010). UPF3B is a component of the NMD (nonsense mediated decay), the RNA surveillance pathway that degrades aberrant mRNAs harboring premature translation termination codons (PTCs). The pathway acts on PTCs originated from genes that are mutated or harboring errors occurring during pre-mRNA synthesis or maturation, and it acts during translation of an mRNA containing a PTC. Furthermore, the expression of several genes is regulated by NMD. UPF3B is a nuclear protein able to shuttle between nucleus and cytoplasm; it represents the link between the NMD machinery and the EJC (exon junction complex) core-complex (e.g., Y14/Magoh heterodimer, the DEAD-box protein eIF4AIII, and Barentsz), that is assembled in the presence of RNA and ATP and follows the mRNA during nuclear export. UPF3B contains an RRM (RNA recognition motif) mediating its interaction with UPF2, another member of the NMD machinery (Buchwald et al., 2010; Gehring et al., 2003). The impact of UPF3B mutants was studied in patients by evaluating the level of NMD targets. Indeed, the expression of the mRNA encoded by the gene *growth arrest and DNA damage-inducible beta* (*GADD45B*)—coding for a protein regulating growth and apoptosis—was increased in individuals carrying UFP3B PTCs compared to controls. Conversely, no effect was observed in the expression of other mRNAs: *sm5 homolog nonsense mediated mRNA decay factor* (*SMG5*) and *pantothenate kinase 2* (*PANK2*), suggesting that the absence of UFP3B has a specific effect on a subset of NMD targets (Tarpey et al., 2007). A more recent analysis indicates that the absence of UPF3B deregulates 5% of the transcriptome (526 genes). Among these genes, *ARHGAP24* encodes a GTPase-activating protein for ras-related p21-rac and a phorbol ester receptor, is predominantly expressed in neurons, plays an important role in neuronal signal-transduction mechanisms, and is considered a good candidate to explain the cognitive deficit in *UPF3B*-null patients. Indeed, overexpression of *ARHGAP24* in mouse primary hippocampal neurons causes a significant reduction of axonal length and an increased level of cell death (Nguyen et al., 2011).

UPF3A is a paralog of UFP3B that may compensate for its absence and mitigate the effect of its mutations (Chan et al., 2009). To date, an animal model for UFP3B does not exist, and the full list of mRNAs whose deregulation is caused by mutation in UPF3 in humans is unknown. However, the relevance of NMD in neuronal differentiation and function has been proved in mouse. Indeed, it was shown that the inactivation of this pathway via miR-128 (having as target the NMD component, UPF1) modulates the expression of a subset of genes involved in neurogenesis, neuron differentiation, and morphology of neuronal projections (Bruno et al., 2011). In *Drosophila*, the important role of the NMD pathway in maintaining synapse architecture, neurotransmission, and synaptic vesicle cycling was also shown (Long et al., 2010).

VCX—Variable charges X-linked family of proteins

VCX/Y proteins family is composed by four X-linked members (namely, VCX-A, VCX-B, VCX-B1, VCX-C) clustered on the Xp22.3 chromosomal region and two Y-linked members encoded by genes located on the chromosome Yq11 (VCY1and VCY1A; MIM 300533). The X-linked members of this family are located in a region that is critical for ID (Ballabio et al., 1989; Turner et al., 2003). Furthermore, a deletion analysis by a phenotype–genotype correlation comparing patients affected or not by ID, resulted in the identification of VCXA as the only gene of this family involved in this phenotype. Indeed, VCXA was retained in patients with normal intelligence; however, in patients with cognitive deficit, the deletion included also the neuroligin 4 (NLGN4X) gene, that was later associated with ID and autism. A more recent analysis studying patients with X/Y translocations confirmed the main role of VCXA in ID. Point mutations in VCXA associated with ID were never shown, but a cognitive handicap was in all cases associated with rearrangement of the VCX region, leading to the suggestion that altered expression of some VCX genes should be present in brain or during development to get an ID phenotype (Fukami et al., 2000; Van Esch et al., 2005). Patients affected by ID associated with VCXA are characterized by mild to moderate ID, hyperactive behavior, delayed growth (Van Esch et al., 2005).

Members of this family share a high level of homology with an identity greater than 95% in their N-terminal region, where the RGG box domain is located, indicating that they represent a family of RNA-binding proteins (Fukami et al., 2000; Jiao et al., 2006). These genes are not conserved during evolution. Indeed, for instance, their orthologs in mouse have not been found so far (Jiao et al., 2009) and in primates they are present only in chimpanzee, gorilla, and macaque (ENSEMBL-NCBI). The X-linked members of the family appear to be expressed in most tissues, while the two Y copies have a prevalent expression in testes. In particular, these genes are only present in humans (Jiao et al., 2009). The precise dosage of these genes has an effect on neuronal function. Indeed, deletions involving more than one member are associated with nonsyndromic forms of ID (Hosomi et al., 2007; Van Esch et al., 2005), suggesting also in this case a compensation effect of the different paralogs, as known for other forms of ID. Given the lack of point mutations in VCX genes, the impossibility to generate animal models for this family of genes represents an obvious difficulty in order to dissect the role of each member in learning and memory. To date, only the function of VCX-A has been studied. First results have shown that VCX-A is a noncanonical cap-binding protein able to bind 5′cap region in mRNA and modulate mRNA decapping. Half-life of each mRNA is a crucial parameter in the expression of a gene and is regulated through the 5′ cap and 3′ polyadenylated tail. Both these structures are important for mRNA stability and translation since their removal results in exonucleolytic decay of the mRNA (Jiao et al., 2006). Furthermore, the role of VCX-A in translation (being also associated to polyribosomes) was also shown (Zou et al., 2003; Jiao et al., 2009). This protein probably acts by inhibiting the translation of a subset of mRNAs by binding their 5′ end. In particular, so far, two targets of VCX-A have been identified, the mRNAs coding the glutamate receptor interacting protein 1 (GRIP1) and the calcium/calmodulin dependent protein kinase I (CAMKI), two proteins that are well-known regulators of dendrite branching. Indeed, in neurons, VCX-A displays a positive effect on the neurite outgrowth, indicating that this protein has a specific neuronal role that, due to the human specificity, could be linked to higher eukaryotic brain functions, thus explaining the intellectual deficit when absent (Jiao et al., 2009).

Summary

It seems clear that some of these RNA-binding proteins act at different levels during mRNA metabolism. As their function is not unique, we can speculate that when they bind an mRNA, they escort it through different aspects of its life. This consideration points out the importance to identify the RNA targets of these proteins, which obviously contribute to the understanding of molecular defect/s causing the ID. The identification of target mRNAs is not an easy task as the past 20years of studies on FMRP have clearly demonstrated. FMRP is a protein which can interact with RNA nonspecifically. This complicates the analysis of its targets since binding assays in the presence of competitor RNA are needed, as it has been shown (Bechara et al., 2009; Schaeffer et al., 2001). Indeed, simplification of this protocol created in the past artifacts for FMRP binding, complicating the interpretation of the cellular role of FMRP (Iacoangeli et al., 2008). Moreover, nonspecific *in vitro* affinity for RNA led to some general conclusions on the function of FMRP (being only a repressor) that in specific cases have been proved not to be true (Bechara et al., 2009; Li et al., 2001). Furthermore, even if hundreds of putative targets of FMRP are now known, only for a few of them functional significance of their interaction with FMRP is known and the mechanism of action of FMRP in the regulation of translation, for instance, is not completely unraveled yet (Bassell and Warren, 2008; Melko and Bardoni, 2010). This fact underlines the difficulty to combine molecular biology, cell biology, and neuroscience experiments.

It is also possible that the proteins we described here have some common target RNAs whose deregulation (at different steps of their metabolism) may result in ID. A broad analysis has been performed to date only for FMRP and only limited data exist for the other RNA-binding proteins linked to ID, however, the mRNA of *Fmr1* appeared to be a target of FMRP itself and of FMR2P, since they both bind with high-affinity the G-quadruplex RNA (Melko and Bardoni, 2010). A link between FXS and Down syndrome (DS) is provided by the fact that mRNAs encoding *SOD1* and *APP* (*Amyloid precursor protein*), whose genes are located in the Down syndrome critical region on chromosome 21 (HSA21, see also below) are modulated by FMRP, even if only APP is overexpressed in FXS brain (Melko and Bardoni, 2010). Recently, also *DSCAM* (Down syndrome cell adhesion molecule) RNA was described as target of FMRP, as well as *AFF3* and *AFF4* mRNAs, that we proposed as possible candidates for ID-causing genes (Darnell et al., 2011).

On the other hand, other forms of ID are caused by proteins that are in the same pathway, so it would not be surprising that the same happens for this group of proteins. The recent identification of several proteins involved in translation and splicing whose functional absence causes ID (Kalscheuer et al., 2011; Najmabadi et al., 2011) will help to address this question only if their functional characterization will be performed in depth.

Another amazing aspect is that most of these proteins are part of a family, whose members share a high level of homology. With the exception of the VCX family that is located in a chromosomal region known to have a high level of instability resulting in deletions, duplications, and X/Y translocations (Ballabio et al., 1989; Van Esch et al., 2005) in all the other cases it seems that the existence of paralogs has a functional relevance since these proteins have a redundant function and some of them can compensate for the absence of another one mitigating the effect of a mutation. In the same case, patients display only nonsyndromic forms of intellectual disability and this could be due to the fact that the concerned processes and pathways are particularly critical in brain, as we explained at the beginning of this review.

Indirect role on RNA metabolism

DYRK1A—DUAL specific tyrosine regulated kinase 1A

DS-(MIM 190685) is an abnormal chromosomal condition characterized by the presence of an extra copy of genetic material on HSA21. DS is associated with some impairment of cognitive ability (IQ 50) and physical growth, and a particular set of facial phenotype such as epicanthic fold of the eyelid, upslanting palpebral fissures, microgenia, and macroglossia. DS is considered the most common form of ID (1: 700/1000 newborns).

Among the genes that are located in the DS critical region (DSCR), DYRK1A is a good candidate to explain ID in these patients (see below). DYRK1A protein belongs to a family of protein kinases called dual-specificity tyrosine-regulated-kinases (DYRKs) due to the ability to phosphorylate serine/threonine and tyrosine residues. In particular, its ability to autophosphorylate at the tyrosine residue 321 is known. Members of this family are conserved from yeast (Yak1p, Pomp1p, YakA) to humans. In yeast, the DYRK proteins are associated with growth control and development (Becker et al., 1998). Mutations in the *minibrain* gene of *Drosophila*, the ortholog of DYRK1A, reduce adult brain size. This effect is particularly evident at the level of optic lobes due to an altered proliferation in the primordial neuroepithelia of this region at larval stage (Tejedor et al., 1995). Like in *Drosophila*, Dyrk1a+/− heterozygous mice display reduced size of specific brain regions (Fotaki et al., 2002). Moreover, the inactivation of DYRK1A interferes with neurite formation, suggesting a complex role of this protein in neural proliferation and in neuronal differentiation (Gockler et al., 2009). Transgenic mice overexpressing Dyrk1a (TgDyrk1A), as well as chromosome 16 partial trisomy mouse model (Ts65Dn) have hyperactivity, impaired neuromotor development, and spatial learning (Altafaj et al., 2001; Martínez de Lagrán et al., 2004). This phenotype is reversible by inactivating Dyrk1A with a specific inhibitory RNA against *Dyrk1A*, delivered by bilateral intrastriatal injections of an adeno-associated virus type 2 vector (AAVshDyrk1A; Ortiz-Abalia et al., 2008), strongly suggesting that *DYRK1A* overexpression has a relevant contribution in the DS intellectual disability and motor anomalies phenotype. Furthermore, *DYRK1A* is highly expressed in brain and, in particular, in those regions rich in neurons. The DYRK1A protein is endowed with several domains: a NLS (in the N-terminal region), a protein kinase domain (a proline-direct-kinase), a leucine zipper motif (involved in protein–protein interaction), a PEST domain (involved in protein degradation), and a 13-consecutive-histidine repeat. The protein is localized both in the nucleus and cytoplasm in neurons and, interestingly, it has been observed also in neurites and at the synapse. In the nucleus, DYRK1A is accumulated in nuclear speckles where it is driven by a 13-consecutive-histidine repeat (Alvarez et al., 2003; Herrmann and Mancini, 2001). DYRK1A has been shown to be involved in many pathways, notably transcription, translation, and signal transduction. Furthermore, starting from its subnuclear localization, the role of DYRK1A in pre-mRNA splicing has been investigated. First, Alvarez et al. observed that overexpression of DYRK1A interferes with the organization/assembly of nuclear speckles, as also recently described for AFF proteins, inducing a redistribution of SC35 in the nucleoplasm (Melko et al., 2011). Since a kinase inactive mutant and also DYRK1B, a paralog of DYRK1A, lack this activity (Alvarez et al., 2003), we can conclude that this effect is linked to specific substrates of DYRK1A. Some studies have highlighted that DYRK1A interacts with cyclin L2 and stimulates its phosphorylation when overexpressed. Cyclin L2 is a mobile element of nuclear speckles, thus DYRK1A might regulate the dynamics of nuclear speckles via its interaction with this cyclin (de Graaf et al., 2004). Furthermore, another substrate of DYRK1A is the SF3bI/SAP155 protein, a component of the U2 small nuclear ribonucleoprotein particle (snRNP) that is known to be phosphorylated during splicing catalysis but also during mitosis. Indeed, overexpression of DYRK1A (as well as DYRK1B) determines an enhanced phosphorylation of Thr434 in endogenous SF3bI in cultured cell lines (de Graaf et al., 2006). Taken together, all these data confirm the role of DYRK1A in splicing modulation, and they also seem to give indications concerning the effect of the overexpression of DYRK1A during brain development. In the same direction, some interesting results have been obtained by interrogating microarrays carrying 252 transcripts encoding for RNAs and proteins related to splicing by RNA obtained from cortex of TgDyrk1A mice. Indeed, the overexpression of DYRK1A resulted in altered levels of different components of the splicing machinery, such as helicases, snRNPs, hnRNPs, and SR proteins (Toiber et al., 2010). This finding strongly suggests a role of DYRK1A as a general regulator of splicing and allows understanding other results previously shown by other laboratories. Importantly, these abnormalities in splicing should affect the alternative splicing of isoforms of proteins whose function is important for brain development and function and whose altered level of expression could explain some DS phenotypes. For this reason, Toiber et al. tested the expression level of different isoforms of transcripts, whose relevance in DS is known. For instance, in DS fetal cerebral cortex the level of TRKBT.1 (RNA and its encoded protein) is reduced. This variant of TRKB is a dominant negative form, mediating inositol-1,4,5-trisphosphate-dependent calcium release and, when reduced, results in the increased activity of TRKB. These results were confirmed also in the adult brain of TgDyrk1A mice. Furthermore, the AChE-S isoform (involved in synaptic transmission and neurite growth) was downregulated in these mice, while the level of AChE-R isoform (involved in distress responses) was significantly increased. On the same path, two variants of Neuroligin were differently expressed in the brain of TgDyrk1A mice when compared with control mice. Taken together, these results show the impact of overexpression of DYRK1A on the unbalanced expression of alternatively spliced synaptic proteins that can influence the synaptic function in animal models and DS patients (Toiber et al., 2010). Other laboratories worked on the identification of the pre-mRNAs that are concerned by the over-function of DYRK1A. This aspect is very important since we can expect that this "cascade effect" may unravel the physiopathology of this disease. In this sense, two articles published by Liu and coworkers showing a role of DYRK1A in the regulation of alternative splicing of *Tau* open up avenues for new treatment of DS (Ittner and Gotz, 2011; Shi et al., 2008). Tau is a protein associated to microtubules and plays a role in their polymerization and stabilization. The function of tau has an important impact in the organization of the cytoskeleton in neurons and, in particular, in regulation of axonal transport. In neurons, the hyperphosphorylation and accumulation of Tau into neurofibrillary tangles (NFTs) was found to characterize some neurodegenerative disorders, called taupathies, among them Alzheimer disease (Ittner and Gotz, 2011). The *Tau* gene is subject to alternative splicing and, in particular, the inclusion of exon 10 (E10) determines two main group of tau isoforms, endowed with three (3R) or four (4R)-microtubule-binding repeats. Interestingly, it was suggested that maintaining an equal level of 3R- and 4R-tau is critical for normal neuronal physiology. In this context, the role of the alternative splicing machinery appears critical. Indeed, alternative splicing of *tau* E10 is regulated by splicing factor 2 (ASF), a member of the serine/arginine rich (SR) protein family, whose role is very important in splicing. ASF interaction with *tau* pre-mRNA is necessary to get the inclusion of E10 in the mature mRNA. DYRK1A directly phosphorylates ASF at the level of three serine residues (227, 234, and 238) and these modifications drive the protein into nuclear speckles, excluding ASF from the tau transcription loci and also prevent the inclusion of E10 in the tau mRNA (Shi et al., 2008; Fig. 4). A study by Weigel et al. has confirmed these data, showing that the overexpression of DYRK1A may alter both phosphorylation of tau and alternative splicing factor (ASF). These authors used two-dimensional electrophoresis to reveal modification of ASF phosphorylation in DS/AD and AD compared to controls (Wegiel et al., 2011).

Two more recent studies provided insight into the mechanism of action of DYRK1A, showing an implication of SC35 and 9G8, both targets of DYRK1A, in the modulation of tau E10 alternative splicing (Ding et al., 2011; Qian et al., 2011). Indeed, SC35 was coprecipitated with tau pre-mRNA and shown to bind to an ESE located on the E10 of tau pre-RNA, promoting the inclusion of this exon in the mature mRNA. DYRKIA has been reported to be able to phosphorylate SC35 *in vitro*. Indeed, overexpression of DYRK1A suppresses the ability of SC35 to act on the E10 ESE, resulting in a reduced inclusion of this exon in tau mRNA (Qian et al., 2011). On the same

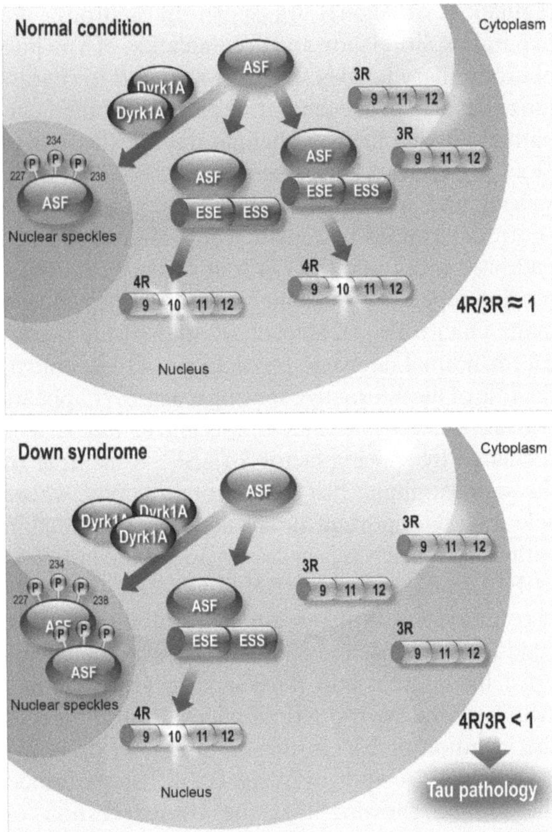

Fig. 4. Model for the role of DYRK1A in the regulation of the alternative splicing of tau exon 10 in normal and pathological conditions (Down syndrome). ASF binds an ESE sequence in the exon 10 of tau pre-mRNA promoting its inclusion in the mature tau mRNA corresponding to the 4R-tau. Overexpression of DYRK1A phosphorylates ASF inducing its retention in nuclear speckles promoting the exclusion of exon 10 from tau mRNA and promoting 3R-tau expression. In this condition, the ratio of 4R/3R tau isoforms is reduced and nonphysiological. (For color version of this figure, the reader is referred to the Web version of this chapter.)

track, the splicing factor 9G8 inhibits tau E10 inclusion, while its phosphorylation via DYRK1A promotes the inclusion of E10 in the mature tau mRNA (Ding et al., 2011). Thus, the overexpression of DYRK1A and the over phosphorylation of ASF causes an imbalance of 3R- and 4R-tau in DS brain. Interestingly enough, the presence of an Alzheimer-type neurofibrillary degeneration was described in the brain of DS patients. This pathology was attributed to the overexpression of the *APP* gene, that is located on the critical DS region on the HSA21, like *DYRK1A* (Moncaster et al., 2010), however, these recent findings implicate also Tau in the neurofibrillary degeneration observed in DS patients. Thus, we can propose that both forms of neurodegeneration can contribute to the initiation or to the acceleration of NFTs generation and, as anticipated, this would be important in the design of a treatment. Interestingly, some molecules have been identified that can modulate splicing selectively targeting DYRK1A/CLK1 kinases (Giraud et al., 2011; Rosenthal et al., 2011), opening new avenues for treatment of DS different from those already proposed (Gardiner et al., 2010 and also see Chapter 1).

CDKL5—Cyclin-dependent kinase-like-5

Mutations in cyclin-dependent kinase-like 5 (CDKL5) have been identified in patients with Rett syndrome and X-linked infantile spasm (OMIM 300203). Rett syndrome is a neurodevelopmental disorder that occurs almost exclusively in females and is characterized by arrested development between 6 and 18months of age, regression of acquired skills, loss of speech, stereotypical movements (classically of the hands), microcephaly, seizures, and intellectual disability. Early infantile epileptic encephalopathy is a severe form of epilepsy and is characterized by frequent tonic seizures or spasms beginning in infancy with a specific EEG finding of suppression-burst patterns, characterized by high-voltage bursts alternating with almost flat suppression phases (Ohtahara et al., 1976; Weaving et al., 2004).

CDKL5 is a large protein endowed with a serine/threonine domain that shares high homology with both mitogen-activated protein kinase (MAPK) and cyclin-dependent kinase (CDK) families (Tao et al., 2004). The important role of this protein in brain was predictable observing its expression that is neuron-specific in brain and very abundant in cortex and hippocampus but is not detectable in glia (Rusconi et al., 2008). The impact of the absence of CDKL5 depends on its targets and their role in brain development and function. Indeed, the inactivation of CDKL5 in rat brain by *in utero* electroporation results in delayed neuronal migration and dendritic arborization. In cortical neurons, it inhibits neurite growth and dendritic arborization, while its overexpression has an opposite effect (Chen et al., 2010). Concerning its role in RNA metabolism, CDKL5 has been reported to be localized in nuclear speckles and to influence splicing by having a role in the assembly/biogenesis of these nuclear domains, probably by preventing the recycling of some splicing factors that are its targets. Interestingly, abnormal morphology and number of nuclear speckles were observed in cell lines obtained by patients carrying a mutated *CDKL5* gene. It was also shown that CDKL5 affects the splicing of an E1A minigene, suggesting a general role of this protein in splicing modulation (Ricciardi et al., 2009). However, the possibility exists that specific pre-mRNAs are targeted by the functional absence of CDKL5 by a cascade effect as described for DYRK1A. Recently, the distribution of CDKL5 in postmitotic neurons was studied. It was shown that the endogenous protein is localized both in the nucleus and in the cytoplasm, even if a shuttling activity was not observed in these conditions. However, CDKL5 can rapidly diffuse from nucleus to cytoplasm through a CRM1-dependent nuclear export process, upon glutamate stimulation. The cytoplasmic localization of CDKL5 suggests also a potential role of this protein on cytoplasmic RNA metabolism that will be interesting to explore (Rusconi et al., 2011). In this sense, a transcriptome analysis of cell lines obtained from CDKL5 mutation patients revealed alteration in the mRNA level of 36 genes (16 upregulated and 20 downregulated). Among these genes, mitogen-activated protein kinase kinase kinase 5 (MAP3K5) seems a good candidate to explain mechanisms underpinning ID and epilepsy in these patients (Nectoux et al., 2011). MAP3K5, by activating MAP kinases, mediates signaling pathways involved in both differentiation and survival of neuronal cells. Indeed, MAP3K5-null mice exhibit impairment of long-term recognition memory, in addition to hyperactivity in a novel environment, and superior motor coordination (Kumakura et al., 2010). It would be interesting to determine the mechanism of action of CDKL5 on the regulation of the expression level of MAP3K5.

Summary

The section “Indirect role on RNA metabolism” of this review describes two proteins (DYRK1A and CDKL5) that seem to have a similar function and a similar mechanism of action on the pathophysiology of two IDs. However, both proteins are likely to have many substrates in neurons and thus it will be difficult to discriminate the effect produced by their action on RNA metabolism or on other pathways. Also in this case, further studies are needed and only a complete identification of mRNAs, whose splicing is altered in the two IDs will allow making some solid conclusions. Interestingly, some factors involved in splicing are localized also in dendrites (Racca et al., 2010), it is then possible that also these two kinases have a similar localization to modulate the activity of their substrates and thus to be component of RNA granules.

On the other hand, other neuronal proteins may have a similar subnuclear localization and function as DYRK1A and CDKL5 and cause ID when mutated. For instance, the AFF members *AFF3* and *AFF4*, the two paralogs of *FMR2*, could be good candidates genes, as discussed in this review.

Closing remarks

In conclusion, it is interesting to mention that other macromolecules involved in RNA metabolism are represented by ncRNAs, a class of

RNA involved in different cellular processes and very abundant in neurons (Kosik, 2006). They have been reported to play a role in neurogenesis and memory formation (Mercer et al., 2008; Shi et al., 2010), thus their potential implication in ID is obvious. To date, only the *FMR3* noncoding RNA is known to be involved in a form of ID (see the section "FMR2P: FRAXE mental retardation 2 protein" Gecz, 2000). It seems clear that due to new technologies, the exploration of ncRNA as cause of disease and, in particular, ID will be a future direction of research in the field.

Acknowledgments

We thank Marius Südol and Amjad Farooq for the kind gift of Fig. 3. The authors are indebted with Enzo Lalli for critical reading of the chapter and with Frederic Brau and Franck Aguila for help. This work was supported by INSERM, CNRS LIA NEOGENEX, ANR (Neuro16-06, Blanc06), by the Cure-FXS project, under the FP7 E-Rare programme (ANR 2009), Fondation pour la Recherche Médicale, Fondation Jérôme Lejeune.

References

Addington, A. M., Gauthier, J., Piton, A., Hamdan, F. F., Raymond, A., Gogtay, N., et al. (2011). A novel frameshift mutation in UPF3B identified in brothers affected with childhood onset schizophrenia and autism spectrum disorders. *Molecular Psychiatry*, *16*, 238–239.

Allemand, E., Batsché, E., & Muchardt, C. (2008). Splicing, transcription, and chromatin: A ménage à trois. *Current Opinion in Genetics & Development*, *18*, 145–151.

Altafaj, X., Dierssen, M., Baamonde, C., Martí, E., Visa, J., Guimerà, J., et al. (2001). Neurodevelopmental delay, motor abnormalities and cognitive deficits in transgenic mice overexpressing Dyrk1A (minibrain), a murine model of Down's syndrome. *Human Molecular Genetics*, *10*, 1915–1923.

Alvarez, M., Estivill, X., & de la Luna, S. (2003). DYRK1A accumulates in splicing speckles through a novel targeting signal and induces speckle disassembly. *Journal of Cell Science*, *116*, 3099–3107.

Aschrafi, A., Cunningham, B. A., Edelman, G. M., & Vanderlish, P. W. (2005). The fragile X mental retardation protein and group I metabotropic glutamate receptors regulate level of mRNA granules in brain. *Proceedings of the National Academy of Sciences of the United States of America*, *102*, 2180–2185.

Ballabio, A., Bardoni, B., Carrozzo, R., Andria, G., Bick, D., Campbell, L., et al. (1989). Contiguous gene syndromes due to deletions in the distal short arm of the human X chromosome. *Proceedings of the National Academy of Sciences of the United States of America*, *86*, 10001–10005.

Bardoni, B., Davidovic, L., Bensaid, M., & Khandjian, E. W. (2006). The fragile X syndrome: Exploring its molecular basis and seeking a treatment. *Expert Reviews in Molecular Medicine*, *8*, 1–16.

Bassell, G. J., & Warren, S. T. (2008). Fragile X syndrome: Loss of local mRNA regulation alters synaptic development and function. *Neuron*, *60*, 201–214.

Bechara, E. G., Didiot, M. C., Melko, M., Davidovic, L., Bensaid, M., Martin, P., et al. (2009). A novel function for fragile X mental retardation protein in translational activation. *PLoS Biology*, *7*, e16.

Becker, W., Weber, Y., Wetzel, K., Eirmbter, K., Tejedor, F. J., & Joost, H. G. (1998). Sequence characteristics, subcellular localization, and substrate specificity of DYRK-related kinases, a novel family of dual specificity protein kinases. *The Journal of Biological Chemistry*, *273*, 25893–25902.

Bensaid, M., Melko, M., Bechara, E. G., Davidovic, E. G., Berretta, A., Catania, M. V., et al. (2009). FRAXE-associated mental retardation protein (FMR2) is an RNA-binding protein with high affinity for G-quartet RNA forming structure. *Nucleic Acids Research*, *37*, 1269–1279.

Bork, P., & Sudol, M. (1994). The WW domain: A new signaling site in dystrophin? *Trends in Biochemical Sciences*, *19*, 531–533.

Bruno, I. G., Karam, R., Huang, L., Bhardwaj, A., Lou, C. H., Shum, E. Y., et al. (2011). Identification of a microRNA that activates gene expression by repressing nonsense-mediated RNA decay. *Molecular Cell*, *42*, 500–510.

Buchwald, G., Ebert, J., Basquin, C., Sauliere, J., Jayachandran, U., Bono, F., et al. (2010). Insights into the recruitment of the NMD machinery from the crystal structure of a core EJC-UPF3b complex. *Proceedings of the National Academy of Sciences of the United States of America*, *107*, 10050–10055.

Carson, M. (1991). Ribbons 2.0. *Journal of Applied Crystallography*, *24*, 958–961.

Cavallaro, S., Paratore, S., Fradale, F., de Vrij, F. M., Willemsen, R., & Oostra, B. A. (2008). Genes and pathways differentially expressed in the brains of Fxr2 knockout mice. *Neurobiology of Disease*, *32*, 510–520.

Chan, W. K., Bhalla, A. D., Le Hir, H., Nguyen, L. S., Huang, L., Gecz, J., et al. (2009). A UPF3-mediated regulatory switch that maintains RNA surveillance. *Nature Structural & Molecular Biology*, *16*, 747–753.

Chen, H. I., & Sudol, M. (1995). The WW domain of Yes-associated protein binds a novel proline-rich ligand that differs from the consensus established for SH3-binding modules. *Proceedings of the National Academy of Sciences of the United States of America*, *92*, 7819–7823.

Chen, Q., Zhu, Y. C., Yu, J., Miao, S., Zheng, J., Xu, L., et al. (2010). CDKL5, a protein associated with Rett syndrome, regulates neuronal morphogenesis via Rac1 signaling. *The Journal of Neuroscience*, *30*, 12777–12786.

Cougot, N., Bhattacharyya, S. N., Tapia-Arancibia, L., Bordonne, R., Filipowicz, W., Bertrand, E., et al. (2008). Dendrites of mammalian neurons contain specialized P-body-like structures that respond to neuronal activation. *The Journal of Neuroscience*, *28*, 13793–13804.

Darnell, J. C. Van, Driesche, S. J., Zhang, C., Hung, K. Y. S., Mele, A., Fraser, C. E., et al. (2011). FMRP stalls ribosomal translocation on mRNAs linked to synaptic function and autism. *Cell*, *146*, 247–261.

Darnell, J. C., Fraser, C. E., Mostovetsky, O., Stefani, G., Jones, T. A., Eddy, S. E., et al. (2005). Kissing complex RNAs mediate interaction between the fragile-X mental retardation protein KH2 domain and brain polyribosomes. *Genes & Development*, *19*, 903–918.

Davidovic, L., Jaglin, X. H., Lespagnol-Bestel, A.-M., Tremblay, S., Simonneau, M., Bardoni, B., et al. (2007). The fragile X mental retardation protein is a molecular adaptor between the neurospecific KIF3C kinesin and dendritic RNA granules. *Human Molecular Genetics*, *16*, 3047–3058.

Davidovic, L., Navratil, V., Bonaccorso, C. M., Catania, M. V., Bardoni, B., & Dumas, M. E. (2011). A metabolomic and systems biology perspective on the brain of the fragile X syndrome mouse model. *Genome Research*, *21*, 2190–2202.

Davidovic, L., Sacconi, S., Bechara, E. G., Delplace, S., Allegra, M., Desnuelle, C., et al. (2008). Alteration of expression of muscle specific isoforms of the fragile X related protein 1 (FXR1P) in facioscapulohumeral muscular dystrophy patients. *Journal of Medical Genetics*, *45*, 679–685.

de Diego-Otero, Y., Romero-Zerbo, Y., el Bekay, R., Decara, J., Sanchez, L., Rodriguez-de Fonseca, F., et al. (2009). Alpha-tocopherol protects against oxidative stress in the fragile X knockout mouse: An experimental therapeutic approach for the Fmr1 deficiency. *Neuropsychopharmacology*, *34*, 1011–1026.

de Graaf, K., Czajkowska, H., Rottmann, S., Packman, L. C., Lilischkis, R., Luscher, B., et al. (2006). The protein kinase DYRK1A phosphorylates the splicing factor SF3b1/SAP155 at Thr434, a novel in vivo phosphorylation site. *BMC Biochemistry*, *7*, 7.

de Graaf, K., Hekerman, P., Spelten, O., Herrmann, A., Packman, L. C., Bussow, K., et al. (2004). Characterization of cyclin L2, a novel cyclin with an arginine/serine-rich domain: Phosphorylation by DYRK1A and colocalization with splicing factors. *The Journal of Biological Chemistry*, *279*, 4612–4624.

Dickson, B. J., van der Straten, A., Dominguez, M., & Hafen, E. (1996). Mutations modulating Raf signaling in *Drosophila* eye development. *Genetics*, *142*, 163–171.

Dictenberg, J. B., Swanger, S. A., Antar, L. N., Singer, R. H., & Bassell, G. J. (2008). A direct role for FMRP in activity-dependent dendritic mRNA transport links filopodial-spine morphogenesis to fragile X syndrome. *Developmental Cell*, *14*, 926–939.

Didiot, M.-C., Subramanian, M., Flatter, E., Mandel, J. L., & Moine, H. (2009). Cells lacking the fragile X mental retardation protein (FMRP) have normal RISC activity but exhibit altered stress granle assembly. *Molecular Biology of the Cell*, *19*, 428–437.

Ding, S., Shi, J., Qian, W., Iqbal, K., Grundke-Iqbal, I., Gong, C. X., et al. (2011). Regulation of alternative sploicing of tau exon 10 by 9G8 and Dyrk1A. *Neurobiology of Aging*, 2011 Jan 5. [Epub ahead of print].

Dolzhanskaya, N., Sung, Y. J., Conti, J., Currie, J. R., & Denman, R. B. (2003). The fragile X mental retardation protein interacts with U-rich RNAs in a yeast three-hybrid system. *Biochemical and Biophysical Research Communications*, *305*, 434–441.

Edbauer, D., Neilson, J. R., Foster, K. A., Wang, C. F., Seeburg, D. P., Batterton, M. N., et al. (2010). Regulation of synaptic structure and function by FMRP-associated microRNAs miR-125b and miR-132. *Neuron*, *65*, 373–384.

el Bekay, R., Romero-Zerbo, Y., Decara, J., Sanchez-Salido, L., Del Arco-Herrera, I., Rodríguez-de Fonseca, F., et al. (2007). Enhanced markers of oxidative stress, altered antioxidants and NADPH-oxidase activation in brains from Fragile X mental retardation 1-deficient mice, a pathological model for Fragile X syndrome. *The European Journal of Neuroscience*, *26*, 3169–3180.

Fotaki, V., Dierssen, M., Alcantara, S., Martinez, S., Marti, E., Casas, C., et al. (2002). Dyrk1A haploinsufficiency affects viability and causes developmental delay and abnormal brain morphology in mice. *Molecular and Cellular Biology*, *22*, 6636–6647.

Fukami, M., Kirsch, S., Schiller, S., Richter, A., Benes, V., Franco, B., et al. (2000). A member of a gene family on Xp22.3, VCX-A, is deleted in patients with X-linked non-specific mental retardation. *American Journal of Human Genetics*, *67*, 563–573.

Gardiner, K., Herault, Y., Lott, I., Antonarakis, S. E., Reeves, R. H., & Dierssen, M. (2010). Down syndrome: From understanding the neurobiology of treatment. *The Journal of Neuroscience*, *30*, 14943–14945.

Gecz, J. (2000). FMR3 is a novel gene associated with FRAXE CpG island and transcriptionally silent in FRAXE full mutations. *Journal of Medical Genetics*, *37*, 782–784.

Gecz, J., Bielby, S., Sutherland, G. R., & Mulley, J. C. (1997a). Gene structure and subcellular localization of FMR2, a member of a new family of putative transcription activators. *Genomics*, *44*, 201–213.

Gecz, J., Gedeon, A. K., Sutherland, G. R., & Mulley, J. C. (1996). Identification of the gene FMR2, associated with FRAXE mental retardation. *Nature Genetics*, *13*, 105–108.

Gecz, J., Oostra, B. A., Hockey, A., Carbonell, P., Turner, G., Haan, E. A., et al. (1997b). FMR2 expression in families with FRAXE mental retardation. *Human Molecular Genetics*, *6*, 435–441.

Gecz, J., Shoubridge, C., & Corbett, M. (2009). The genetics landscape of intellectual disability arising from chromosome X. *Trends in Genetics*, *25*, 308–316.

Gehring, N. H., Neu-Yilik, G., Schell, T., Hentze, M. W., & Kulozik, A. E. (2003). Y14 and hUpf3b form an NMD-activating complex. *Molecular Cell*, *11*, 939–949.

Giraud, F., Alves, G., Debiton, E., Nauton, L., Thèry, V., Durieu, E., et al. (2011). Synthesis, protein kinase inhibitory potencies, and in vitro antiproliferative activities of meridianin derivatives. *Journal of Medicinal Chemistry*, *54*, 4479–4489.

Gockler, N., Jofre, G., Papadopoulos, C., Soppa, U., Tejedor, F. J., & Becker, W. (2009). Harmine specifically inhibits protein kinase DYRK1A and interferes with neurite formation. *The FEBS Journal*, *276*, 6324–6337.

Gross, C., Yao, X., Pong, D. L., Jeromin, A., & Bassell, G. J. (2011). Fragile X mental retardation protein regulates protein expression and mRNA translation of the potassium channel Kv4.2. *The Journal of Neuroscience*, *31*, 5693–5698.

Gu, Y., McIlwain, K. L., Weeber, E. J., Yamagata, T., Xu, B., Antalffy, B. A., Reyes, C., Yuva-Paylor, L., Armstrong, D., Zoghbi, H., Sweatt, J. D., Paylor, R., & Nelson, D. L. (2002). Impaired conditioned fear and enhanced long-term potentiation in *Fmr2* knock-out mice. *The Journal of Neuroscience*, *22*, 2753–2763.

Gu, Y., Shen, Y., Gibbs, R. A., & Nelson, D. L. (1996). Identification of FMR2, a novel gene associated with FRAXE CCG repeat and CpG island. *Nature Genetics*, *13*, 109–113.

Hagerman, R., Hoem, G., & Hagerman, P. (2010). Fragile X and autism: Intertwined at the molecular level leading to targeted treatments. *Molecular Autism*, *1*, 12.

Herrmann, C. H., & Mancini, M. A. (2001). The Cdk9 and cyclin T subunits of TAK/P-TEFb localize to splicing factor-rich nuclear speckle regions. *Journal of Cell Science*, *114*, 1491–1503.

Hoeft, F., Carter, J. C., Lightbody, A. A., Cody Hazlett, H., Piven, J., & Reiss, A. L. (2010). Region-specific alterations in brain development in one- to three-year-old boys with fragile X syndrome. *Proceedings of the National Academy of Sciences of the United States of America*, *107*, 9335–9339.

Hosomi, N., Oiso, N., Fukai, K., Hanada, K., Fujita, H., & Ishii, M. (2007). Deletion of distal promoter of VCXA in a patient with X-linked ichthyosis associated with borderline mental retardation. *Journal of Dermatological Science*, *45*, 31–36.

Iacoangeli, A., Rozhdestvensky, T. S., Dolzhanskaya, N., Tournier, B., Schutt, J., Brosius, J., et al. (2008). On BC1 RNA and the fragile X mental retardation protein. *Proceedings of the National Academy of Sciences of the United States of America*, *105*, 734–739.

Ito, H., Yoshimura, N., Kurosawa, M., Ishii, S., Nukina, N., & Okazawa, H. (2009). Knock-down of PQBP1 impairs anxiety-related cognition in mouse. *Human Molecular Genetics*, *18*, 4239–4254.

Ittner, L. M., & Gotz, J. (2011). Amyloid-beta and tau—A toxic pas de deux in Alzheimer's disease. *Nature Reviews. Neuroscience*, *12*, 65–72.

Jiao, X., Chen, H., Chen, J., Herrup, K., Firestein, B. L., & Kiledjian, M. (2009). Modulation of neuritogenesis by a protein implicated in X-linked mental retardation. *The Journal of Neuroscience*, *29*, 12419–12427.

Jiao, X., Wang, Z., & Kiledjian, M. (2006). Identification of an mRNA-decapping regulator implicated in X-linked mental retardation. *Molecular Cell*, *24*, 713–722.

Kalscheuer, V. M., Freude, K., Musante, L., Jensen, L. R., Yntema, H. G., Gecz, J., et al. (2003). Mutations in the polyglutamine binding protein 1 gene cause X-linked mental retardation. *Nature Genetics*, *35*, 313–315.

Kalscheuer, V. M., Hu, H., Haas, S. A., Chelly, J., Van Esch, H., Raynaud, M., et al. (2011). Next-generation sequencing in 248 families with X-linked intellectual disability. In: *Presented at the 12th International Congress of Human Genetics/61st Annual Meeting of The American Society of Human Genetics, October 13, 2011, Montreal, Canada.*

Kato, Y., Nagata, K., Takahashi, M., Lian, L., Herrero, H., Sudol, M., et al. (2004). Common mechanism of ligand recognition by Group-II/III WW domains—Redefining their functional classification. *The Journal of Biological Chemistry*, *279*, 318.

Knight, S. J. L., Flannery, A. V., Hist, M. C., Campbell, L., Christodouluo, Z., Phelps, S. R., et al. (1993). Trinucleotide repeat amplification and hypermethylation of a CpG island in FRAXE mental retardation. *Cell*, *74*, 127–134.

Kosik, K. (2006). The neuronal microRNA system. *Nature Reviews. Neuroscience*, *7*, 911–920.

Krichevsky, A. M., & Kosik, K. S. (2001). Neuronal RNA granules: A link between RNA localization and stimulation-dependent translation. *Neuron*, *32*, 683–696.

Kumakura, K., Nomura, H., Toyoda, T., Hashikawa, K., Noguchi, T., Takeda, K., et al. (2010). Hyperactivity in novel environment with increased dopamine and impaired novelty preference in apoptosis signal-regulating kinase 1 (ASK1)-deficient mice. *Neuroscience Research*, *66*, 313–320.

Kunde, S. A., Musante, L., Grimme, A., Fischer, U., Müller, E., Wanker, E. E., et al. (2011). The X-chromosome-linked intellectual disability protein PQBP1 is a component of neuronal RNA granules and regulates the appearance of stress granules. *Human Molecular Genetics*, *20*, 4916–4931.

Laumonnier, F., Shoubridge, C., Antar, C., Nguyen, L. S., Van Esch, H., Kleefstra, T., et al. (2010). Mutations of the UPF3B gene, which encodes a protein widely expressed in neurons, are associated with nonspecific mental retardation with or without autism. *Molecular Psychiatry*, *15*, 767–776.

Li, Z., Zhang, Y., Ku, L., Wilkinson, K. D., Warren, S. T., & Feng, Y. (2001). The fragile X mental retardation protein inhibits translation via interacting with mRNA. *Nucleic Acids Research*, *29*, 2276–2283.

Llorian, M., Beullens, M., Lesage, B., Nicolaescu, E., Beke, L., Landuyt, W., et al. (2005). Nucleocytoplasmic shuttling of the splicing factor SIPP1. *The Journal of Biological Chemistry*, *280*, 38862–38869.

Long, A. A., Mahapatra, C. T., Woodruff, E. A., 3rd, Rohrbough, J., Leung, H. T., Shino, S., et al. (2010). The nonsense-mediated decay pathway maintains synapse architecture and synaptic vesicle cycle efficacy. *Journal of Cell Science*, *123*, 3303–3315.

Lu, R., wang, H., Liang, Z., O'Donnell, W. T., Warren, S. T., & Feng, Y. (2004). The fragile X protein controls microtubule-associated protein 1B translation and microtubule stability in brain neuron development. *Proceedings of the National Academy of Sciences of the United States of America*, *101*, 15201–15206.

Lubs, H., Abidi, F. E., Echeverri, R., Holloway, L., Meindl, A., Stevenson, R. E., et al. (2006). Golabi-Ito-Hall syndrome results from a missense mutation in the WW domain of the PQBP1 gene. *Journal of Medical Genetics*, *43*, e30.

Lukong, K. E., Chang, K. W., Khandjian, E. W., & Richard, S. (2008). RNA-binding proteins in human genetic disease. *Trends in Genetics*, *24*, 416–425.

Martin, K. C., & Ephrussi, A. (2009). mRNA localization: Gene expression in the spatial dimension. *Cell*, *136*, 719–730.

Martínez de Lagrán, M., Altafaj, X., Gallego, X., Martí, E., Estivill, X., Sahún, I., et al. (2004). Motor phenotypic alterations in TgDyrk1a transgenic mice implicate DYRK1A in Down syndrome motor dysfunction. *Neurobiology of Disease*, *15*, 132–142.

Marti-Renom, M. A., Stuart, A. C., Fiser, A., Sanchez, R., Melo, F., & Sali, A. (2000). Comparative protein structure modeling of genes and genomes. *Annual Review of Biophysics and Biomolecular Structure*, *29*, 291–325.

Mazroui, R., Huot, M. E., Tremblay, S., Boilard, N., Labelle, Y., & Khandjian, E. W. (2003). Fragile X mental retardation protein determinants required for its association with polyribosomal mRNPs. *Human Molecular Genetics*, *12*, 3087–3096.

Melko, M., & Bardoni, B. (2010). The role of G-quadruplex in RNA metabolism: Involvement of FMRP and FMR2P. *Biochimie*, *92*, 919–926.

Melko, M., Douguet, D., Bensaid, M., Zongaro, S., Verheggen, C., Gecz, J., et al. (2011). Functional characterization of the AFF (AF4/FMR2) family of RNA-binding proteins: Insights into the molecular pathology of FRAXE intellectual disability. *Human Molecular Genetics*, *20*, 1873–1885.

Mercer, T. R., Dinger, M. E., Mariani, J., Kosik, K. S., Mehler, M. F., & Mattick, J. S. (2008). Noncoding RNAs in long-term memory formation. *The Neuroscientist*, *14*, 434–445.

Mientjes, E. J., Willemsen, R., Kirkpatrick, L. L., Nieuwenhuizen, I. M., Hoogeveen-Westerveld, M., Verweij, M., et al. (2004). Fxr1 knockout mice show a striated muscle phenotype: Implications for Fxr1p function in vivo. *Human Molecular Genetics*, *13*, 1291–1302.

Moncaster, J. A., Pineda, R., Moir, R. D., Lu, S., Burton, M. A., Ghosh, J. G., et al. (2010). Alzheimer's disease amyloid-beta links lens and brain pathology in Down syndrome. *PLoS One*, *5*, e10659.

Muddashetty, R. S., Nalavadi, V. C., Gross, C., Yao, X., Xing, L., Laur, O., et al. (2011). Reversible inhibition of PSD-95 mRNA translation by miR-125a, FMRP phosphorylation, and mGluR signaling. *Molecular Cell*, *42*, 673–688.

Najmabadi, H., Hu, H., Garshasbi, M., Zemojtel, T., Abedini, S. S., Chen, W., et al. (2011). Deep sequencing reveals 50 novel genes for recessive cognitive disorders. *Nature*, *478*, 57–63.

Nectoux, J., Fichou, Y., Cagnard, N., Bahi-Buisson, N., Nusbaum, P., Letourneur, F., et al. (2011). Transcriptome profiling of UPF3B/NMD-deficient lymphoblastoid cells from patients with various forms of intellectual disability. *Molecular Psychiatry*, *89*, 193–202.

Nguyen, L. S., Jolly, L., Shoubridge, G., Chan, W. C., Huang, L., Laumonnier, F., et al. (2011). Transcriptome profiling of UPF3B/NMD-deficient lymphoblastoid cells from patients with various forms of intellectual disability. *Molecular Psychiatry*, doi:10.1038/mp. 2011.163 [Epub ahead of print].

Ohtahara, S., Yamatogi, Y., & Ohtsuka, Y. (1976). Prognosis of the Lennox syndrome-long-term clinical and electroencephalographic follow-up study, especially with special reference to relationship with the West syndrome. *Folia Psychiatrica et Neurologica Japonica*, *30*, 275–287.

Okazawa, H., Rich, T., Chang, A., Lin, X., Waragai, M., Kajikawa, M., et al. (2002). Interaction between mutant ataxin-1 and PQBP-1 affects transcription and cell death. *Neuron*, *34*, 701–713.

Okuda, T., Hattori, H., Takeuchi, S., Shimizu, J., Ueda, H., Palvimo, J. J., et al. (2003). PQBP-1 transgenic mice show a late-onset motor neuron disease-like phenotype. *Human Molecular Genetics*, *12*, 711–725.

Ortiz-Abalia, J., Sahun, I., Altafaj, X., Estivill, X., Dierssen, M., & Fillat, C. (2008). Targeting Dyrk1A with AAVshRNA attenuates motor alterations in TgDyrk1A, a mouse model of Down syndrome. *American Journal of Human Genetics*, *83*, 479–488.

Qian, W., Liang, H., Shi, J., Jin, N., Grundke-Iqbal, I., Iqbal, K., Gong, C. X., & Liu, F. (2011). Regulation of the alternative splicing of tau exon 10 by SC35 and Dyrk1A. *Nucleic Acids Research*, *39*, 6161–6171.

Racca, C., Gardiol, A., Eom, T., Ule, J., Triller, A., & Darnell, R. B. (2010). The neuronal splicing factor nova co-localizes with target RNAs in the dendrite. *Frontiers in Neural Circuits*, *4*, 5.

Ramakers, G. J. (2002). Rho proteins, mental retardation and the cellular basis of cognition. *Trends in Neurosciences*, *25*, 191–199.

Ricciardi, S., Kilstrup-Nielsen, C., Bienvenu, T., Jacquette, A., Landsberger, N., & Broccoli, V. (2009). CDKL5 influences RNA splicing activity by its association to the nuclear speckle molecular machinery. *Human Molecular Genetics*, *18*, 4590–4602.

Romero-Zerbo, Y., Decara, J., el Bekay, R., Sanchez-Salido, L., Del Arco-Herrera, I., de Fonseca, F. R., et al. (2009). Protective effects of melatonin against oxidative stress in Fmr1 knockout mice: A therapeutic research model for the fragile X syndrome. *Journal of Pineal Research*, *46*, 224–234.

Rosenthal, A. S., Tanega, C., Shen, M., Mott, B. T., Bougie, J. M., Nguyen, D. T., et al. (2011). An inhibitor of the Cdc2-like kinase 4 (Clk4). In: *Probe Reports from the NIH Molecular Libraries Program [Internet]*, Bethesda, MD: National Center for Biotechnology Information (US) 2010-2010 Mar 29 [updated 2011 Mar 3].

Rusconi, L., Kilstrup-Nielsen, C., & Landsberger, N. (2011). Extrasynaptic N-Methyl-D-aspartate (NMDA) receptor stimulation induces cytoplasmic translocation of CDKL5 kinase and its proteasomal degradation. *The Journal of Biological Chemistry*, *286*, 36550–36558.

Rusconi, L., Salvatoni, L., Giudici, L., Bertani, I., Kilstrup-Nielsen, C., Broccoli, V., et al. (2008). CDKL5 expression is modulated during neuronal development and its subcellular distribution is tightly regulated by the C-terminal tail. *The Journal of Biological Chemistry*, *283*, 30101–30111.

Schaeffer, C., Bardoni, B., Mandel, J.-L., Ehresmann, B., Ehresmann, C., & Moine, H. (2001). The fragile X mental retardation protein binds specifically to its mRNA via a purine quartet motif. *The EMBO Journal*, *20*, 4803–4813.

Schütt, J., Falley, K., Richter, D., Kreienkampt, H.-J., & Kindler, S. (2009). Fragile X mental retardation protein regulates the levels of scaffold proteins and glutamte receptors in postsynaptic densities. *The Journal of Biological Chemistry*, *284*, 25479–25487.

Schwanhausser, B., Busse, D., Li, N., Dittmar, G., Schuchhardt, J., Wolf, J., et al. (2011). Global quantification of mammalian gene expression control. *Nature*, *473*, 337–342.

Shi, J., Zhang, T., Zhou, C., Chohan, M. O., Gu, X., Wegiel, J., et al. (2008). Increased dosage of Dyrk1A alters alternative splicing factor (ASF)-regulated alternative splicing of tau in Down syndrome. *The Journal of Biological Chemistry*, *283*, 28660–28669.

Shi, Y., Zhao, X., Hsieh, J., Wichterle, H., Impey, S., Banerjee, S., et al. (2010). MicroRNA regulation of neural stem cells and neurogenesis. *The Journal of Neuroscience*, *30*, 14931–14936.

Sittler, A., Devys, D., Weber, C., & Mandel, J. L. (1996). Alternative splicing of exon 14 determines nuclear or cytoplasmic localisation of fmr1 protein isoforms. *Human Molecular Genetics*, *5*, 95–102.

Stevenson, R. E., Bennett, C. W., Abidi, F., Kleefstra, T., Porteous, M., Simensen, R. J., et al. (2005). Renpenning syndrome comes into focus. *American Journal of Medical Genetics*, *134*, 415–421.

Subramanian, M., Rage, F., Tabet, R., Flatter, E., Mandel, J. L., & Moine, H. (2011). G-quadruplex RNA structure as a signal for neurite mRNA targeting. *EMBO Reports*, *12*, 697–704.

Südol, M., & Harvey, K. (2010). The modularity in the Hippo signaling pathway. *Trends in Biochemical Sciences*, *35*, 627–633.

Swanger, S. A., & Bassell, G. J. (2011). Making and breaking synapses through local mRNA regulation. *Current Opinion in Genetics & Development*, *214*, 414–421.

Takahashi, K., Yoshina, S., Masashi, M., Ito, W., Inoue, T., Shiwaku, H., et al. (2009). Nematode homologue of PQBP1, a mental retardation causative gene, is involved in lipid metabolism. *PLoS One*, *4*, e4104.

Tamura, T., Horiuchi, D., Chen, Y. C., Sone, M., Miyashita, T., Saitoe, M., et al. (2010). Drosophila PQBP1 regulates learning acquisition at projection neurons in aversive olfactory conditioning. *The Journal of Neuroscience*, *30*, 14091–14101.

Tang, A. H., Neufeld, T. P., Rubin, G. M., & Müller, H.-A. J. (2001). Transcriptional regulation of cytoskeletal functions and segmentation by a novel maternal pair-rule gene, lilliputian. *Development*, *128*, 801–813.

Tao, J., Van Esch, H., Hagedorn-Greiwe, M., Hoffmann, K., Moser, B., Raynaud, M., et al. (2004). Mutations in the X-linked cyclin-dependent kinase-like 5 (CDKL5/STK9) gene are associated with severe neurodevelopmental retardation. *American Journal of Human Genetics*, *75*, 1149–1154.

Tapia, V. E., Nicolaescu, E., McDonald, C. B., Musi, V., Oka, T., Inayoshi, Y., et al. (2010). Y65C missense mutation in the WW domain of the globali-ito-hall syndrome protein PQBP1 affects its binding activity and deregulates pre-mRNA splicing. *The Journal of Biological Chemistry*, *285*, 19391–19401.

Tarpey, P. S., Raymond, F. L., Nguyen, L. S., Rodriguez, J., Hackett, A., Vandeleur, L., et al. (2007). Mutations in UPF3B, a member of the nonsense-mediated mRNA decay complex, cause syndromic and nonsyndromic mental retardation. *Nature Genetics*, *39*, 1127–1133.

Tejedor, F., Zhu, X. R., Kaltenbach, E., Ackermann, A., Baumann, A., Canal, I., et al. (1995). minibrain: A new protein kinase family involved in postembryonic neurogenesis in Drosophila. *Neuron*, *14*, 287–301.

Terribilini, M., Lee, J. H., Yan, C., Jernigan, R. L., Honovar, V., & Dobbs, D. (2006). Prediction of RNA binding sites in proteins from amino acids sequence. *RNA*, *12*, 1450–1462.

Toiber, D., Garikoitz, A., Ben-Ari, S., Toran, N., Soreq, H., & Dierssen, M. (2010). Engineerinh *DIRK1A* overdosage yields Down syndrome-characteristic cortical splicing aberrations. *Neurobiology of Disease*, *40*, 348–359.

Turner, G., Gedeon, A., Kerr, B., Bennett, R., Mulley, J., & Partington, M. (2003). Syndromic form of X-linked mental retardation with marked hypotonia in early life, severe mental handicap, and difficult adult behavior maps to Xp22. *American Journal of Medical Genetics. Part A*, *117A*, 245–250.

Ule, J. (2008). Ribonucleoprotein complexes in neurological diseases. *Current Opinion in Neurobiology*, *18*, 516–523.

Urano, A., Endoh, M., Wada, T., Morikawa, Y., Itoh, M., Kataoka, Y., et al. (2005). Infertility with defective spermiogenesis in mice lacking AF5q31, the target of chromosomal translocation in human infant leukemia. *Molecular and Cellular Biology*, *25*, 6834–6845.

Van Esch, H., Hollanders, K., Badisco, L., Melotte, C., Van Hummelen, P., Vermeesch, J. R., et al. (2005). Deletion of VCX-A due to NAHR plays a major role in the occurrence of mental retardation in patients with X-linked ichthyosis. *Human Molecular Genetics*, *14*, 1795–1803.

Vanderzwan-Butler, C. J., Prazak, L. M., & Gergen, J. P. (2007). The HMG-box protein Lilliputian is required for Runt-dependent activation of the pair-rule gene fushi-tarazu. *Developmental Biology*, *301*, 350–360.

Van't Padje, S., Chaudhry, B., Severijnen, L. A., van der Linde, H. C., Mientjes, E. J., Oostra, B. A., et al. (2009). Reduction in fragile X related 1 protein causes cardiomyopathy and muscular dystrophy in zebrafish. *The Journal of Experimental Biology*, *212*, 2564–2570.

Vasudevan, S., & Steitz, J. A. (2007). AU-rich-element-mediated upregulation of translation by FXR1 and Argonaute 2. *Cell*, *128*, 1105–1118.

Wang, Q., Gu, Y., Ferguson, J. M., Chen, Q., Boatwright, S., Gardiner, J., et al. (2003). Cytogenetic analysis of obsessive–compulsive disorder (OCD): Identification of a FRAXE fragile site. *American Journal of Medical Genetics. Part A*, *118A*, 25–28.

Waragai, M., Junn, E., Kajikawa, M., Takeuchi, S., Kanazawa, I., Shibata, M., et al. (2000). PQBP-1/Npw38, a nuclear protein binding to the polyglutamine tract, interacts with U5-15kD/dim1p via the carboxyl-terminal domain. *Biochemical and Biophysical Research Communications*, *273*, 592–595.

Weaving, L. S., Christodoulou, J., Williamson, S. L., Friend, K. L., McKenzie, O. L., Archer, H., et al. (2004). Mutations of CDKL5 cause a severe neurodevelopmental disorder with infantile spasms and mental retardation. *American Journal of Human Genetics*, *75*, 1079–1093.

Wegiel, J., Kaczmarski, W., Barua, M., Kuchna, I., Nowicki, K., Wang, K. C., et al. (2011). Link between DYRK1A overexpression and several-fold enhancement of neurofibrillary degeneration with 3-repeat tau protein Down syndrome. *Journal of Neuropathology and Experimental Neurology*, *70*, 3–50.

Whitman, S. A., Cover, C., Yu, L., Nelson, D. L., Zarnescu, D. C., & Gregorio, C. C. (2011). Desmoplakin and Talin2 are novel mRNA targets of fragile X-related protein-1 in cardiac muscle whitman; Dsp and Tln2 are mRNA targets of FXR1 in the heart. *Circulation Research*, *109*, 262–271.

Wittwer, F., van der Straten, A., Keleman, K., Dickson, B. J., & Hafen, E. (2001). Lilliputian: an AF4/FMR2-related protein that controls cell identity. *Development*, *128*, 791–800.

Yoshimura, N., Horiuchi, D., Shibata, M., Saitoe, M., Qi, M. L., & Okazawa, H. (2006). Expression of human PQBP-1 in Drosophila impairs long-term memory and induces abnormal courtship. *FEBS Letters*, *580*, 2335–2340.

Zhang, J., Fang, Z., Jud, C., Vansteensel, M. J., Kaasik, K., Lee, C. C., et al. (2008). Fragile X-related proteins regulate mammalian circadian behavioral rhythms. *American Journal of Human Genetics*, *83*, 43–52.

Zhang, Y. Z., Lindblom, T., Carnahan, R., Chang, A., Gould, K., Sudol, M., et al. (2000). Evidence that dim1 associates with proteins involved in pre-mRNA splicing, and delineation of residues essential for dim1 interactions with hnRNP F and Npw38/PQBP-1. *Gene*, *257*(1), 33–43.

Zhong, J., Chuang, S. C., Bianchi, R., Zhao, W., Paul, G., Thakkar, P., et al. (2010). Regulatory BC1 RNA and the fragile X mental retardation protein: Convergent functionality in brain. *PLoS One*, *5*, e15509.

Zou, S. W., Zhang, J. C., Zhang, X. D., Miao, S. Y., Zong, S. D., Sheng, Q., et al. (2003). Expression and localization of VCX/Y proteins and their possible involvement in regulation of ribosome assembly during spermatogenesis. *Cell Research*, *13*, 171–177.

M. Dierssen and R. de la Torre (Eds.)
Progress in Brain Research, Vol. 197
ISSN: 0079-6123

CHAPTER 4

Aberrant epigenetic landscape in intellectual disability

J. V. Sanchez-Mut, D. Huertas and M. Esteller*

Cancer Epigenetics and Biology Program (PEBC), Bellvitge Biomedical Research Institute, L'Hospitalet, Barcelona, Catalonia, Spain

Abstract: In recent decades, epigenetics has emerged as a broad-ranging regulatory layer that modulates the whole genome and transcriptome. It largely determines the firing of transcription start sites, the splicing processes, and the binding of transcription factors, among many other processes. Its wide spectrum of action has provided us with the keys to new doors to investigate many diseases, including intellectual disability syndromes.

The involvement of epigenetic factors in Rett syndrome is already well established, and its involvement in alpha-thalassemia/mental retardation-X-linked and Rubinstein–Taybi syndromes is also being elucidated. Down syndrome is not an exception, and the most recent reports suggest that epigenetic factors may play a crucial role in its etiology and also have the potential to provide new panels of biomarkers and tailored treatments.

Keywords: epigenetics; brain function; Down syndrome; DNA methylation; histone modification.

General introduction to epigenetics

Epigenetics (from the Greek *epi-*: "over, above"; *genetikos*: "genitive, genesis, origin") is a discipline devoted to the study of heritable traits resulting from changes in a chromosome that do not alter the DNA sequence (Berger et al., 2009). It is generally understood as the way by which the environment modifies the phenotype without changing the genotype and, likewise, is influenced by diet (Chan and Shea, 2006), cigarette smoking, social behavior, lifestyle, and aging (Fraga et al., 2005).

Epigenetics plays an important part in genomic regulation. It controls the whole genome and transcriptome by modifying the chromatin structure. The idea that the chromatin is merely the scaffold of the DNA sequence has been abandoned. In fact, it has come to be realized that it is a crucial component of genome regulation. Through two main molecular mechanisms—the direct methylation of DNA (DNAm) and the modification of the histones, the proteins that package DNA—epigenetics controls the binding

*Corresponding author.
Tel.: +34-93-2607253; Fax: +34-93-2607219
E-mail: mesteller@iconcologia.net

DOI: 10.1016/B978-0-444-54299-1.00004-2

of transcription factors (Choy et al., 2010), the firing of transcription start sites (TSS) (Elango and Yi, 2008), and the splicing process (Choi, 2010). Epigenetics is closely regulated in development (Palini et al., 2011) and is responsible for the tissue-specific gene expression patterns, the silencing of transposable elements, gene imprinting, and X chromosome inactivation.

DNA methylation

In somatic cells, DNAm consists of the addition of a methyl group at cytosines followed by guanines (CpG dinucleotides), which are underrepresented in the genome and tend to concentrate at specific clusters, the so-called CpG islands (CGIs). Despite this, around 80% of all CpGs are distributed throughout the genome. DNAm has been considered a repressive mark, since hypermethylation of promoter CGIs was associated with tissue-specific gene silencing and allele-specific inactivation of the X chromosome (Straussman et al., 2009). Nevertheless, ongoing investigations are revealing that reality is much more complex. In general, CGIs are predominantly unmethylated and non-CGI CpG sites are methylated. Very few CGI promoters are methylated in normal cells, and their hypermethylation is more related to tumorigenesis than to normal development. Noticeably, around 70% of annotated genes, including almost all housekeeping genes and a fraction of tissue-specific and regulatory genes, are associated with CGIs in their promoter regions (Deaton and Bird, 2011), reinforcing the importance of CGIs in gene regulation. Consequently, the role of CGI in promoter regions cannot be associated only with its regulation by DNAm. The representation of the CpG content of the promoters is bimodally distributed, suggesting that the division in CGI-containing promoters and non-CGI-containing promoters is not random (Sharif et al., 2010). CGI promoters are usually associated with widely and strongly expressed genes with a low evolutionary rate and relaxed TSS firing. Conversely, non-CGI-promoters are mainly associated with weaker and more restricted gene expression and TSS firing, exhibiting higher evolutionary rates (Tang and Epstein, 2007). In keeping with this, CGI promoters tend to lack TATA box elements and are enriched in SP1, NRF1, E2F, and ETS transcription binding motifs (Deaton and Bird, 2011).

Chromatin structure determines the accessibility to DNA and nucleosomal occupancy can constitute a physical barrier for polymerase and transcription factors. Likewise, a low nucleosomal occupancy of TSS is associated with active gene expression, whereas its occlusion is associated with gene repression (Schones et al., 2008). In this context, CGIs represent an unstable platform for nucleosomal occupancy, providing an accessible space for transcription machinery and suggesting a reasonable explanation for its importance in CGI-promoter regulation (Ramirez-Carrozzi et al., 2009).

DNAm cannot be considered independently of its localization (Fig. 1a). DNAm at TSS is associated with gene silencing. DNAm on transcription factor binding motifs is associated with its accessibility (Choy et al., 2010). When occurring on gene bodies, it is associated with gene activity (Zemach et al., 2010) and, specifically on exons, with exon inclusion in the final spliced transcript (Choi, 2010). Consequently, nucleosomal positioning on exons determines its recognition, supporting the idea that DNAm and nucleosomal positioning are associated (Schwartz et al., 2009; Tilgner et al., 2009).

DNAm is carried out by the DNA methyltransferases (DNMTs) DNMT1, DNMT3A, and DNMT3B. DNMT1 is required for maintaining DNAm patterns during semiconservative DNA replication, whereas DNTM3A and DNMT3B are responsible for the establishment of *de novo* DNAm profiles during development. Interestingly, there is a tight correlation between nucleosome positioning and DNAm (Chodavarapu et al., 2010), and the fact that DNMT3A/B are strongly anchored to nucleosomes (Jeong et al., 2009) links DNAm and nucleosomal positioning. DNMT depletion causes general hypomethylation in a passive and

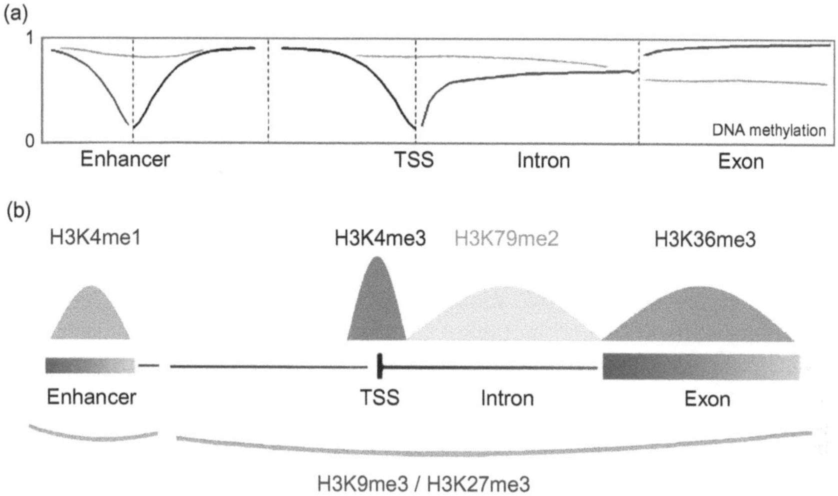

Fig. 1. Schematic representation of the main epigenetic marks. (a) Specific histone marks are enriched in specific contexts. In the upper part are shown the main activation histone marks. In the bottom part are shown the main repressive histone marks. (a) DNA methylation shows specific patterns in different contexts. The black line represents the pattern of DNA methylation of an active gene. The red line represents the DNA methylation profile of a repressed gene. (For interpretation of the references to color in this figure legend, the reader is referred to the Web version of this chapter.)

replication-dependent manner (Rhee et al., 2002), but whether DNA is actively demethylated independently of replication is under debate. Several proteins have been reported to exhibit DNA demethylase activity; however, 10–11 translocation proteins (TET) are shoo-in. Conversely to DNA methylation, DNA demethylation orchestrates different enzymatic activities in series. TET proteins convert 5-methyl-cytosine (5mC) to 5-hydroxymethyl-cytosine (5hmC). Then, cytidine deaminases process the 5hmC to 5-hydroxymethyluracil (5hmU), and finally, 5hmU is turned to cytosine by the base excision repair system (Guo et al., 2011).

Nowadays, the role of TET proteins in active DNA demethylation is supported by a plethora of evidence, but it has remained an enigma that has just been deciphered thanks to the trail of 5hmC. As we have aforementioned, CpGs tend to be methylated throughout the entire genome except for the TSS and CGIs. Strikingly, TET proteins preferentially bind these regions that are also associated with high levels of 5hmC (Xu et al., 2011) and gene activation (Ficz et al., 2011). It has been postulated that DNMTs tend to methylate all CpGs in the genome, and that the maintenance of unmethylated CpG status is either a consequence of the blocking of DNMT activity or results from active DNA demethylation (Deaton and Bird, 2011). In this context, the knockdown of Tet1 and Tet2 induces gene repression, which is subsequently associated with promoter hypermethylation (Ficz et al., 2011). Otherwise, during early development, the paternal pronucleus in zygote is dramatically reprogrammed, the preestablished DNAm pattern being erased. This process results in a large decrease in 5mC levels, which is correlated with *TET3* expression, and an increase in 5hmC levels (Iqbal et al., 2011). A similar challenge is provided by the reprogramming of differentiated cells into stem cells, and as expected, the conversion of fibroblasts into iPSCs is again associated with increased 5hmC levels and requires *TET1* and *TET2* expression (Koh et al., 2011). Whether 5hmC is just an intermediate or actively participates in chromatin regulation remains an intriguing matter that will be clarified in future studies.

Histone modifications

The histone octamer—two dimers of H2.A–H2.B and one tetramer of H3–H4—forms a globular structure from which the N-terminal tails of histones protrudes. It provides an unstructured platform where the majority of histone modifications take place. A large number of histone modifications have been reported and occur as a consequence of different sets of antagonistic complexes. The two main histone modifications, the acetylation and methylation of lysines, result from the balanced activity of histone acetyltransferases (HATs) and deacetylases (HDACs), and histone methyltransferases (HMTs) and demethylases (HDMTs), respectively. Phosphorylations, ADP-ribosylations, ubiquitylations, sumoylations, adenylations, and others are fulfilled by similar sets of complexes. The list of histone modifications is ever-growing. In addition, several amino acids are susceptible to being modified in different ways. Lysines can be acetylated; mono-, di-, or trimethylated; and mono- or polyubiquitylated; arginines can be deiminated; mono-, symmetrically or asymmetrically dimethylated; and mono- or poly-ADP-ribosylated; serines and threonines can be *N*-acetylglucosaminated and phosphorylated; glutamates can be mono- or poly-ADP ribosylated; prolines can be isomerized; and tyrosines can be phosphorylated. Altogether, histone modifications constitute a huge regulatory system that has been scarcely understood and requires deeper research to characterize their importance to cell homeostasis fully.

One of the most thoroughly studied histone modifications is the acetylation (ac) of lysines (K). Most of them—H3K9, H3K14, H3K18, H3K27, H3K56, H4K5, H4K8, H4K12, and H4K16—are located on histone tails, and their acetylation is associated with gene activation (Wang et al., 2008). Noticeably, some histone modifications take place inside the nucleosome core, such as H3K56ac. Three families of HATs—GNAT, MYST, and CBP/p300—carry out the acetylation at multiple sites, and the specificity of these modifications depends on their association with large multiprotein complexes (Bannister and Kouzarides, 2011). For instance, GCN5—a member of the GNAT family—acetylates H3K9 and H3K14 when it is associated with Ada2b but acetylates H4K5 and H4K12 when it is associated with ADA2A (Pankotai et al., 2005). The classical and SIRT2-related families of HDACs antagonize the effects of HATs and, consequently, are considered transcriptional repressors. HDACs have low substrate specificity by themselves, and as occurs with HATs, the noncatalytic subunits of the complexes are responsible for their specificity (Yang and Seto, 2007).

Another well-known histone modification is the methylation of lysines, which can be mono-, di-, or trimethylated. Conversely to acetylation, the meaning of each modification depends on which amino acid is modified and on the number of methyl groups added. As example, H3K9 monomethylation (H3K9me1) and H4K20me1 are associated with active genes, whereas H3K9 trimethylation (H3K9me3) and H4K20me3 are associated with repressed genes (Barski et al., 2007; Wang et al., 2008). Also, H3K4me3 and H3K4me1 represent specific marks enriched at the TSS and the enhancers of active genes, respectively (Heintzman et al., 2007).

The two classes of lysine HMTs—SET domain and non-SET domain-containing HMTs—are highly specific enzymes that modify a particular lysine at a specific degree of methylation. Three members of the SET HMTs carry out the methylation of H4K20 in a specific degree way. The monomethylation of H4K20 is exclusively carried out by PR-SET7, the dimethylation is carried out by SUV420H1, and the trimethylation is carried out by SUV420H2. Other histone methylations show similar specificity. H3K79 methylation is entirely carried out by DOT1-A non-SET HMT, and H3K27 methylation is solely carried out by EZH2. On the other hand, the two classes of HDMTs—LSD1- and Jumonji-related HDMTs—also display a high level of substrate specificity. LSD1 demethylates monomethylated and dimethylated H3K4 and H3K9, but not their trimethylations. The JMJD2/KDM4 subfamily specifically demethylates di- and trimethylated H3K9 and H3K36, but not the monomethylated form (for a further review, see Bannister and Kouzarides, 2011).

Considering that lysines can be targeted by several different modifications, and given that some changes would preclude others and *vice versa*, the complexity of the *histone code* is striking. Besides, some histone modifications can directly regulate chromatin structure altering the electric charge of histones, such as acetylations and phosphorylations. Others, such as ubiquitylations, necessarily have to alter the conformation of nucleosome as a consequence of its large size. However, although these effects cannot be disregarded, its effect on chromatin structure is most likely associated with its significance for the cell machinery. For instance, H3K4me3 is associated with active TSS, correlates with gene expression, and would be read by the cell machinery as an indicator of "active gene start." Also, the H3K36me3 mark is associated with gene bodies, especially in exons that are included in the final spliced transcript (Hon et al., 2009), and interpreted by the splicing machinery as an "exon inclusion mark." H3K79me2, which is enriched in the first introns (Huff et al., 2010), is considered as an "intron mark," and H3K4me1 indicates the location of an active enhancer (Fig. 1a).

Cross talk

A worthwhile analogy is that of the epigenetic code with a language, in which the letters would be the DNAm and histone modification patterns; the writers the DNMTs, HATs, and HMTs; and the erasers the TET proteins, HDACs, and HDMTs. This language would also need its readers. Three families of proteins read 5mC in nonoverlapping target regions: methyl-CpG binding domain proteins (MBD) (MeCP2, MBD1, MBD2, MBD3, and MBD4), SET- and Ring finger-associated proteins (SRA) (UHRF1 and UHRF2), and KAISO family proteins (Kaiso, ZBTB4, and ZBTB38). Acetylated histones are read by bromodomains and the tandem PHD domains containing proteins. Finally, methylated histones are read by PHD, chromo, WD40, Tudor, double/tandem Tudor, MBT, Ankyrin Repeats, zf-CW, and PWWP domains containing proteins. Similar to HATs, acetylation readers can bind multiple acetylated histones. They display certain specificity, DPB3B prefers acetylated H3 to H4, and work cooperatively in heavy complexes that fine-tune their interactions. Unlike acetylation, methylation readers normally recognize methylation states on concrete lysines. PDP1 binds to H4K20me1, whereas CRB2 recognizes H4K20me2, although, rather as with the acetylation readers, they are also included in heavy complexes (reviewed in Yun et al., 2011).

There is intense cross talk between epigenetic modifications. DNAm and histone modifications are intimately related and work together in genomic and transcriptomic regulation. MECP2 silences transcription by recruiting the histone deacetylase HDAC1, which links DNAm and histone deacetylation (Suzuki et al., 2003). UHRF1 interacts with the three DNMTs and the H3K9 methyltransferase G9A, providing a link between DNAm and H3K9 methylation (Meilinger et al., 2009; Rottach et al., 2010). KAISO protein also interacts with HDAC3, which links the DNAm and histone deacetylation in its specific context (Yoon et al., 2003). Otherwise, histone acetylase GCN5 interacts with H3K4me3, reinforcing the "message of activation" and linking histone methylation and histone acetylation. In this way, polymerase II transcription factor TFIID binds more strongly to H3K4me3 when it is flanked by H3K9ac and H3K14ac, indicating a direct connection between these modifications and active transcription (Vermeulen et al., 2010).

All the languages need their "speakers," and in the epigenetic language, these correspond to the ATP-remodeling complexes. Remodeling complexes mobilize the nucleosomes and regulate the presence and position of nucleosomes on DNA in an ATP-dependent manner. This nucleosome reorganization is carried out by the SWI/SNF,

NURD, and ISWI families of ATP-remodeling complexes. The SWI/SNIF family encompasses a series of complexes composed of 11 subunits encoded by 20 genes, with 288 different possible assemblies. An alternative ATPase, BRG or BRM, in combination with a specific BAF60 protein, BAF60A, BAF60B, or BAF60C, is contained in each complex. In addition, several specialized subunits are included in each complex. BAF53B is exclusively expressed in neurons, and null mice die prenatally with reduced neurite outgrowth and synapse formation (Wu et al., 2007). The NURD family of complexes, with 48 possible combinations, and the ISWI family of complexes, with seven possible combinations both present a similar modular structure (Wu et al., 2009). In general, remodeling complexes are polymorphic readers of epigenetic marks that interact with many chromatin proteins and transcription factors (Reddy et al., 2010). A combination of strong recognition, weak contacts, and transitory interactions regulate chromatin-remodeling complexes which, in turn, are also regulated by subunit exchange and posttranslational modifications (Simone, 2006). At the same time, the huge chromatin-remodeling complexes can embrace different nucleosomes, increasing their interactions and the possibilities of major fine-tuning regulations.

In summary, epigenetics can interpret the genome, showing or hiding specific functional regions (transcription factor binding sites, TSS, exons, etc.) and providing a kind of comprehensive "working genome" to the cell machinery. The cell machinery is unable to function with all the information included in the genome at the same time, and epigenetics could summarize what is important for each cell type at each particular moment.

Epigenetic involvement in brain functions

The human brain is a highly specialized system that orchestrates the function of hundreds of billions of cells (Azevedo et al., 2009). Different cell types, with their specific epigenetic patterns, are organized into distinct structures that display specific transcriptome and characteristic DNAm profiles (Xin et al., 2010). The nervous system represents an immensely complex structure in which the transcriptome machinery faces its major challenge. Three of four genes are expressed in the brain (Johnson et al., 2009), and it expresses more alternatively spliced transcripts (Yeo et al., 2004) and miRNAs (Cao et al., 2006) than any other tissue. As we have seen above, epigenetics is crucial to gene regulation, controlling transcription and splicing. Consequently, epigenetic regulation has also been of cardinal importance for brain function.

DNAm and histone modifications have been associated with neurodevelopment and brain functions such as synaptic plasticity, learning, and memory. Contrary to expectation, neurons express high levels of DNMT1 and DNMT3A (Feng et al., 2005). This is surprising because neurons are considered a differentiated state of post-mitotic cells for which further differentiation or maintenance of DNAm patterns would not be necessary. The presence of DNMTs, together with the high levels of MBD1 and MeCP2, suggests that DNAm has a specific role in neuron function. Furthermore, *TET1* is highly expressed in brain (Lein et al., 2007) and 5hmC levels in the frontal cortex are four times higher than in stem cells (Jin et al., 2011). Consequently, given the idea that DNAm has a role in brain function, the ablation of DNMT1 and DNTM3A in post-mitotic neurons perturbs learning and memory (Feng et al., 2010), and MECP2 mutations (discussed below) cause intellectual disability. In addition, DNMT1 mutations cause hereditary sensory neuropathy with dementia (Klein et al., 2011) and the ablation of DNMT3A impairs neurogenesis in adult brain (Wu et al., 2010).

One of the most intriguing breakthroughs in neuroscience has been the discovery that neuronal activity induces active DNA demethylation at specific loci (Ma et al., 2009). As expected, this DNA demethylation is dependent on TET1 and cytidine deaminase APOBEC1 (Guo et al., 2011) and, noticeably, changes in DNAm associated with learning can persist for more than 1 month (Miller et al., 2010). Intuitively, a parallel between epigenetics and memory arises with the idea that epigenetics might constitute a scaffold for memory. Neuronal activity modifies gene expression, which is essential for long-term memory, and necessarily must be associated with epigenetic chromatin changes. These changes are not confined to DNAm and also include histone modifications. In this way, neuronal activity also increases histone acetylation (Levenson et al., 2004). *HDAC2* overexpression decreases dendritic spine density, synapse number, synaptic plasticity, and memory formation. Conversely, HDAC2 inhibition results in increased synapse number and memory facilitation, highlighting the role of histone acetylation in learning and memory formation (Guan et al., 2009). In keeping with this, mutations in the CREB-binding protein (CBP), which regulates the function of CREB/p300 HAT, cause intellectual disability that can be ameliorated by HDAC inhibitors (Alarcón et al., 2004). Although it continues to be less well studied, histone methylation is also known to participate in brain function. Likewise, neuronal ablation of GLP and G9A HMTs induces learning defects (Kramer et al., 2011), and contextual fear conditioning induces changes in H3K4me3 and H3K9me3 (Gupta et al., 2010).

Overall, it is clear that epigenetics plays a crucial role in the nervous system, and although its involvement is not fully understood, it offers a comprehensive framework of pathways and mechanisms that underpins its function. The complexity of the nervous system makes it especially sensitive to epigenetic disturbances, and consequently, many mental disorders are caused by mutations in the epigenetic machinery (Urdinguio et al., 2009). We will describe some observations that will give us an insight into the general epigenetic landscape of mental disorders (Fig. 2).

Epigenetic alterations in several intellectual disability

Alpha-thalassemia/mental retardation, X-linked syndrome

X-linked alpha-thalassemia mental retardation (ATRX) syndrome (OMIM 301040) is associated in males with profound mental deficiency, facial

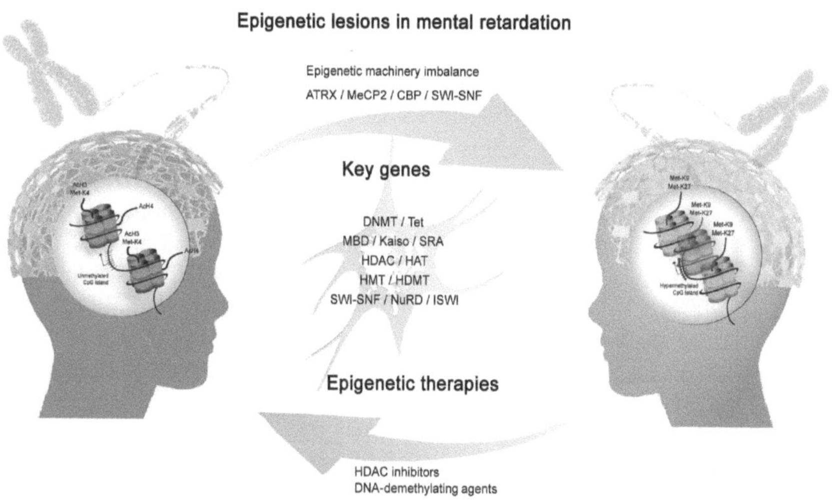

Fig. 2. Schematic representation of the epigenetic landscape of intellectual disability. (For color version of this figure, the reader is referred to the Web version of this chapter.)

dimorphism, genital abnormalities, and alpha-thalassemia. Female carriers exhibit a less severe phenotype and are usually physically and intellectually normal, although they show subtle signs of language limitation, or alpha-thalassemia.

This syndrome is X-linked recessive and arises from mutations of the *ATRX* gene (Gibbons et al., 1995, 2008). ATRX is an ATPase member of the SWI/SNF family that simultaneously acts as an epigenetic reader and speaker (Eisen et al., 1995). The helicase/ATPase domain, in which most syndrome-causing mutations occur, is the catalytic unit of the protein, and the ADD (ATRX–DNMT3–DNMT3L) domain allows the specific recognition of the status of lysine methylation thanks to the PHD finger domain (Ooi et al., 2007; Otani et al., 2009).

ATRX protein is predominantly associated with heterochromatin—regions of silenced chromatin enriched with H3K9me3 and deprived of H3K4me3—where it directs H3.3 deposition on the telomeres through the interaction with Daxx, a histone chaperone (Goldberg et al., 2010; Lewis et al., 2010). ATP-dependent chromatin remodelers are involved in the assembly of different histone variants in the nucleosome, which induces changes in the chromatin structure and constitutes an extra-epigenetic mechanism of regulation.

ATRX is recruited to heterochromatin thanks to the recognition of unmodified H3K4 and dimethylated or trimethylated H3K9 by the ADD domain (Eustermann et al., 2011; Iwase et al., 2011). ATRX also interacts with heterochromatin protein 1 (HP1); in fact, *Suv39H1/H2* HMTs double knockouts, which are responsible for H3K9me2/3, cause its delocalization from heterochromatin (Kourmouli et al., 2005; Lechner et al., 2005; Thiru et al., 2004).

It seems likely that one of the functions of the ATRX protein is mainly associated with the establishment and organization of constitutive heterochromatin. Thus, as expected, *ATRX* mutations or low *ATRX* gene expression has been linked to an altered pattern of DNA methylation of several repetitive elements (Gibbons et al., 2000) and telomere and centromere defects (De la Fuente et al., 2004; Wong et al., 2010). ATRX is also involved in the maintenance of X chromosome inactivation in women (Baumann and De la Fuente, 2009) and localizes at the nuclear matrix during interphase, where it may participate in transcriptional regulation (Bérubé et al., 2000).

Additionally, ATRX interacts with known epigenetic regulators, such as EZH2 methyltransferase (Cardoso et al., 1998) and MECP2, which might help control the expression of downstream genes (Nan et al., 2007) (Table 1 summarizes the epigenetic alterations of the syndromes discussed).

Rett syndrome

Rett syndrome (RTT; OMIM 312750) is an X-linked neurodevelopmental disorder that mainly affects females and leads to postnatal lethality in males. RTT offspring develop without symptoms until 6–18 months of life. Afterward, they gradually lose mental skills and develop brain hypotrophy and severe intellectual disability (Esteller, 2007). RTT is caused by mutations in *MECP2* (Amir et al., 1999) and affects females who are heterozygous and mosaic for the expression of *MECP2* mutations.

MECP2 is a methyl-CpG-binding protein of 50kDa with four functional domains: a methyl-binding domain (MBD), an RNA binding domain (arginine–glycine repeat RNA binding domain) (Jeffery and Nakielny, 2004), a transcriptional repression domain (TRD), and an RNA *splicing* binding factor region (WW group II binding domain). It also has an NLS (nuclear localization signal) region. MECP2 is able to recognize a unique methylated CpG and binds methylated DNA through the MBD domain (Lewis et al., 1992) and recruits the HDAC complex Sin3a through the TRD domain (Jones et al., 1998; Nan et al., 1998). MECP2 is also involved in the regulation of alternative *splicing* of mRNA through its association with YB-1 (Y-box-binding protein), which is involved in alternative *splicing* (Young et al., 2005).

Table 1. Summary of the main epigenetic alterations of intellectual disability syndromes

Disease	Epigenetic alteration	Gene/complex	Mechanism	Neuronal consequences
ATRX	Chromatin remodelers	SWI/SNF	Mutation	Reduced neuronal maturation
Rett syndrome	Epigenetic readers	MeCP2	Mutation	Impairment neuronal functions
	Chromatin remodelers	SWI–SNF*	Imbalance	Reduced neuronal maturation
	Epigenetic writers	CREB–CBP*/ Suv39H*	Imbalance	Abnormal histone acetylation patterns
	Epigenetic erasers	Sin3a*	Imbalance	Abnormal histone acetylation patterns
Rubinstein–Taybi syndrome	Epigenetic writers	CBP	Mutation	Reduced histone acetylation
Down syndrome	Chromatin remodelers	SWI/SNF*	Imbalance	Reduced neuronal maturation
	Epigenetic writers	CBP	Imbalance	Abnormal histone acetylation patterns
		DNMT3L*	Overexpression	Abnormal DNA methylation patterns
	Epigenetic readers	MeCP2	Downregulation	Impairment neuronal functions
	Epigenetic erasers	SIRT1*	Imbalance	Abnormal histone acetylation patterns

Asterisk remarks some of the gene/complex altered, including the mechanism by which could be affected, that are not yet demonstrated. Those have been included in the table because several evidences suggest that must be affected (see text).

Initial studies found abundant MECP2 binding sites in genomic chromatin and MECP2 seemed capable of long-range repression (Nan et al., 1997). It was subsequently found to be involved in large-scale reorganization of pericentromeric heterochromatin during cellular differentiation (Brero et al., 2005). Although MECP2 shows three times more affinity for methylated DNA, it is able to bind to unmethylated DNA (Yasui et al., 2007). Even though *MECP2* is expressed in a wide range of tissues, the major features of RTT have already been reported as being mainly caused by the neuronal deficiency of MECP2 (Chen et al., 2001). In fact, the chromatin-containing MECP2 in the brain is thought to be different than that in all other tissues (Skene et al., 2010). In neuronal nuclei, MECP2 is nearly as abundant as histone H1, and the two compete to occupy internucleosomal regions. MECP2 preferentially associates with methylated regions, displacing H1 (Ghosh et al., 2010). Likewise, MECP2 deficiency results in global changes in neuronal chromatin structure, including elevated histone acetylation and a doubling of histone H1 (Skene et al., 2010).

Although it was thought that MECP2 might function as a global repressor, further studies have suggested that it acts as a transcriptional regulator coupling DNA methylation and transcriptional modulation. MECP2 binds to methylated DNA, HDAC SIN3A, retinoid and thyroid hormone receptors (SMRT) (Stancheva et al., 2003), HP1(Agarwal et al., 2007), HMT SUV39H (Fuks et al., 2003), the SWI/SNF Brahma complex (Harikrishnan et al., 2005), and ATRX (Kernohan et al., 2010). Notably, MECP2 does not completely co-fractionate with these partners, suggesting that it is not a component of a stably assembled complex (Klose and Bird, 2004). Unexpectedly, data from recent studies have revealed that

MECP2 can also act as gene activator that directly binds the promoter of some genes in association with the transcriptional activator CREB1 (cAMP response element binding protein 1) (Chahrour et al., 2008).

The complex role of MECP2 indicates that it cannot be considered a classical transcriptional repressor. To advance our basic knowledge of the role of MECP2 in brain development and to understand the molecular bases of RTT pathophysiology, several different mouse models have been generated. MECP2-deficient mice have lower brain weight, smaller neurons, reduced exploratory activity (Chen et al., 2001; Guy et al., 2001, Jentarra et al., 2010), cognitive deficits, and impaired synaptic plasticity (Collins et al., 2004). Importantly, postnatal reexpression of *Mecp2* partially rescues the phenotype (Guy et al., 2007), and contrary to the classical idea of the neuron role MECP2, the reexpression of *Mecp2* on astrocytes also mitigates the RTT phenotype (Lioy et al., 2011).

Given that MECP2 affects gene expression through chromatin conformational changes that are secondary to histone modifications, MECP2 dysfunction found in RTT should be associated with a different pattern of histone markers. Several studies have already examined the relationship between DNA methylation, MECP2, and histone modifications but have given different results. For instance, established cell lines from two RTT patients showed an increase in the acetylation of histone H4 at lysine 16 (Wan et al., 2001) but no difference in H3 acetylation. However, there was no difference in the global acetylation of H3 or H4 in nine RTT-patient T-cell clones (Balmer et al., 2002). On the other hand, untransformed cells bear differences in methylation and acetylation of H3 without any change in H4 modifications (Kaufmann et al., 2005). Using mice harboring, a truncated version of MECP2 revealed increased acetylation of histone H3 and an absence of changes of histone H4 in the brain of transgenic animals when compared with their normal counterparts (Shahbazian et al., 2002). Finally, global changes in H3 and H4 histone modification patterns were not detected when comparing cortex, midbrain, and cerebellum in wild-type and *Mecp2* knockout mice (Urdinguio et al., 2007). This apparent contradiction suggests that MeCP2 may act as a specific gene regulator with restricted epigenetic changes instead of global effects with opposite trends.

A different approach has used microarray and chromatin immunoprecipitation technologies to interrogate the downstream consequences of *MECP2* mutations. In general, these global analyses have revealed subtle differences in RNA expression, chromatin enrichment, and miRNAs profiles. However, despite these small differences, a set of target genes and miRNAs have been identified (Guy et al., 2011; Urdinguio et al., 2008, 2010).

It is still uncertain if MECP2 acts merely as a methyl-CpG binding protein, whether it is a multifunctional protein or mainly a transcriptional repressor. Neither do we know whether MECP2 acts at specific sites within the genome or more globally, nor the cell types in which it is functionally important. And finally, taking into account recent observations, it is also remains to be determined whether MECP2 is a regulator of brain development or if it is necessary for maintenance of the function at the neuronal and glial levels. Thus, the role of MECP2 in the etiology of RTT is not completely characterized.

Rubinstein–Taybi syndrome

Rubinstein–Taybi syndrome (RTS; OMIM 180849) is an autosomal dominant disorder characterized by intellectual disability and physical abnormalities associated with *de novo* mutations in *CBP* and *EP300* genes (Bartholdi et al., 2007). These genes share a highly related structure and participate in gene regulation through their HAT activity (Kalkhoven, 2004). RTS is characterized by the loss of function of CBP/P300 and its deficiency in mouse models results in similar phenotypes. The ablation of one CBP allele in mouse models results in deficits in long-term memory and cognitive impairments comparable to what is observed in RTS (Alarcón et al., 2004). At the molecular level, these mice have reduced HAT activity, decreased acetylation of specific histone proteins, and impaired CBP-dependent gene expression. As we have aforesaid, histone acetylation is important for synaptic plasticity, learning, and memory formation and explains how these skills are affected in RTS (Gaub et al., 2010). Interestingly, CBP deficiency induces *HMT* expression and, consequently, regulates histone methylation patterns spreading the downstream effects and increases the number of genes affected (Lee et al., 2008).

Epigenetic alterations in Down syndrome

Down syndrome (DS) is considered the most frequent cause of intellectual disability. Neuropathological outcomes of DS include defects in neurogenesis, neuronal differentiation, and brain hypotrophy (Lepagnol-Bestel et al., 2009). DS is mainly caused by the complete trisomy of HSA21 although an extra copy of the Down syndrome critical region (DSCR) is sufficient to cause DS.

HSA21 contains 225 genes and 59 pseudogenes, and around 100 of these are located in the DSCR (Hattori et al., 2000). Although by itself, DSCR cannot recapitulate the complete DS features, it is sufficient to induce the majority of DS traits. Single-gene mouse models interrogating the specific DSCR genes have greatly contributed to our knowledge of the disease. Interestingly, some of them have important epigenetic consequences.

DYRK1A is considered one of the major effectors in DS pathology, and its overexpression causes learning and memory impairment and also contributes to the Alzheimer's disease pathology by phosphorylating APP and TAU proteins (Park et al., 2009). Interestingly, with respect its epigenetics, DYRK1A interacts with the SWI/SNIF complex. As mentioned above, the SWI/SNIF complex is a chromatin-remodeling complex that participates in the arrangement of the chromatin structure. It is generally associated with gene expression and affects explicit subsets of genes according to its interactors. The SWI/SNF complex induces the expression of downstream genes such as *JARID1D* (HDMTs), *L3MBTL2* (HMT), and *NCOR* (HDAC interactor) (Lepagnol-Bestel et al., 2009). DYRK1A also phosphorylates HDAC SIRT1 and CREB, regulating the transcription of many genes through histone acetylation (Guo et al., 2010; Park et al., 2009). Thus, DYRK1A is completely involved in the epigenetic machinery, suggesting an epigenetic mediation of the DYRK1A effect on DS gene deregulation. Interestingly, DYRK1A is not the only DSCR protein that interacts with the SWI/SNF complex. The bromodomain and WD repeat domain containing 1 (BRWD1) and the Runt-related transcription factor 1 (RUNX1) are also associated with the SWI/SNF complex (Bakshi et al., 2010; Huang et al., 2003). Likewise, three DSCR proteins interact with SWI/SNF, reinforcing the possible importance of this complex in the etiopathology of DS. Indeed, BRG1 containing the SWI/SNF complex is critical during neurodevelopment. *BRG1* is strongly expressed in telencephalon, cerebellum, and hindbrain regions and induces brain abnormalities when it is deregulated (Dorfman et al., 2003). BRG1 inhibition also causes neurogenesis and neuronal differentiation impairment (Eroglu et al., 2006), thereby exhibiting clear parallels between its deregulation and DS features. Furthermore, BRG1 positively regulates the T-cell transcription factor, which is involved in non-solid tumors and could contribute to the elevated incidence of leukemia in DS (Dorfman et al., 2003). Noticeably, ATRX syndrome is caused by mutations of the SWI/SNF complex, reinforcing the existence of some analogies between the two diseases. Unfortunately, the involvement of SWI/SNF in DS has not been investigated, and it is not known whether these similarities are meaningful. DSCR also includes some chromatin constitutive genes. Chromatin assembly factor 1, subunit B (CHAF1B) is a constitutive protein of chromatin that participates in nucleosome assembly. CHAF1B interacts with MBD1 and HP1 and participates in transcription repression through DNAm and H3K9me3 (Reese et al., 2003). Nucleosome-binding high-mobility group domain 1 (HMGN1) is also a constitutive protein of chromatin and regulates H3K14 acetylation through CBP/P300 HAT activity (Ueda et al., 2006). Again, there is a convergence on epigenetic effectors. Another DSCR protein, the erythroblastosis virus E26 oncogene homolog 2 (ETS2), also regulates CBP/P300 activity (Sun et al., 2006). As already mentioned, the CBP/P300 mutation causes RTS, and histone acetylation is crucial for learning and memory formation. In summary, it seems that CBP/P300 dysfunction may be at least partially responsible for the cognitive disabilities observed in DS.

Overall, the evidence suggests that DSCR genes can induce global epigenetic changes that affect the expression of genes located inside and outside the HSA21. Indeed, studies profiling mRNA expression have shown that many genes outside of HSA21 are deregulated and that some genes on HSA21 exhibit dose-independent expression (Mao et al., 2005). Despite only one high-throughput study on the DNA methylation patterns has been published, the first differentially methylated genes in DS have already been identified. Interestingly, most of these are located outside of HSA21 (Kerkel et al., 2010). Unfortunately, to date, no reports investigating histone profiles in DS have been published and their involvement in DS etiopathology remain undetermined.

It is important to mention that miR-99a, let-7c, miR-125b-2, miR-155, and miR-802 are located in HSA21 and, consequently, are overexpressed in DS. Interestingly, mir-155 and miR-802 target *MECP2* (Kuhn et al., 2010), which is downregulated in DS (Nagarajan et al., 2006). As we have said, MeCP2 mutation causes RTT, which is also associated with intellectual disability. Notably, miR-802 is located in the DSCR region, and it suggests an independent connection between DSCR genes and epigenetic effectors. Furthermore, H2A histone family member Z pseudogene 1 (H2AFZP) is located on HSA21. It is not known whether this pseudogene is functional, but its homolog H2AZ participates in gene regulation (Barski et al., 2007). Remarkably, DNA 5mC methyltransferase 3-like (DNMT3L), a noncatalytic member of DNMTs family, is also located in HSA21. DNMT3L binds nucleosomes when they are not targeted by H3K4 methylation and stimulates DNMT3A and DNMT3B methylation activity (Ooi et al., 2007). Once more, this suggests that epigenetic factors play an important role in DS etiopathogenesis.

In recent years, epigenetics has been come to be considered a new paradigm for many diseases. It is of cardinal importance in regulating gene expression, and consequently, it is involved in all the processes involved in gene expression regulation. The alteration of the epigenetic machinery appears to be a common process in many neurological diseases that recurrently affect the same pathways. DS is not an exception, and several lines of evidences suggest that epigenetics contributes greatly to DS. Interrogating the epigenetic landscape of DS offers new opportunities to gain better insight into the etiopathology of DS.

Epigenetic drugs and treatment

We have described intellectual disability disorders caused by mutations in genes encoding epigenetic players. Thus, it is expected that epigenetic deregulation takes place in the pathogenesis of these brain disorders, with the appearance of changes in the activity and expression of epigenetic modifiers. In order to regulate and restore epigenetic status in neurological disorders, epigenetic drugs can be used. Currently, there are two types of epigenetic drugs that are extensively used, DNA-demethylating drugs and HDAC inhibitors. Although DNA-demethylating drugs might be suitable for several neurodegenerative and neurodevelopmental diseases, more effort has been devoted to research with histone deacetylase inhibitors. By reducing deacetylation, HDAC inhibitors increase the acetylation levels of histones, causing the upregulation of genes, which may act in molecular pathways involved in neuroprotection. HDAC inhibitors can be classified into four main chemical families: short-chain fatty acids (sodium butyrate, phenylbutyrate, and valproic acid (VPA)), hydroxamic acids (trichostatin A and suberoylanilide hydroxamic acid, SAHA), epoxyketones (trapoxin), and benzamides. Of these, the most widely studied are sodium butyrate, phenylbutyrate, VPA, trichostatin A, and SAHA.

The studies done in the CBP-negative allele mouse model of RTS have shown that SAHA and trichostatin A are capable of reversing the late phase of long-term potentiation impairment and long-term memory deficits observed in these mice (Alarcón et al., 2004; Korzus et al., 2004). But, even though the intellectual disability syndromes described in this chapter are the consequence of mutations in epigenetic enzymes, only RTS has been reported as being treated by epigenetic drugs in mouse models, giving promising results with HDACs inhibitor (Vecsey et al., 2007). This syndrome is probably the most likely condition to be treated with the current epigenetic drugs because HDAC inhibitors can induce an increase in histone acetylation that presumably compensates for the absence of CBP HAT activity in the CBP-negative mutant mice. However, Rett and ATRX syndromes are linked with mutations in epigenetic enzymes that are mainly associated with the maintenance of repressive chromatin and, at first sight, neither HDACs nor DNMT inhibitors appear to be likely candidates, and so far they have not been tested in these syndromes. Indeed, it has recently been shown that the knockdown of HDAC1 and HDAC2 in hippocampus neurons can replicate some of the imbalance between excitation and inhibition in the brains of *Mecp2* KO mice, particularly after the loss of HDAC2 (Nelson et al., 2011). Otherwise, DS is not linked with mutations on epigenetic regulators, and therefore, the proof-of-principle for epigenetic treatments has not yet been determined.

VPA is a common drug used to treat epilepsy and bipolar mood disorder. By using a microarray, it has been shown that VPA upregulates genes involved in neuronal excitability and downregulates those involved in neuronal inhibition in rat cortical primary cultures (Fukuchi et al., 2009). These results suggest that VPA may mediate the imbalance between excitation and inhibition, which is an alteration present in RTT and DS mice models (Best et al., 2007; Chao et al., 2010). In addition, VPA treatment increases the levels of BDNF, brain-derived neurotrophic factor (Fukuchi et al., 2009), which is a crucial factor associated with memory and cognition abilities (Lubin, 2011), and that is also downregulated in DS (Dorsey et al., 2002). In this way, a study evaluating the use of VPA in a *MECP2* knockdown system demonstrated a partial restoration of *MECP2* expression levels and an increase in levels of the BRM/SNF2 component of the SWI/SNF complex, as well as of BDNF (Vecsler et al., 2010), which is an important cofactor in RTT and DS. Regrettably, VPA has been used as an antiepileptic drug in the treatment of RTT with important secondary effects, such as risk of bone fracture (Leonard et al., 2010) and the induction hyperammonemic encephalopathy (Rath et al., 2005).

The maintenance of DNA methylation is a process that depends on the availability of micronutrients and vitamins, which serve as essential cofactors for all the enzymes required in this pathway. Those cofactors are folate, vitamin B6, vitamin B12, and S-adenosyl methionine, which are required for the DNMT transfer reaction of methyl groups from S-adenosyl methionine to cytosine residues within CpG-rich regions of the genome. Thus, these cofactors may also be considered to be epigenetic drugs since they are essential for DNA methylation. Previous studies in RTT and DS patients suggested various abnormalities in folate (Ormazabal et al., 2005; Pogribna et al., 2001). To evaluate whether folates improve RTT or DS, many studies administrating this compounds have been carried out. However, the vast majority of results indicate that folinic acid supplementation has no significant effect on the course of the disease (Hagebeuk et al., 2011; Salman, 2002; Temudo et al., 2009).

Despite the broad effects of epigenetic treatments, they have been successful for several diseases. No epigenetic treatments have yet been tested in DS but, since many epigenetic effectors are located within DSCR, this is not an unreasonable avenue to pursue. Many reports show that HDAC inhibitor treatments induce the expression of the SWI/SNF components BRM and BRG1 (Glaros et al., 2007; Vecsler et al., 2010), and as mentioned above, DYK1A, BRWD1, and RUNX1 interact with these proteins. Otherwise, DNA methylation profiles could also be altered in DS as a consequence, at least in part, of MECP2 deficiency and DNMT3L overexpression. Indeed, the first set of differentially methylated genes has been identified (Kerkel et al., 2010), and the possibility of using DNMT inhibitors such as 5-Aza-dC to restore these alterations in DS cannot be ignored.

References

Agarwal, N., Hardt, T., Brero, A., Nowak, D., Rothbauer, U., Becker, A., et al. (2007). MeCP2 interacts with HP1 and modulates its heterochromatin association during myogenic differentiation. *Nucleic Acids Research, 35*, 5402–5408.

Alarcón, J. M., Malleret, G., Touzani, K., Vronskaya, S., Ishii, S., Kandel, E. R., et al. (2004). Chromatin acetylation, memory, and LTP are impaired in CBP+/- mice: A model for the cognitive deficit in Rubinstein-Taybi syndrome and its amelioration. *Neuron, 42*(6), 947–959.

Amir, R. E., Van den Veyver, I. B., Wan, M., Tran, C. Q., Francke, U., & Zoghbi, H. Y. (1999). Rett syndrome is caused by mutations in X-linked MECP2, encoding methyl-CpG-binding protein 2. *Nature Genetics, 23*, 185–188.

Azevedo, F. A., Carvalho, L. R., Grinberg, L. T., Farfel, J. M., Ferretti, R. E., Leite, R. E., et al. (2009). Equal numbers of neuronal and nonneuronal cells make the human brain an isometrically scaled-up primate brain. *The Journal of Comparative Neurology, 513*(5), 532–541.

Bakshi, R., Hassan, M. Q., Pratap, J., Lian, J. B., Montecino, M. A., van Wijnen, A. J., et al. (2010). The human SWI/SNF complex associates with RUNX1 to control transcription of hematopoietic target genes. *Journal of Cellular Physiology, 225*(2), 569–576.

Balmer, D., Arredondo, J., Samaco, R. C., & LaSalle, J. M. (2002). MECP2 mutations in Rett syndrome adversely affect lymphocyte growth, but do not affect imprinted gene expresión in blood or brain. *Human Genetics, 110*, 545–552.

Bannister, A. J., & Kouzarides, T. (2011). Regulation of chromatin by histone modifications. *Cell Research, 21*(3), 381–395.

Barski, A., Cuddapah, S., Cui, K., Roh, T. Y., Schones, D. E., Wang, Z., et al. (2007). High-resolution profiling of histone methylations in the human genome. *Cell, 129*(4), 823–837.

Bartholdi, D., Roelfsema, J. H., Papadia, F., Breuning, M. H., Niedrist, D., Hennekam, R. C., et al. (2007). Genetic heterogeneity in Rubinstein-Taybi syndrome: Delineation of the phenotype of the first patients carrying mutations in EP300. *Journal of Medical Genetics, 44*(5), 327–333.

Baumann, C., & De La Fuente, R. (2009). ATRX marks the inactive X chromosome (Xi) in somatic cells and during imprinted X chromosome inactivation in trophoblast stem cells. *Chromosoma, 18*, 209–222.

Berger, S. L., Kouzarides, T., Shiekhattar, R., & Shilatifard, A. (2009). An operational definition of epigenetics. *Genes & Development, 23*(7), 781–783.

Bérubé, N. G., Smeenk, C. A., & Picketts, D. J. (2000). Cell cycle-dependent phosphorylation of the ATRX protein correlates with changes in nuclear matrix and chromatin association. *Human Molecular Genetics, 9*, 539–547.

Best, T. K., Siarey, R. J., & Galdzicki, Z. (2007). Ts65Dn, a mouse model of Down syndrome, exhibits increased GABAB-induced potassium current. *Journal of Neurophysiology, 97*(1), 892–900.

Brero, A., Easwaran, H. P., Nowak, D., Grunewald, I., Cremer, T., Leonhardt, H., et al. (2005). Methyl CpG-binding proteins induce large-scale chromatin reorganization during terminal differentiation. *The Journal of Cell Biology, 169*, 733–743.

Cao, X., Yeo, G., Muotri, A. R., Kuwabara, T., & Gage, F. H. (2006). Noncoding RNAs in the mammalian central nervous system. *Annual Review of Neuroscience, 29*, 77–103.

Cardoso, C., Timsit, S., Villard, L., Khrestchatisky, M., Fontes, M., & Colleaux, L. (1998). Specific interaction between the XNP/ATR-X gene product and the SET domain of the human EZH2 protein. *Human Molecular Genetics, 7*, 679–684.

Chahrour, M., Jung, S. Y., Shaw, C., Zhou, X., Wong, S. T., Qin, J., et al. (2008). MeCP2, a key contributor to neurological disease, activates and represses transcription. *Science, 320*, 1224–1229.

Chan, A., & Shea, T. B. (2006). Supplementation with apple juice attenuates presenilin-1 overexpression during dietary and genetically-induced oxidative stress. *Journal of Alzheimer's Disease, 10*, 353–358.

Chao, H. T., Chen, H., Samaco, R. C., Xue, M., Chahrour, M., Yoo, J., et al. (2010). Dysfunction in GABA signalling mediates autism-like stereotypies and Rett syndrome phenotypes. *Nature, 468*(7321), 263–269.

Chen, R. Z., Akbarian, S., Tudor, M., & Jaenisch, R. (2001). Deficiency of metil-CpG binding protein-2 in CNS neurons result in a Rett-like phenotype in mice. *Nature Genetics, 27*, 327–331.

Chodavarapu, R. K., Feng, S., Bernatavichute, Y. V., Chen, P. Y., Stroud, H., Yu, Y., et al. (2010). Relationship between nucleosome positioning and DNA methylation. *Nature*, *466*(7304), 388–392.

Choi, J. K. (2010). Contrasting chromatin organization of CpG islands and exons in the human genome. *Genome Biology*, *11*(7), R70 (Epub 2010 July 5).

Choy, M. K., Movassagh, M., Goh, H. G., Bennett, M. R., Down, T. A., & Foo, R. S. (2010). Genome-wide conserved consensus transcription factor binding motifs are hyper-methylated. *BMC Genomics*, *11*, 519.

Collins, A. L., Levenson, J. M., Vilaythong, A. P., Richman, R., Armstrong, D. L., Noebels, J. L., et al. (2004). Mild overexpression of MeCP2 causes a progressive neurological disorder in mice. *Human Molecular Genetics*, *13*, 2679–2689.

De La Fuente, R., Viveiros, M. M., Wigglesworth, K., & Eppig, J. J. (2004). ATRX, a member of the SNF2 family of helicase/ATPases, is required for chromosome alignment and meiotic spindle organization in metaphase II stage Mouse oocytes. *Developmental Biology*, *272*, 1–14.

Deaton, A. M., & Bird, A. (2011). CpG islands and the regulation of transcription. *Genes & Development*, *25*(10), 1010–1022 (Review).

Dorfman, D. M., Greisman, H. A., & Shahsafaei, A. (2003). Loss of expression of the WNT/beta-catenin-signaling pathway transcription factors lymphoid enhancer factor-1 (LEF-1) and T cell factor-1 (TCF-1) in a subset of peripheral T cell lymphomas. *The American Journal of Pathology*, *162*(5), 1539–1544.

Dorsey, S. G., Bambrick, L. L., Balice-Gordon, R. J., & Krueger, B. K. (2002). Failure of brain-derived neurotrophic factor-dependent neuron survival in mose trisomy 16. *The Journal of Neuroscience*, *22*, 2571–2578.

Eisen, J. A., Sweder, K. S., & Hanawalt, P. C. (1995). Evolution of the SNF2 family of proteins: Subfamilies with distinct sequences and functions. *Nucleic Acids Research*, *23*, 2715–2723.

Elango, N., & Yi, S. V. (2008). DNA methylation and structural and functional bimodality of vertebrate promoters. *Molecular Biology and Evolution*, *25*(8), 1602–1608.

Eroglu, B., Wang, G., Tu, N., Sun, X., & Mivechi, N. F. (2006). Critical role of Brg1 member of the SWI/SNF chromatin remodeling complex during neurogenesis and neural crest induction in zebrafish. *Developmental Dynamics*, *235*(10), 2722–2735.

Esteller, M. (2007). Rett syndrome: The first forty years: 1966–2006. *Epigenetics*, *2*, 1.

Eustermann, S., Yang, J. C., Law, M. J., Amos, R., Chapman, L. M., Jelinska, C., et al. (2011). Combinatorial readout of histone H3 modifications specifies localization of ATRX to heterochromatin. *Nature Structural & Molecular Biology*, *18*, 777–782.

Feng, J., Chang, H., Li, E., & Fan, G. (2005). Dynamic expression of de novo DNA methyltransferases Dnmt3a and Dnmt3b in the central nervous system. *Journal of Neuroscience Research*, *79*(6), 734–746.

Feng, J., Zhou, Y., Campbell, S. L., Le, T., Li, E., Sweatt, J. D., et al. (2010). Dnmt1 and Dnmt3a maintain DNA methylation and regulate synaptic function in adult forebrain neurons. *Nature Neuroscience*, *13*(4), 423–430.

Ficz, G., Branco, M. R., Seisenberger, S., Santos, F., Krueger, F., Hore, T. A., et al. (2011). Dynamic regulation of 5-hydroxymethylcytosine in mouse ES cells and during differentiation. *Nature*, *473*(7347), 398–402.

Fraga, M. F., Ballestar, E., Paz, M. F., Ropero, S., Setien, F., Ballestar, M. L., et al. (2005). Epigenetic differences arise during the lifetime of monozygotic twins. *Proceedings of the National Academy of Sciences of the United States of America*, *102*(30), 10604–10609.

Fuks, F., Hurd, P. J., Wolf, D., Nan, X., Bird, A. P., & Kouzarides, T. (2003). The methyl-CpG-binding protein MeCP2 links DNA methylation to histone methylation. *The Journal of Biological Chemistry*, *278*, 4035–4040.

Fukuchi, M., Nii, T., Ishimaru, N., Minamino, A., Hara, D., Takasaki, I., et al. (2009). Valproic acid induces up- or down-regulation of gene expression responsible for the neuronal excitation and inhibition in rat cortical neurons through its epigenetic actions. *Neuroscience Research*, *65*, 35–43.

Gaub, P., Tedeschi, A., Puttagunta, R., Nguyen, T., Schmandke, A., & Di Giovanni, S. (2010). HDAC inhibition promotes neuronal outgrowth and counteracts growth cone collapse through CBP/p300 and P/CAF-dependent p53 acetylation. *Cell Death & Differentiation*, *17*(9), 1392–1408.

Ghosh, R. P., Horowitz-Scherer, R. A., Nikitina, T., Shlyakhtenko, L. S., & Woodcock, C. L. (2010). MeCP2 binds cooperatively to its substrate and competes with histone H1 for chromatin binding sites. *Molecular and Cellular Biology*, *30*, 4656–4670.

Gibbons, R. J., Brueton, L., Buckle, V. J., Burn, J., Clayton-Smith, J., Davison, B. C., et al. (1995). Clinical and hematologic aspects of the X-linked alpha-thalassemia/mental retardation syndrome (ATR-X). *American Journal of Medical Genetics*, *55*, 288–299.

Gibbons, R. J., Wada, T., Fisher, C. A., Malik, N., Mitson, M. J., Steensma, D. P., et al. (2008). Mutations in the chromatin-associated protein ATRX. *Human Mutation*, *29*, 796–802.

Gibbons, R. J., McDowell, T. L., Raman, S., O'Rourke, D. M., Garrick, D., Ayyub, H., et al. (2000). Mutations in ATRX, encoding a SWI/SNF-like protein, cause diverse changes in the pattern of DNA methylation. *Nature Genetics*, *24*(4), 368–371.

Glaros, S., Cirrincione, G. M., Muchardt, C., Kleer, C. G., Michael, C. W., & Reisman, D. (2007). The reversible epigenetic silencing of BRM: Implications for clinical targeted therapy. *Oncogene*, *26*, 7058–7066.

Goldberg, A. D., Banaszynski, L. A., Noh, K. M., Lewis, P. W., Elsaesser, S. J., Stadler, S., et al. (2010).

Distinct factors control histone variante H3.3 localization at specific genomic regions. *Cell*, *140*, 678–691.

Guan, J. S., Haggarty, S. J., Giacometti, E., Dannenberg, J. H., Joseph, N., Gao, J., et al. (2009). HDAC2 negatively regulates memory formation and synaptic plasticity. *Nature*, *459*(7243), 55–60.

Guo, J. U., Su, Y., Zhong, C., Ming, G. L., & Song, H. (2011). Hydroxylation of 5-methylcytosine by TET1 promotes active DNA demethylation in the adult brain. *Cell*, *145*(3), 423–434 (Epub 2011 April 14).

Guo, X., Williams, J. G., Schug, T. T., & Li, X. (2010). DYRK1A and DYRK3 promote cell survival through phosphorylation and activation of SIRT1. *The Journal of Biological Chemistry*, *285*(17), 13223–13232 (Epub 2010 February 18).

Gupta, S., Kim, S. Y., Artis, S., Molfese, D. L., Schumacher, A., Sweatt, J. D., et al. (2010). Histone methylation regulates memory formation. *The Journal of Neuroscience*, *30*(10), 3589–3599.

Guy, J., Cheval, H., Selfridge, J., & Bird, A. (2011). The role of MeCP2 in the brain. *Annual Review of Cell and Developmental Biology*, *27*, 631–652.

Guy, J., Gan, J., Selfridge, J., Cobb, S., & Bird, A. (2007). Reversal of neurological defects in a mouse model of Rett syndrome. *Science*, *315*, 1143–1147.

Guy, J., Hendrich, B., Holmes, M., Martin, J. E., & Bird, A. (2001). A mouse Mecp2-null mutation causes neurological symptoms that mimic Rett syndrome. *Nature Genetics*, *27*, 322–326.

Hagebeuk, E. E., Koelman, J. H., Duran, M., Abeling, N. G., Vyth, A., & Poll-The, B. T. (2011). Clinical and electroencephalographic effects of folinic acid treatment in Rett syndrome patients. *Journal of Child Neurology*, *26*(6), 718–723.

Harikrishnan, K. N., Chow, M. Z., Baker, E. K., Pal, S., Bassal, S., Brasacchio, D., et al. (2005). Brahma links the SWI/SNF chromatin-remodeling complex with MeCP2-dependent transcriptional silencing. *Nature Genetics*, *37*, 254–264.

Hattori, M., Fujiyama, A., Taylor, T. D., Watanabe, H., Yada, T., Park, H. S., et al. (2000). Chromosome 21 mapping and sequencing consortium. The DNA sequence of human chromosome 21. *Nature*, *405*(6784), 311–319.

Heintzman, N. D., Stuart, R. K., Hon, G., Fu, Y., Ching, C. W., Hawkins, R. D., et al. (2007). Distinct and predictive chromatin signatures of transcriptional promoters and enhancers in the human genome. *Nature Genetics*, *39*(3), 311–318.

Hon, G., Wang, W., & Ren, B. (2009). Discovery and annotation of functional chromatin signatures in the human genome. *PLoS Computational Biology*, *5*(11), e1000566.

Huang, H., Rambaldi, I., Daniels, E., & Featherstone, M. (2003). Expression of the Wdr9 gene and protein products during mouse development. *Developmental Dynamics*, *227* (4), 608–614.

Huff, J. T., Plocik, A. M., Guthrie, C., & Yamamoto, K. R. (2010). Reciprocal intronic and exonic histone modification regions in humans. *Nature Structural and Molecular Biology*, *17*(12), 1495–1499.

Iqbal, K., Jin, S. G., Pfeifer, G. P., & Szabó, P. E. (2011). Reprogramming of the paternal genome upon fertilization involves genome-wide oxidation of 5-methylcytosine. *Proceedings of the National Academy of Sciences of the United States of America*, *108*(9), 3642–3647.

Iwase, S., Xiang, B., Ghosh, S., Ren, T., Lewis, P. W., Cochrane, J. C., et al. (2011). ATRX ADD domain links an atypical histone methylation recognition mechanism to human mental-retardation syndrome. *Nature Structural and Molecular Biology*, *18*, 769–776.

Jeffery, L., & Nakielny, S. (2004). Components of the DNA methylation system of chromatin control are RNA-binding proteins. *The Journal of Biological Chemistry*, *279*, 49479–49487.

Jentarra, G. M., Olfers, S. L., Rice, S. G., Srivastava, N., Homanics, G. E., Blue, M., et al. (2010). Abnormalities of cell packing density and dendritic complexity in the MeCP2 A140V mouse model of Rett syndrome/X-linked mental retardation. *BMC Neuroscience*, *11*, 19.

Jeong, S., Liang, G., Sharma, S., Lin, J. C., Choi, S. H., Han, H., et al. (2009). Selective anchoring of DNA methyltransferases 3A and 3B to nucleosomes containing methylated DNA. *Molecular and Cellular Biology*, *29*(19), 5366–5376.

Jin, S. G., Wu, X., Li, A. X., & Pfeifer, G. P. (2011). Genomic mapping of 5-hydroxymethylcytosine in the human brain. *Nucleic Acids Research*, *39*(12), 5015–5024 (Epub 2011 March 4).

Johnson, M. B., Kawasawa, Y. I., Mason, C. E., Krsnik, Z., Coppola, G., Bogdanović, D., et al. (2009). Functional and evolutionary insights into human brain development through global transcriptome analysis. *Neuron*, *62*(4), 494–509.

Jones, P. L., Veenstra, G. J., Wade, P. A., Vermaak, D., Kass, S. U., Landsberger, N., et al. (1998). Methylated DNA and MeCP2 recruit histone deacetylase to repress transcription. *Nature Genetics*, *19*, 187–191.

Kalkhoven, E. (2004). CBP and p300: HATs for different occasions. *Biochemical Pharmacology*, *68*, 1145–1155.

Kaufmann, W. E., Jarrar, M. H., Wang, J. S., Lee, Y. J., Reddy, S., Bibat, G., et al. (2005). Histone modifications in Rett syndrome lymphocytes: A preliminary evaluation. *Brain & Development*, *27*, 331–339.

Kerkel, K., Schupf, N., Hatta, K., Pang, D., Salas, M., Kratz, A., et al. (2010). Altered DNA methylation in leukocytes with trisomy 21. *PLoS Genetics*, *6*(11), e1001212.

Kernohan, K. D., Jiang, Y., Tremblay, D. C., Bonvissuto, A. C., Eubanks, J. H., Mann, M. R., et al. (2010). ATRX partners with cohesin and MeCP2 and contributes to developmental silencing of imprinted genes in the brain. *Developmental Cell*, *18*, 191–202.

Klein, C. J., Botuyan, M. V., Wu, Y., Ward, C. J., Nicholson, G. A., Hammans, S., et al. (2011). Mutations in DNMT1 cause hereditary sensory neuropathy with dementia and hearing loss. *Nature Genetics*, *43*(6), 595–600.

Klose, R. J., & Bird, A. P. (2004). MeCP2 behaves as an elongated monomer that does not stably associate with the Sin3a chromatin remodeling complex. *The Journal of Biological Chemistry*, *279*, 46490–46496.

Koh, K. P., Yabuuchi, A., Rao, S., Huang, Y., Cunniff, K., Nardone, J., et al. (2011). Tet1 and Tet2 regulate 5-hydroxymethylcytosine production and cell lineage specification in mouse embryonic stem cells. *Cell Stem Cell*, *8* (2), 200–213.

Korzus, E., Rosenfeld, M. G., & Mayford, M. (2004). CBP histone acetyltransferase activity is a critical component of memory consolidation. *Neuron*, *42*, 961–972.

Kourmouli, N., Sun, Y. M., van der Sar, S., Singh, P. B., & Brown, J. P. (2005). Epigenetic regulation of mammalian pericentric heterochromatin in vivo by HP1. *Biochemical and Biophysical Research Communications*, *337*, 901–907.

Kramer, J. M., Kochinke, K., Oortveld, M. A., Marks, H., Kramer, D., de Jong, E. K., et al. (2011). Epigenetic regulation of learning and memory by Drosophila EHMT/G9a. *PLoS Biology*, *9*(1), e1000569.

Kuhn, D. E., Nuovo, G. J., Terry, A. V., Jr., Martin, M. M., Malana, G. E., Sansom, S. E., et al. (2010). Chromosome 21-derived microRNAs provide an etiological basis for aberrant protein expression in human Down syndrome brains. *The Journal of Biological Chemistry*, *285*(2), 1529–1543.

Lechner, M. S., Schultz, D. C., Negorev, D., Maul, G. G., & Rauscher, F. J. 3rd (2005). The mammalian hetrochromatin protein 1 binds diverse nuclear proteins through a common motif that targets the chromoshadow domain. *Biochemical and Biophysical Research Communications*, *331*, 929–937.

Lee, J., Hagerty, S., Cormier, K. A., Kim, J., Kung, A. L., Ferrante, R. J., et al. (2008). Monoallele deletion of CBP leads to pericentromeric heterochromatin condensation through ESET expression and histone H3 (K9) methylation. *Human Molecular Genetics*, *17*(12), 1774–1782.

Lein, E. S., Hawrylycz, M. J., Ao, N., Ayres, M., Bensinger, A., Bernard, A., et al. (2007). Genome-wide atlas of gene expression in the adult mouse brain. *Nature*, *445*(7124), 168–176.

Leonard, H., Downs, J., Jian, L., Bebbington, A., Jacoby, P., Nagarajan, L., et al. (2010). Valproate and risk of fracture in Rett syndrome. *Archives of Disease in Childhood*, *95*, 444–448.

Lepagnol-Bestel, A. M., Zvara, A., Maussion, G., Quignon, F., Ngimbous, B., Ramoz, N., et al. (2009). DYRK1A interacts with the REST/NRSF-SWI/SNF chromatin remodelling complex to deregulate gene clusters involved in the neuronal phenotypic traits of Down syndrome. *Human Molecular Genetics*, *18*(8), 1405–1414.

Levenson, J. M., O'Riordan, K. J., Brown, K. D., Trinh, M. A., Molfese, D. L., & Sweatt, J. D. (2004). Regulation of histone acetylation during memory formation in the hippocampus. *The Journal of Biological Chemistry*, *279*(39), 40545–40559.

Lewis, P. W., Elsaesser, S. J., Noh, K. M., Stadler, S. C., & Allis, C. D. (2010). Daxx is an H3.3-specific histone chaperone and cooperates with ATRX in replication-independent chromatin assembly at telomeres. *Proceedings of the National Academy of Sciences of the United States of America*, *107*, 14075–14078.

Lewis, J. D., Meehan, R. R., Henzel, W. J., Maurer-Fogy, I., Jeppesen, P., Klein, F., et al. (1992). Purification, sequence, and cellular localization of a novel chromosomal protein that binds to methylated DNA. *Cell*, *69*, 905–914.

Lioy, D. T., Garg, S. K., Monaghan, C. E., Raber, J., Foust, K. D., Kaspar, B. K., et al. (2011). A role for glia in the progression of Rett's syndrome. *Nature*, *475*(7357), 497–500.

Lubin, F. D. (2011). Epigenetic gene regulation in the adult mammalian brain: Multiples role in memory formation. *Neurobiology of Learning and Memory*, *96*, 68–78.

Ma, D. K., Jang, M. H., Guo, J. U., Kitabatake, Y., Chang, M. L., Pow-Anpongkul, N., et al. (2009). Neuronal activity-induced Gadd45b promotes epigenetic DNA demethylation and adult neurogenesis. *Science*, *323*(5917), 1074–1077.

Mao, R., Wang, X., Spitznagel, E. L., Jr., Frelin, L. P., Ting, J. C., Ding, H., et al. (2005). Primary and secondary transcriptional effects in the developing human Down syndrome brain and heart. *Genome Biology*, *6*(13), R107.

Meilinger, D., Fellinger, K., Bultmann, S., Rothbauer, U., Bonapace, I. M., Klinkert, W. E., et al. (2009). Np95 interacts with de novo DNA methyltransferases, Dnmt3a and Dnmt3b, and mediates epigenetic silencing of the viral CMV promoter in embryonic stem cells. *EMBO Reports*, *10*(11), 1259–1264 (Epub 2009 October 2).

Miller, C. A., Gavin, C. F., White, J. A., Parrish, R. R., Honasoge, A., Yancey, C. R., et al. (2010). Cortical DNA methylation maintains remote memory. *Nature Neuroscience*, *13*(6), 664–666.

Nagarajan, R. P., Hogart, A. R., Gwye, Y., Martin, M. R., & LaSalle, J. M. (2006). Reduced MeCP2 expression is frequent in autism frontal cortex and correlates with aberrant MECP2 promoter methylation. *Epigenetics*, *1*(4), e1–e11.

Nan, X., Campoy, F. J., & Bird, A. (1997). MeCP2 is a transcriptional repressor with abundant binding sites in genomic chromatin. *Cell*, *88*, 471–481.

Nan, X., Hou, J., Maclean, A., Nasir, J., Lafuente, M. J., Shu, X., et al. (2007). Interaction between chromatin proteins MECP2 and ATRX is disrupted by mutations that cause inherited mental retardation. *Proceedings of the National Academy of Sciences of the United States of America*, *104*, 2709–2714.

Nan, X., Ng, H. H., Johnson, C. A., Laherty, C. D., Turner, B. M., Eisenman, R. N., et al. (1998). Transcriptional repression by the metil-CpG-binding protein MeCP2 involves a histone deacetylase complex. *Nature*, *393*, 386–389.

Nelson, E. D., Bal, M., Kavalali, E. T., & Monteggia, L. M. (2011). Selective impact of MeCP2 and associated histone deacetylases on the dynamics of evoked excitatory neurotransmission. *Journal of Neurophysiology*, *106*, 193–201.

Ooi, S. K., Qiu, C., Bernstein, E., Li, K., Jia, D., Yang, Z., et al. (2007). DNMT3L connects unmethylated lysine 4 of histone H3 to de novo methylation of DNA. *Nature*, *448* (7154), 714–717.

Ormazabal, A., Artuch, R., Vilaseca, M. A., Aracil, A., & Pineda, M. (2005). Cerebrospinal fluid concentrations of folate, biogenic amines and pterins in Rett syndrome: treatment with folinic acid. *Neuropediatrics*, *36*(6), 380–385.

Otani, J., Nankumo, T., Arita, K., Inamoto, S., Ariyoshi, M., & Shirakawa, M. (2009). Structural basis for recognition of H3K4 methylation status by the DNA methyltransferase 3A ATRX-DNMT3-DNMT3L domain. *EMBO Reports*, *10*, 1235–1241.

Palini, S., De Stefani, S., Scala, V., Dusi, L., & Bulletti, C. (2011). Epigenetic regulatory mechanisms during preimplantation embryo development. *Annals of the New York Academy of Sciences*, *1221*, 54–60. doi:10.1111/j.1749-6632.2010.05937.x.

Pankotai, T., Komonyi, O., Bodai, L., Ujfaludi, Z., Muratoglu, S., Ciurciu, A., et al. (2005). The homologous Drosophila transcriptional adaptors ADA2a and ADA2b are both required for normal development but have different functions. *Molecular and Cellular Biology*, *25*(18), 8215–8227.

Park, J., Oh, Y., & Chung, K. C. (2009). Two key genes closely implicated with the neuropathological characteristics in Down syndrome: DYRK1A and RCAN1. *BMB Reports*, *42*(1), 6–15.

Pogribna, M., Melnyk, S., Pogribny, I., Chango, A., Yi, P., & James, S. J. (2001). Homocysteine metabolism in children with Down syndrome: In vitro modulation. *The American Journal of Human Genetics*, *69*(1), 88–95.

Ramirez-Carrozzi, V. R., Braas, D., Bhatt, D. M., Cheng, C. S., Hong, C., Doty, K. R., et al. (2009). A unifying model for the selective regulation of inducible transcription by CpG islands and nucleosome remodeling. *Cell*, *138*(1), 114–128.

Rath, A., Naryanan, T. J., Chowdhary, G. V., & Murthy, J. M. (2005). Valproate-induced hyperammonemic encephalopathy with normal liver function. *Neurology India*, *53*, 226–228.

Reddy, B. A., Bajpe, P. K., Bassett, A., Moshkin, Y. M., Kozhevnikova, E., Bezstarosti, K., et al. (2010). Drosophila transcription factor Tramtrack69 binds MEP1 to recruit the chromatin remodeler NuRD. *Molecular and Cellular Biology*, *30*(21), 5234–5244.

Reese, B. E., Bachman, K. E., Baylin, S. B., & Rountree, M. R. (2003). The methyl-CpG binding protein MBD1 interacts with the p150 subunit of chromatin assembly factor 1. *Molecular and Cellular Biology*, *23*(9), 3226–3236.

Rhee, I., Bachman, K. E., Park, B. H., Jair, K. W., Yen, R. W., Schuebel, K. E., et al. (2002). DNMT1 and DNMT3b cooperate to silence genes in human cancer cells. *Nature*, *416* (6880), 552–556.

Rottach, A., Frauer, C., Pichler, G., Bonapace, I. M., Spada, F., & Leonhardt, H. (2010). The multi-domain protein Np95 connects DNA methylation and histone modification. *Nucleic Acids Research*, *38*(6), 1796–1804.

Salman, M. (2002). Systematic review of the effect of therapeutic dietary supplements and drugs on cognitive function in subjects with Down syndrome. *European Journal of Paediatric Neurology*, *6*(4), 213–219.

Schones, D. E., Cui, K., Cuddapah, S., Roh, T. Y., Barski, A., Wang, Z., et al. (2008). Dynamic regulation of nucleosome positioning in the human genome. *Cell*, *132*(5), 887–898.

Schwartz, S., Meshorer, E., & Ast, G. (2009). Chromatin organization marks exon-intron structure. *Nature Structural and Molecular Biology*, *16*(9), 990–995.

Shahbazian, M., Young, J., Yuva-Paylor, L., Spencer, C., Antalffy, B., Noebels, J., et al. (2002). Mice with truncated MeCP2 recapitulate many syndrome features and display hyperacetylation of histone H3. *Neuron*, *35*, 243–254.

Sharif, J., Endo, T. A., Toyoda, T., & Koseki, H. (2010). Divergence of CpG island promoters: A consequence or cause of evolution? *Development, Growth & Differentiation*, *52*(6), 545–554.

Simone, C. (2006). SWI/SNF: The crossroads where extracellular signaling pathways meet chromatin. *Journal of Cellular Physiology*, *207*(2), 309–314.

Skene, P. J., Illingworth, R. S., Webb, S., Kerr, A. R., James, K. D., Turner, D. J., et al. (2010). Neuronal MeCP2 is expressed at near histone-octamer levels and globally alters the chromatin state. *Molecular Cell*, *37*, 457–468.

Stancheva, I., Collins, A. L., Van den Veyver, I. B., Zoghbi, H., & Meehan, R. R. (2003). A mutant form of MeCP2 protein associated with human Rett syndrome cannot be displaced from methylated DNA by notch in Xenopus embryos. *Molecular Cell*, *12*, 425–435.

Straussman, R., Nejman, D., Roberts, D., Steinfeld, I., Blum, B., Benvenisty, N., et al. (2009). Developmental programming of CpG island methylation profiles in the human genome. *Nature Structural and Molecular Biology*, *16*(5), 564–571.

Sun, H. J., Xu, X., Wang, X. L., Wei, L., Li, F., Lu, J., et al. (2006). Transcription factors Ets2 and Sp1 act synergistically with histone acetyltransferase p300 in activating human interleukin-12 p40 promoter. *Acta Biochimica et Biophysica Sinica*, *38*(3), 194–200.

Suzuki, M., Yamada, T., Kihara-Negishi, F., Sakurai, T., & Oikawa, T. (2003). Direct association between PU.1 and

MeCP2 that recruits mSin3A-HDAC complex for PU.1-mediated transcriptional repression. *Oncogene*, *22*(54), 8688–8698.

Tang, C. S., & Epstein, R. J. (2007). A structural split in the human genome. *PLoS One*, *2*(7), e603.

Temudo, T., Rios, M., Prior, C., Carrilho, I., Santos, M., Maciel, P., et al. (2009). Evaluation of CSF neurotransmitters and folate in 25 patients with Rett disorder and effects of treatment. *Brain & Development*, *31*(1), 46–51.

Thiru, A., Nietlispach, D., Mott, H. R., Okuwaki, M., Lyon, D., Nielsen, P. R., et al. (2004). Structural basis of HP1/PXVXL motif peptide interactions and HP1 localisation to heterochromatin. *The EMBO Journal*, *23*, 489–499.

Tilgner, H., Nikolaou, C., Althammer, S., Sammeth, M., Beato, M., Valcárcel, J., et al. (2009). Nucleosome positioning as a determinant of exon recognition. *Nature Structural & Molecular Biology*, *16*(9), 996–1001.

Ueda, T., Postnikov, Y. V., & Bustin, M. (2006). Distinct domains in high mobility group N variants modulate specific chromatin modifications. *The Journal of Biological Chemistry*, *281*(15), 10182–10187.

Urdinguio, R. G., Fernandez, A. F., Lopez-Nieva, P., Rossi, S., Huertas, D., Kulis, M., et al. (2010). Disrupted microRNA expression caused by Mecp2 loss in a mouse model of Rett syndrome. *Epigenetics*, *5*, 656–663.

Urdinguio, R. G., Lopez-Serra, L., Lopez-Nieva, P., Alaminos, M., Diaz-Uriarte, R., Fernandez, A. F., et al. (2008). Mecp2-null mice provide new neuronal targets for Rett syndrome. *PLoS One*, *3*, e3669.

Urdinguio, R. G., Pino, I., Ropero, S., Fraga, M. F., & Esteller, M. (2007). Histone H3 and H4 modification profiles in a Rett syndrome mouse model. *Epigenetics*, *2*, 11–14.

Urdinguio, R. G., Sanchez-Mut, J. V., & Esteller, M. (2009). Epigenetic mechanisms in neurological diseases: Genes, syndromes, and therapies. *Lancet Neurology*, *8*, 1056–1072.

Vecsey, C. G., Hawk, J. D., Lattal, K. M., et al. (2007). Histone deacetylase inhibitors enhance memory and synaptic plasticity via CREB: CBP-dependent transcriptional activation. *The Journal of Neuroscience*, *27*, 6128–6140.

Vecsler, M., Simon, A. J., Amariglio, N., Rechavi, G., & Gak, E. (2010). MeCP2 deficiency downregulates specific nuclear proteins that could be partially recovered by valproic acid in vitro. *Epigenetics*, *5*, 61–67.

Vermeulen, M., Eberl, H. C., Matarese, F., Marks, H., Denissov, S., Butter, F., et al. (2010). Quantitative interaction proteomics and genome-wide profiling of epigenetic histone marks and their readers. *Cell*, *142*(6), 967–980.

Wan, M., Zhao, K., Lee, S. S., & Francke, U. (2001). MECP2 truncating mutations cause histone hyperacetylation in Rett syndrome. *Human Molecular Genetics*, *10*, 1085–1092.

Wang, Z., Zang, C., Rosenfeld, J. A., Schones, D. E., Barski, A., Cuddapah, S., et al. (2008). Combinatorial patterns of histone acetylations and methylations in the human genome. *Nature Genetics*, *40*(7), 897–903.

Wong, L. H., McGhie, J. D., Sim, M., Anderson, M. A., Ahn, S., Hannan, R. D., et al. (2010). ATRX interacts with H3.3 in maintaining telomere structural integrity in pluripotent embryonic stem cells. *Genome Research*, *20*, 351–360.

Wu, H., Coskun, V., Tao, J., Xie, W., Ge, W., Yoshikawa, K., et al. (2010). Dnmt3a-dependent nonpromoter DNA methylation facilitates transcription of neurogenic genes. *Science*, *329*(5990), 444–448.

Wu, J. I., Lessard, J., & Crabtree, G. R. (2009). Understanding the words of chromatin regulation. *Cell*, *136*(2), 200–206.

Wu, J. I., Lessard, J., Olave, I. A., Qiu, Z., Ghosh, A., Graef, I. A., et al. (2007). Regulation of dendritic development by neuron-specific chromatin remodeling complexes. *Neuron*, *56*(1), 94–108.

Xin, Y., Chanrion, B., Liu, M. M., Galfalvy, H., Costa, R., Ilievski, B., et al. (2010). Genome-wide divergence of DNA methylation marks in cerebral and cerebellar cortices. *PLoS One*, *5*(6), e11357.

Xu, Y., Wu, F., Tan, L., Kong, L., Xiong, L., Deng, J., et al. (2011). Genome-wide regulation of 5hmC, 5mC, and gene expression by Tet1 hydroxylase in mouse embryonic stem cells. *Molecular Cell*, *42*(4), 451–464.

Yang, X. J., & Seto, E. (2007). HATs and HDACs: From structure, function and regulation to novel strategies for therapy and prevention. *Oncogene*, *26*(37), 5310–5318.

Yasui, D. H., Peddada, S., Bieda, M. C., Vallero, R. O., Hogart, A., Nagarajan, R. P., et al. (2007). Integrated epigenomic analyses of neuronal MeCP2 reveal a role for long-range interaction with active genes. *Proceedings of the National Academy of Sciences of the United States of America*, *104*, 19416–19421.

Yeo, G., Holste, D., Kreiman, G., & Burge, C. B. (2004). Variation in alternative splicing across human tissues. *Genome Biology*, *5*(10), R74.

Yoon, H. G., Chan, D. W., Reynolds, A. B., Qin, J., & Wong, J. (2003). N-CoR mediates DNA methylation-dependent repression through a methyl CpG binding protein Kaiso. *Molecular Cell*, *12*(3), 723–734.

Young, J. I., Hong, E. P., Castle, J. C., Crespo-Barreto, J., Bowman, A. B., Rose, M. F., et al. (2005). Regulation of RNA splicing by the methylation-dependent transcriptional repressor methyl-CpG binding protein 2. *Proceedings of the National Academy of Sciences of the United States of America*, *102*, 17551–17558.

Yun, M., Wu, J., Workman, J. L., & Li, B. (2011). Readers of histone modifications. *Cell Research*, *21*(4), 564–578.

Zemach, A., McDaniel, I. E., Silva, P., & Zilberman, D. (2010). Genome-wide evolutionary analysis of eukaryotic DNA methylation. *Science*, *328*(5980), 916–919.

M. Dierssen and R. de la Torre (Eds.)
Progress in Brain Research, Vol. 197
ISSN: 0079-6123

CHAPTER 5

Pathways to cognitive deficits in Down syndrome

Xiaolu Sturgeon[†,‡,1], Thanh Le[§,1], Md. Mahiuddin Ahmed[†] and Katheleen J. Gardiner[†,‡,¶,#,*]

[†] *Department of Pediatrics, University of Colorado Denver, Denver, CO, USA*
[‡] *Computational Biosciences Program, University of Colorado Denver, Denver, CO, USA*
[§] *Department of Computer Science and Engineering, University of Colorado Denver, Denver, CO, USA*
[¶] *Intellectual and Developmental Disabilities Research Center, University of Colorado Denver, Denver, CO, USA*
[#] *Human Medical Genetics and Neuroscience Programs, University of Colorado Denver, Denver, CO, USA*

Abstract: Major efforts in Down syndrome (DS) research have been directed at the identification and functional characterization of genes encoded by human chromosome 21 (HSA21). In parallel with this, tissue samples and cell lines derived from individuals with DS have been examined for abnormalities in gene expression and cellular morphology, and mouse models of DS have been characterized for abnormalities at the molecular, cellular, electrophysiological, and behavioral level. One goal of such investigations has been the identification of effective targets for pharmacotherapies that can prevent or correct the abnormalities and, by extension to human clinical trials, prevent or lessen aspects of the cognitive deficits seen in people with DS. Because it is caused by an extra copy of an entire chromosome, DS has been considered by some as too complicated a genetic perturbation to be amenable to postnatal pharmacological interventions. However, recent data from experiments with one mouse model, the Ts65Dn, have clearly demonstrated that several pharmacological interventions can indeed rescue DS-relevant learning and memory deficits. Extension of mouse data to successful human clinical trials will be aided by understanding the molecular basis of successful drug treatments, that is, how increased expression of HSA21 genes perturbs molecular mechanisms that are targeted and rescued by specific drugs. Here, we review information on HSA21 genes, their expression and their likely contributions to the DS phenotype. We then describe results of a bioinformatics effort that integrates information on genes known to cause intellectual disability when mutated, the pathways in which these genes function, and how these pathways are impacted by HSA21 encoded proteins. This pathway approach to the molecular basis of ID in

*Corresponding author.
Tel.: +1-303-724-0572; Fax: +1-303-724-3838
E-mail: katheleen.gardiner@ucdenver.edu

[1]These authors contributed equally to this work.

DOI: 10.1016/B978-0-444-54299-1.00005-4

DS aids in understanding why some drug therapies have been successful in the Ts65Dn and in predicting whether these same drugs are likely to be successful in treating ID in DS. These data can be used to design new experiments and interpret information for prediction of additional targets for effective drug treatments.

Keywords: partial trisomy; protein expression; protein interactions; mouse model; pharmacotherapy; posttranslational modification.

Introduction

It was in 1959 that Jerome Lejeune recognized that Down syndrome (DS) was associated with an extra, third copy of human chromosome 21 (HSA21; Lejeune et al., 1959). Since then, one hypothesis has been that intellectual disability (ID) in DS is caused by the increased expression, due to gene dosage, of normal but trisomic HSA21 genes. Accordingly, much basic research has been driven by the logic that understanding the molecular abnormalities that underlie DS-ID will lead to the development of treatments to prevent or ameliorate at least some aspects of the phenotype. Broadly, this effort has involved attempts to link functions of specific HSA21 genes to abnormalities observed in DS or in DS model systems at the molecular, cellular, electrophysiological, or behavioral level. This logic was supported by information on the first gene mapped to HSA21, superoxide dismutase, *SOD1* (Sichitiu et al., 1974; Tan et al., 1973). Elevated levels of reactive oxygen species in DS have been observed, as have the consequences for cellular damage, especially in sensitive neuronal structures of the developing brain (Busciglio and Yankner, 1995; Groner et al., 1994; Lott, 1982). Overexpression of protein SOD1 leading to excessive production of hydrogen peroxide remains an important hypothesis in the search for gene–phenotype correlations (Perluigi et al., 2011).

Early gene identification

Throughout the 1970s and 1980s, gene mapping was a slow and laborious process. By 1990, only 13 HSA21 genes with biochemical or biological associations had been identified (Gardiner et al., 1990). For some genes, arguments for relevance to ID in DS were strong, for example, the amyloid precursor protein gene, APP, also mutated in some forms of familial Alzheimer's disease (AD), suggested a direct link to neurodegeneration seen in DS (Goldgaber et al., 1987; Robakis et al., 1987; St George-Hyslop et al., 1987), and S100B gene, encoding a Ca^{2+}-binding protein, suggested involvement of glial cell activation and neuroinflammation (Allore et al., 1988). However, for other genes, for example, the interferon inducible genes *MX1* and *MX2* (Aebi et al., 1989) and the lens protein gene, α-crystallin, CRYAA (Hawkins et al., 1987), the functional information did not suggest obvious consequences for brain development or function in trisomy.

The pace of HSA21 gene identification increased in the 1990s, when the development of the techniques of exon trapping and cDNA selection provided methods for comprehensive identification of transcribed sequences from defined genomic regions (Chen et al., 1996; Dahmane et al., 1998; Tassone et al., 1995). Together with the expansion of the number of dbEST (database of expressed sequence tags) entries from human tissues, this facilitated a rapid increase in complete cDNA open reading frame (ORF) sequences for newly assigned protein-coding genes on HSA21 (Gardiner and Yaspo, 1998). Using sequence homologies to proteins already studied in *Drosophila* and *Caenorhabditis elegans*, and analysis of mouse models with knockouts or transgenic overexpression of HSA21 cDNAs or their orthologs, newly identified HSA21 proteins were frequently reported with compelling or novel rationales for relevance to the DS cognitive phenotype. Among these were the genes encoding the neural cell adhesion molecule NCAM2 (Paoloni-Giacobino et al., 1997), the pre-mRNA adenosine to inosine deaminase, ADAR2, that edits the coding sequences of some serotonin and glutamate receptors (Villard et al., 1997), the protein kinase DYRK1A (Becker et al., 1998 and see Chapter 3), and the C21ORF5 gene, whose *C. elegans* ortholog results in abnormal patterning in development (Guipponi et al., 2000). Other genes, with equally compelling potential relevance, sometimes passed under the DS interest radar because they were originally identified in the study of a biological process or disease independent of DS. Examples of these include the gene encoding the guanine nucleotide exchange factor TIAM1 with activity specific for RAC; TIAM1 was first and is still largely studied for its role in cancer and metastasis (Michiels et al., 1995; Xu et al., 2010) although it is also involved in neurite outgrowth and dendritic spine formation (Shirazi Fard et al., 2010). The gene encoding the nuclear receptor interacting protein 1, NRIP1, was identified in studies of estrogen and other hormone receptor activity and more recently in metabolic regulation (reviewed in Fritah et al., 2010; White et al., 2008). The U2AF1 gene was identified as encoding a constitutive splicing factor (Zuo and Maniatis, 1996), and the Purkinje cell protein 4 (PCP4), as a modulator of calmodulin activity (Slemmon et al., 2000). While PCP4 is now being investigated for its possible role in ID in DS (Mouton-Liger et al., 2011), NRIP1 and U2AF1, like TIAM1, have not received much attention.

Gene identification from genomic sequence

In 2000, the genomic sequence of HSA21q was published (Hattori et al., 2000). Not considered a draft but high-quality finished sequence, it had few gaps, and annotation for gene content identified 225 genes and gene models. Gene discovery was dependent both upon experimental evidence of transcription, making significant use of dbEST, and on *ab initio* exon predictions. Table 1 lists the

Table 1. Comparison of HSA21q gene annotations in 2000 and in 2011

Hattori et al. (2000)		Sturgeon and Gardiner (2011)			
				mc	ms
Known protein coding	127	RefSeqP	161	157	3
Protein similarity	13	MicroRNA	5	5	0
Domain similarity	17				
Novel	68	Novel; ORF>50 AA	146	139[a]	279
		Novel; ORF<50 AA, repetitive or incomplete	250		
Total	225		562	444	

RefSeqP: protein-coding genes annotated in the Reference Sequence database; mc, mouse transcripts conserved in HSA21; ms, mouse-specific transcripts.

[a]HSA21 transcripts with minimal conservation in orthologous mouse genomic sequence, and no mouse transcripts identified (60% identity over>100nt or complete cDNA).

classes and numbers of genes present in that initial annotation. As dbEST expanded in both human and model organism entries, and when the mouse genome was sequenced, the HSA21 gene catalog was revised and supplemented with experimental validation and information on conservation in mouse (Gardiner et al., 2002, 2003; Reymond et al., 2001, 2002). Data from the most recent annotation of HSA21 are also presented in Table 1 (Sturgeon and Gardiner, 2011). Of the original gene annotations, the large majority remains; in only a small number of cases, two models were merged or a model was subsequently found to be a pseudogene. Notably, the original annotation identified at least partial transcripts for essentially all the classical protein-coding genes currently annotated in the SwissProt and the Reference Sequence (RefSeq) protein databases and showing open reading conservation in mouse.

The most dramatic difference in comparison with the 2000 gene list is the more than twofold increase in the number of genes. This is not due to a change in criteria for gene identification, but rather almost entirely to the expansion of the mRNA and dbEST databases. Although many appear rare, with <3 independent database entries, all new transcripts that are based on dbEST entries are multiexonic with a majority of consensus splice sites and therefore cannot simply be dismissed. Many of the new gene models have nonclassical protein-coding gene characteristics, many span large genomic regions (>100 kb), and many others are antisense or likely antisense to RefSeq protein-coding genes. Another notable feature of the current HSA21 gene catalog is that the majority of transcripts show low or no sequence similarity in mouse genomic sequence. These nonconserved transcripts include ~30–40 from the original annotation and many with further experimental validation (Gardiner et al., 2002; Reymond et al., 2001). Almost all have also been included in the annotation generated by VEGA (vertebrate genome annotation), which speaks to the robustness of the supporting evidence (Wilming et al., 2008). A reciprocal gene annotation of the orthologous regions of mouse chromosomes 16, 17, and 10 shows a similar frequency of apparent mouse-specific transcripts with low or no nucleotide sequence similarity within HSA21. As summarized in Table 1, the only HSA21-encoded transcripts that are highly conserved in orthologous mouse genomic regions are 157 of the 161 classical protein-coding genes and five microRNAs. These facts obviously complicate the development of valid mouse models for complete trisomy of HSA21.

Gene dosage and gene transcription

Gene dosage effects of trisomy have been assumed to result in a 50% increase in the mRNA levels of trisomic genes. With the advent of microarray technology, this assumption could be tested for essentially all HSA21 genes. Numerous experiments have used mRNAs from DS, and DS mouse models to screen whole genome or HSA21-specific oligonucleotide and cDNA arrays (see Vilardell et al., 2011 for a summary and Chapter 2 and 3). Broad conclusions about HSA21 gene overexpression are complicated because not all experiments included a complete set of HSA21 genes. Arrays used in early experiments lacked probes for many HSA21 genes, and all the mouse models of DS are only partial trisomies (Fig. 1; Roubertoux and Carlier, 2010; see also Chapter 9) carrying three copies of only 70–120 of the 161 HSA21 protein-coding genes (or their mouse orthologs). Nevertheless, two conclusions are robust: (i) most but not all trisomic classical protein-coding genes show increased levels of mRNAs and (ii) the specifics of this subset depend on organism and tissue source. In their meta-analysis of microarray data (and some protein data), Vilardell et al. (2011) generated a list of 69 HSA21 genes most commonly identified as overexpressed in trisomy. This is a valuable overview, but does not eliminate the need for investigation of specific biological processes. For example, understanding the effects of trisomy on hippocampal development or adult neurogenesis likely will require experiments specifically addressing these processes.

An additional caution associated with all microarray data is that mRNA levels do not predict protein levels (Maier et al., 2009; Schwanhäusser et al., 2011). Schwanhäusser et al. (2011) recently compared levels in mouse fibroblasts of ~5500 proteins determined by mass spectrometry to their corresponding mRNA levels determined by RNAseq. They concluded that variation in RNA levels contributed ~40% to the variation in protein levels. Inspection of their data for the 42 HSA21 orthologs detected showed that protein to mRNA ratios ranged over almost 20-fold differences, for example, protein to mRNA levels were ~500 for U2AF1 and 19,000 for CCT8. Predicting the outcome at the protein level of increased mRNA expression in trisomy is therefore difficult. Furthermore, it will be of interest to see results of RNAseq applied to lymphocytes from DS and controls, to ascertain more accurately the natural variation in mRNA levels resulting from heterogeneous genetic backgrounds. Given

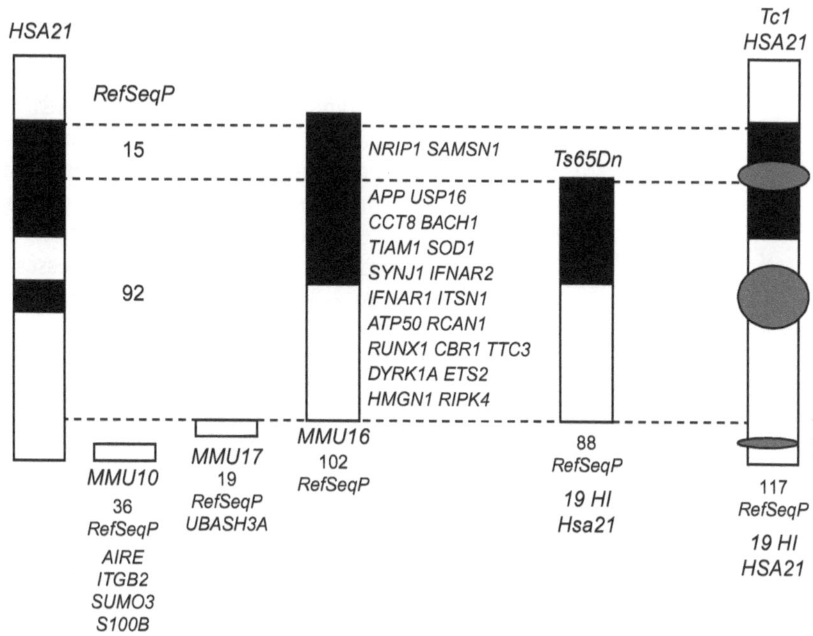

Fig. 1. Distribution of HSA21 protein-coding genes within mouse chromosomes and major mouse models of DS. A schematic of HSA21 is shown at the left. Genes within three regions of HSA21 map to segments of mouse chromosomes 10, 17, and 16 (MMU10, MMU17, and MMU16) as indicated. Mouse models, separately carrying a duplication of each of these regions, have been created (Yu et al., 2010). The Ts65Dn mouse model is trisomic for the telomeric segment of MMU16, and the Tc1 mouse model, in addition to the normal complement of mouse chromosomes, carries a human chromosome 21 in which there are internal deletions (indicated by shaded circles; O'Doherty et al., 2005). The number of HSA21 Reference Sequence database protein-coding genes (RefSeqP) mapping to HSA21 regions and regions in mouse models are indicated. See Fig. 1 for numbers of additional human- and mouse-specific genes. Names and locations of the 27 "high-impact" HSA21 proteins (see Table 7) are indicated (see also chapter 9).

that the popular DS mouse model, Ts65Dn is not maintained on an inbred strain, variation in mRNA levels from a set of these mice is also of interest.

Functional interactions among HSA21 genes

As functional studies have expanded, many examples have been uncovered of HSA21 proteins sharing targets or impacting the same biological processes. Examples of these interactions are shown in Table 2 (Gardiner and Costa, 2006; see also Siddiqui et al., 2008 and references therein). They include DYRK1A (see Chapter 3), a protein kinase that phosphorylates some substrates that are dephosphorylated by the protein phosphatase, calcineurin. Calcineurin activity is regulated in part by the HSA21 protein RCAN1 and also by calmodulin whose activity is modulated, in part, by the HSA21 protein PCP4. Furthermore, the HSA21 protein synaptojanin 1 (SYNJ1) is a target of both DYRK1A and calcineurin. The consequences for activity levels of proteins targeted by both DYRK1A and calcineurin are difficult to predict. In another example, APP and SOD1 must interact functionally because the double transgenic overexpressing both genes displays phenotypic features including protein expression differences, learning and memory deficits, and LTP abnormalities that are not seen in the separate single gene transgenics (Harris-Cerruti et al., 2004). Potential perturbations

Table 2. Functional interactions among HSA21 encoded proteins

HSA21 proteins	Functional interactions
DYRK1A, RCAN1, PCP4, SYNJ1	Phosphorylation levels of SYNJ1, dynamin
APP, SOD1	Learning/memory transgenic mice
SCAF4, DYRK1A, U2AF1[a]	Pre-mRNA splicing
GRIK1, ADAR2[b]	Glutamate receptor functional properties
HSPA13[c], USP25, USP16, TTC3, UBASH3A[a], UBE2G2[b]	Ubiquitin pathway
NRIP1[c], DYRK1A, SUMO3[b]	Activity of transcription factors, ELK, GR
S100B[b], TRPM2[b]	Ca-signaling
SOD1, TRPM2[b]	Oxidative stress

See text for references. [a] and [b], orthologs map to mouse chromosomes 17 and 10, respectively. [c], ortholog maps to mouse chromosome 16 outside the segment trisomic in the Ts65Dn.

of pre-mRNA splicing are predicted in DS by the presence on HSA21 of the constitutive splicing factor U2AF1, the serine–arginine (SR) splicing factor SCAF4 (a.k.a. SFRS15), and again DYRK1A whose targets include additional SR splicing factor proteins. Indeed, overexpression of DYRK1A alone has been shown to alter proportions of splice variants of neuronally relevant transcripts (Toiber et al., 2010). ADAR2 (a.k.a. ADARB1) is a pre-mRNA adenosine deaminase whose targets include the HSA21-encoded kainate receptor subunit, GRIK1; editing of GRIK1 by ADAR2 alters the coding sequence and the functional properties of the receptor. Six HSA21 proteins function in the ubiquitin pathway although specific protein substrates are largely unknown. Phosphorylation by DYRK1A and sumoylation by SUMO3 antagonistically impact activities of important transcription factors such as ELK (Yang et al., 2003), whose activation is required for successful performance in the hippocampal-based task, context fear conditioning (CFC; Sananbenesi et al., 2002), and the glucocorticoid receptor (Davies et al., 2008) involved in stress responses. As last examples, both S100B and the transient receptor potential cation channel, TRPM2, are involved in calcium signaling, and both SOD1 and TRPM2 are involved in responses to oxidative stress (Hecquet et al., 2008).

A good illustration of the need to understand *in vivo* complexities of HSA21 gene interactions comes from phenotype comparisons of a transgenic mouse that overexpresses the single HSA21 ortholog DYRK1A from a genomic (BAC clone) construct (Ahn et al., 2006), and the Ts1Rhr mouse model that overexpresses DYRK1A from an ~3Mb genomic segment containing an additional ~30 protein-coding genes (Olson et al., 2007). The former displays deficits in the Morris water maze and the latter does not. A reasonable interpretation is that functions of one or more of the 30 additional genes trisomic in the Ts1Rhr repress the effects of DYRK1A alone, that is, genomic context matters.

An important use of these data is relevant to modeling DS in mouse. Table 2 includes the chromosomal location of the mouse orthologs (chromosomes 16, 17, or 10) of the genes described above (see also Fig. 1). Notably, 7 of the 19 genes in Table 2 are not trisomic in the popular Ts65Dn model, and six of these map to mouse chromosomes 17 or 10. Of the eight interaction groups, six contain proteins not trisomic in the Ts65Dn. When only one or a few members of a functional interaction group are overexpressed, the phenotypic consequences may not be the same as in a full trisomy. Given current knowledge, it must also be assumed that many additional HSA21-HSA21 protein functional interactions remain to be discovered.

Functional interactions between HSA21 and non-HSA21encoded proteins

Microarray experiments have shown that transcript levels of many non-HSA21 genes are also altered in trisomy (Vilardell et al., 2011 see also Chapter 2). This is not surprising, given that ~18 HSA21-encoded proteins are known to be transcription factors or transcription regulators (Gardiner and Costa, 2006). verexpression of these proteins could result in altered stoichiometry of transcriptional complexes and contribute to enhanced activation or repression of target genes. Additional subtleties can arise from posttranslational modifications (PTMs) of transcription factors, carried out directly or indirectly by HSA21 proteins, for example, direct phosphorylation by DYRK1A and sumoylation by SUMO3, and modulation of dephosphorylation by RCAN1. DYRK1A and SUMO3 compete for an overlapping site of modification of ELK, respectively activating and repressing its activity (Yang et al., 2003), with potential consequences in trisomy for transcription of its target genes. Detecting and interpreting the causes and consequences for trisomy-induced alterations in non-HSA21 gene expression is a challenge, because the number and identity of genes with altered expression differs among experiments, as a result of the differing tissues, developmental ages, and organisms investigated. Standard analysis tools such as Gene Ontology and Gene Set Enrichment Analysis (GSEA) have not provided grand insights. Rather they provide experiment-specific information for which a broad, integrated view is beyond current knowledge. For realistic interpretations, especially if the ultimate goal is understanding the potential for therapeutics, more global understanding is required.

Candidate genes and the Down syndrome critical region

In parallel with efforts at gene identification and characterization, rare cases of DS due to partial trisomy of HSA21 were studied for genotype– phenotype mapping (Rahmani et al., 1989). It was hoped that identification of a consistent reduced segment whose trisomy results in diagnosis of DS would limit the number of HSA21 genes that needed to be considered. This led early on to the designation of an ~3Mb segment within 21q22 as the Down syndrome critical (or chromosomal) region, the DSCR. This was defined as containing genes that are necessary for the phenotype that is diagnostic of DS. While the attraction of the DSCR as a rationale to focus the search for genes relevant to the DS phenotype was clear, there was controversy. The DSCR was identified as a region of minimal overlap among a small set of partial trisomies and had not been observed as a sole trisomic region in any patient with DS. The validity of the concept of a DSCR was questioned by the observations of Korenberg et al. (1994) that a diagnosis of DS was obtained with individuals trisomic for segments of HSA21 that excluded the DSCR. The concept of the DSCR was also not supported by analysis of the Ts1Rhr mouse that is trisomic for the mouse genomic segment orthologous to the DSCR and, as mentioned above, does not show some significant attributes, including learning/memory deficits in the Morris water maze (Olson et al., 2007). Recent more detailed analysis of human DS partial trisomies, with identification at the sequence level of the break points, emphasized that DS features, including ID, can be caused by trisomy of multiple segments of HSA21 not including the DSCR (Korbel et al., 2009). Candidate genes for significant contributions to ID in DS must therefore be considered throughout the long arm and further discussion of a DSCR is unwarranted (see also Chapter 2).

New challenges for candidate gene analysis

Several new challenges confront attempts to link HSA21 genes to features of DS-ID. First, based on data for mouse orthologs from the UniGene database and the Allen Brain Atlas, >90% of HSA21 protein-coding genes are likely expressed in brain. Thus, lack of detectable brain expression is not a useful filter for exclusion of a gene as a candidate. Second, as shown in Table 1 and discussed in Sturgeon and Gardiner (2011), the majority of HSA21 gene models and validated transcripts encode short ORFs (<80 amino acids, and many <50 amino acids) and lack identified functional domain similarities. Methods for discriminating protein coding from functional RNA genes, and discriminating both from potential transcriptional noise, are not established (Dinger et al., 2008). It is also noteworthy that, of the five microRNAs identified on HSA21, only one is represented by transcripts in cDNA sequence databases, suggesting that other RNA transcripts may await discovery. Once non-protein-coding genes are predicted, identifying their targets or assaying their functions remains difficult and laborious. Together, these observations suggest that we lack the tools to evaluate individually the majority of HSA21 genes for potential contributions to DS ID. This also suggests that efforts to distill the genetic basis for DS phenotypic features to a small number of protein-coding genes may be misplaced, based as it is on limited knowledge of individual gene functions. This is an important consideration for choosing targets for therapeutic interventions.

Pathway analysis: A complementary approach

As illustrated by the foregoing, substantial information has been obtained from experiments characterizing the functions of individual HSA21 genes and abnormalities in segmental trisomy mouse models. However, results have been constrained by the model systems used and specifically by the lack of a neurological model of a complete trisomy of HSA21. Undiscovered and unpredicted consequences of increased expression of HSA21 genes present in complete trisomy and absent in mouse models can confound interpretation of learning and memory deficits and responses to drug treatments. In spite of progress made, essential questions in DS research remain: how do the genes of HSA21, through their altered levels of expression, normal functions, and perturbed interactions, combine to produce the DS features, and how can these molecular consequences be manipulated to ameliorate their effects? Additional resources for information on HSA21 genes are contained in publically available, curated databases, in particular, those annotating pathways and protein–protein interactions (PPIs). These data were not obtained in DS-directed experiments and, in the context of DS, have remained largely unexploited.

As a complement to the experimental approaches described above, a bioinformatics effort that mines existing database content can provide a broad, unbiased look at pathway associations of HSA21 genes. A pathway perspective emphasizes that learning and memory, and in particular, deficits in DS, are not driven by single genes, but by complex and dynamic interactions among many protein activities. Such a view is compelled by the data in Table 2. Pathway analysis does not depend on detailed knowledge of the functions of HSA21 genes. Indeed, an experimentally observed (or predicted) pathway perturbation reflects the integration of the inputs, positive and negative, from all HSA21 genes (protein coding and functional RNA) that are overexpressed. The knowledge, and the predictions, gained from pathway analysis can contribute to interpretation of existing experimental results and to the direction of future experiments.

Methods

Protein-coding genes

The RefSeq protein database at the National Center for Biotechnology Information (NCBI) was used as the source for all human protein-coding genes (total ~19,000), and the subsets identified as ID genes, HSA21 protein-coding genes, and their mouse orthologs. All protein-coding genes were identified by ENTREZ Gene ID numbers and are referred to as RefSeq protein-coding genes and RefSeq proteins.

ID genes

An ID gene is defined as a gene for which at least one mutation has been identified that results in ID. Only human genes and mutations are considered. An initial set of ID genes was obtained from Inlow and Restifo (2004). Additional ID genes were obtained by searching entries in the database Online Mendelian Inheritance in Man with the text string "mental retardation" using the number limit to restrict results to genes with a phenotype description for which the molecular basis is known. Each entry was manually reviewed. A gene is classed as an ID gene if at least one individual with an identified mutation in that gene has been described with mental retardation.

Pathway data

Human pathways and pathway components were retrieved from four sources: the Kyoto Encyclopedia of Genes and Genomes (KEGG, http://www.genome.jp/kegg/; Kanehisa and Goto, 2000), Reactome (http://www.reactome.org, Croft et al., 2011), the Pathway Interaction Database (PID, http://pid.nci.nih.gov; Schaefer et al., 2009), and BioCarta (http://www.biocarta.com). For KEGG, proteins not classed within pathway reactions were excluded. In the KEGG, PID, and BioCarta databases, a protein is labeled either as a pathway protein or as a protein within a complex; only pathway proteins were included here.

Protein–protein interaction data

Human pair wise PPI data were retrieved from three data sources: the Human Protein Reference Database (HPRD, http://www.hprd.org; Peri et al., 2004), the Biological General Repository for Interaction Datasets (BioGRID, http://thebiogrid.org; Stark et al., 2006), and the IntAct database (http://www.ebi.ac.uk/intact/; Aranda et al., 2010). For the IntACT database, to restrict PPI to direct pair wise interactions, data obtained through "Spoke" or "Matrix" filters were excluded. PPIs were divided into two confidence levels. A PPI is annotated as level 1 (higher confidence) if it meets at least one of the following criteria: it was identified by an *in vivo* experiment as annotated in HPRD, the two proteins share at least one Gene Ontology (GO) cellular component term (Camon et al., 2003; see also Wu et al., 2010), or the mouse orthologs of the two genes are expressed in at least one common brain region as reported in the Allen Brain Atlas (Jones et al., 2009). For this last, the assumption is made that mouse and human orthologs will share similar expression patterns. Remaining PPIs, including those identified only by *in vitro* experiments and/or through yeast-two-hybrid experiments and lacking further supporting information, were annotated as level 2. This classification does not imply that these PPI are artifacts.

Related databases

The Gene_info file was downloaded from the NCBI (Maglott et al., 2011) and processed to retrieve gene information. Because some pathway data sources use Uniprot accession numbers as the identifiers for proteins, the idmapping_selected file was downloaded from Uniprot and used to create a mapping table for converting Uniprot accession numbers to Entrez Gene ID numbers. The mouse gene expression data were downloaded from the Allen Brain Atlas (Jones et al., 2009) and used as reference for brain region expression of the orthologous human genes. Additional human gene expression data were retrieved from the EST (expressed sequence tags) tissue profiles in the UniGene database (NCBI; Pontius et al. 2003) and filtered for brain expression. The GO database (Camon et al., 2003) was processed for retrieving annotations for each protein under cell components.

Identification of DS-ID, pathway data sets

Extended pathways were generated by annotating each component of a pathway with all primary interaction partners. ID pathways are defined as all pathways for which at least one component is an ID protein. DS pathways are the subset of pathways that have at least one HSA21 protein as a pathway component or as a primary interaction partner with any component. DS-ID pathways are the subset of DS pathways that are also ID pathways. The data processing is shown in Fig. 1.

Data availability

HSA21 and ID gene lists are available at http://gfuncpathdb.ucdenver.edu/iddrc/home.php. Complete datasets of pathways and attributes, including total numbers of components, PPI, and numbers of ID and HSA21 components and PPI are available in Supplementary Tables S1 and S2 (http://www.elsevierdirect.com/companions/9780444542991). Tools found at http://gfuncpathdb.ucdenver.edu/iddrc/hsa21gdpw/home.htm allow searching (i) pathways by name to retrieve lists of components and PPI to components, or the subsets that are ID or HSA21 proteins, and (ii) proteins by name for lists of PPI, and pathway associations as components or as PPI to components. Search tools also allow identification of protein components common among a user-specified set of pathways and pathway associations common to a user-specified set of proteins.

Results of pathway analysis

Pathways and extended pathways

Figure 2 illustrates data collection and integration used to define pathways of predicted relevance to ID in DS. Pathways and their components were obtained from databases (KEGG, Reactome, PID, and BioCarta) that are annotated and curated by experts using data from peer-reviewed literature reports. Each database is largely an independent effort, and each has specific criteria and priorities for annotation, and speed of curation. Thus, there are both redundancies and inconsistencies among the databases. As one example, Reactome, PID,

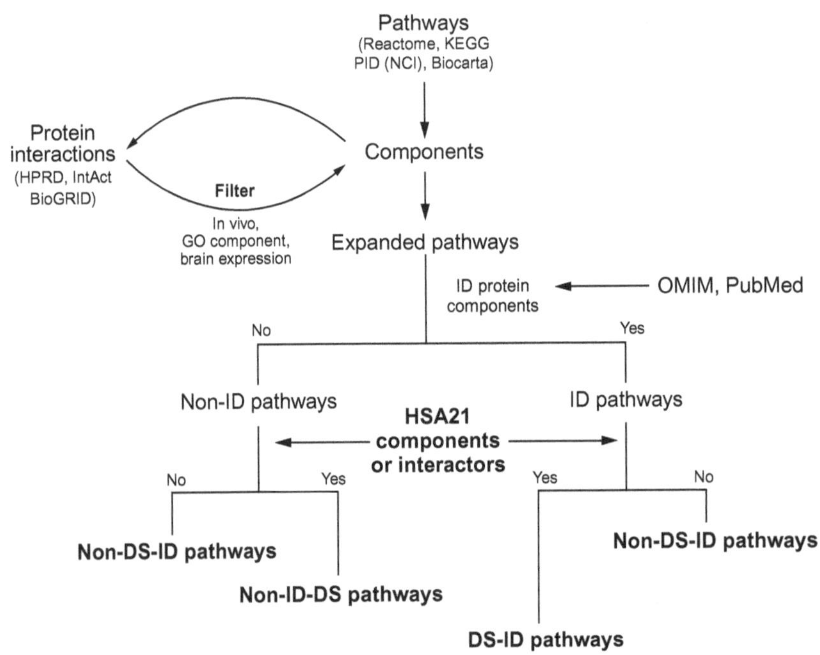

Fig. 2. Data retrieval and integration for identification of DS-ID pathways

and KEGG databases include pathways called signaling by Wnt, Wnt signaling network, and Wnt signaling pathway with 63, 28, and 144 components, respectively. The name overlap is a confound because the pathways have fewer than 20 protein components in common. While biological review can help to understand the reasons behind the differences, automated comparisons are not adequate for reliable decisions on when and how to combine pathways. Because of these ambiguities, we chose not to eliminate redundancies and instead we retain all individual database-specific pathways. As indicated in Table 3(a), a total of 1746 pathways were retrieved. The largest number, 1066, came from Reactome, while KEGG and PID each provided ~200. The total number of pathway protein components varies from ~1000 in BioCarta pathways to ~5000 in Reactome. Although Reactome has annotated five times the number of pathways as KEGG, the latter still includes almost 4500 proteins compared to Reactome's 5000. This is a result of Reactome's protocol of separately annotating subpathways and functionally related pathways, much of which describes additional relationships among proteins without adding new proteins. Comparing all pathway components from the four databases produced a nonredundant set of 7420 RefSeq proteins.

Table 3. Pathway and protein interaction data

(a) Pathways		Numbers of Proteins		
Database	No. Pathways	Pathway components	ID	HSA21
Reactome	1066	5016	206	43
KEGG	211	4467	189	43
PID (NCI-Nature)	215	2137	77	23
BioCarta	254	1053	47	13
Total (nr)	1746*	7420	267	69

(b) Protein–protein interactions (PPI)

Database	No. Pairs	Proteins	ID	HSA21
IntAct	24,439	7501	210	77
HPRD	38,995	9599	288	103
BioGRID	37,762	9068	270	93
Total (nr)	69,534	11,661	328	122
Total (nr, filtered)	52,484	9681	299	99
Totals (path+PPI)		13,364	370	132

*, not non-redundant

The activity of a pathway is impacted, not only by the levels of the individual components but also by the levels of the primary PPI partners of the components. To extend pathways, we retrieved all primary PPI for each component of each pathway from the curated PPI databases, HPRD, BioGRID, and IntAct. Experimental methods to ascertain PPI can be categorized broadly as *in vivo*, *in vitro*, and yeast-two-hybrid, and the potential for the biological relevance of a reported PPI can be judged by the confidence that the two proteins indeed have the possibility to interact *in vivo*, that is, they are least expressed in the same time and place. We separated PPI into two levels. A level 1 PPI is associated with higher confidence because it was identified by an *in vivo* experiment (as annotated in HPRD) or, if not *in vivo*, the possibility that the two proteins interact is increased because either they share a GO term under Cellular Component or the mouse orthologs are both expressed in at least one of the same brain regions as reported in the Allen Brain Atlas. A level 2 PPI was detected *in vitro* or by yeast-two-hybrid experiments and lacks further supporting information for biological validity. Assignment to level 2 reflects only a summary of available information and does not imply these PPI are artifacts. A nonredundant set of almost 70,000 PPI were retrieved, of which more than 52,000 are level 1. These PPI involve >11,000 individual proteins, of which 9681 are present in level 1 interactions. Considering pathway components and PPI together, extended pathway datasets involve more than 13,000 individual proteins, ~60% of the 22,000 annotated RefSeq proteins.

ID genes

We reasoned that functional relationships of HSA21 genes with ID in DS can be discovered by understanding the relationships of HSA21 proteins to molecular mechanisms underlying ID in general. We assume that ID can be caused by perturbations in multiple individual pathways and that each of these pathways can be perturbed independently by mutations in multiple individual genes. The differences among individuals in the specific features of an ID phenotype arise from the specifics of which gene is mutated and which pathways it impacts (and of course, the genetic background of the individual carrying the mutation plus environmental influences, but we restrict discussion here to direct genetic contributions). We started by generating a list of ID genes, where an ID gene is defined as any human gene identified to result in ID when mutated. Mutation information was obtained from the database OMIM, by searching with the text string "mental retardation" and limiting results to phenotypes associated with genes with known mutations. OMIM is an expert curated database and contains information from peer-reviewed primary literature reports. It does not, however, use a controlled vocabulary and therefore all entries retrieved were reviewed manually prior to including the gene as an ID gene. A similar effort was carried out by Inlow and Restifo (2004) where 284 ID genes were identified. Our effort has increased this number to 405 genes. The list includes genes with a relatively common mutation, for example, FXMR1, as well as many where the mutation has been identified in only a single or a few families. The phenotype associated with the mutation may be highly specific for ID, or it may be associated with abnormalities in many other organs or organ systems. Because the goal is to be inclusive, the list not only has genes where ID is seen in essentially all individuals with a mutation in that gene, but also those where ID is reported in only a subset. In the latter case, the lower association with ID could be due to ascertainment (ID not tested for and mild ID not noticed) or to background effects and allelic variation preventing ID expression. We note that ID associated with copy number variations where multiple genes are duplicated or deleted is not included in our analysis unless the ID causative gene or genes within the CNV has been identified.

Inlow and Restifo (2004) divided ID genes into primary and secondary. Broadly, primary ID is caused by mutations in genes involved in synaptic structures or functions, signaling cascades, or transcription of such genes, while secondary ID was defined as due to "systemic toxicity." These latter genes commonly encode enzymes involved in metabolic processes and mutation causes altered, toxic levels of biochemical intermediates. Further expert review is being used to refine these classifications (manuscript in preparation). It is noteworthy that data from UniGene and for mouse orthologs in the Allen Brain Atlas show that few ID genes are expressed specifically in brain or in a restricted subset of brain regions (data not shown) and therefore their mutation may have pleiotropic effects beyond ID. Of the 405 ID genes, two are encoded by HSA21, holocarboxylase synthase (HCLS) and cystathionine beta-synthetase (CBS). Their contributions to ID when overexpressed are not known from human studies. The list of ID genes alone, therefore, does not suggest the role of HSA21 genes in the molecular basis of ID in DS, and the effects of increased HSA21 protein expression must be less direct.

Thirty-one HSA21 proteins interact with one or more of 28 ID proteins (Table 4). APP and AIRE each interact with three ID proteins (CTSD, HSD17B10 and NF1, and CREBBP, SMC1A and SMC3, respectively), and ITSN1, CBR1, and TCP10L interact with two. Conversely, the ID protein CREBBP directly interacts with four HSA21 proteins (AIRE, ETS2, IFNAR2, and GABPA), while the ID protein IKBKG (the regulatory subunit of the inhibitor of kappa B kinase (IKK) complex) interacts with three (GART, ATP5O, and TCP10L). ID proteins CTSD, HRAS, SOX10, and THRB each interact with two HSA21 proteins (APP and CSTB; TIAM1 and ITSN1; SUMO3 and OLIG2; and NRIP1 and PRMT2, respectively). These data suggest that HSA21 proteins could contribute to ID in DS indirectly when their

Table 4. HSA21-ID protein interactions. a) HSA21 proteins are listed with the names of their interacting ID proteins. b) ID proteins are listed with the numbers of their interacting HSA21 proteins

(a) HSA21	ID PPI	HSA21	ID PPI	(b) ID	# HSA21 PPI
APP	CTSD, HSD17B10, NF1	SAMSN1	YWHAE	CTSD	2
IFNAR1	PTPN11	S100B	GFAP	HRAS	2
TIAM1	HRAS	PRMT2	THRB	CREBBP	4
AIRE	CREBBP, SMC1A, SMC3	USP25	MEF2C	THRB	2
RCAN1	RAF1	PFKL	ATN1	SOX10	2
ITSN1	HRAS, SOS1	PCNT	BRAF	IKBKG	3
ETS2	CREBBP	UBE2G2	UBE3A		
HMGN1	RPS6KA3	GART	IKBKG		
NRIP1	THRB	OLIG2	SOX10		
DYRK1A	YWHAE	ATP5O	IKBKG		
CCT8	IGBP1	MX1	FANCA		
IFNAR2	CREBBP	GRIK1	FLNA		
SUMO3	SOX10	DNMT3L	DNMT3B		
UPS16	DBT	CSTB	CTSD		
CBR1	ERCC8, FTSJ1	TCP10L	IKBKG, KIAA1279		
GABPA	CREBBP				

overexpression perturbs the stoichiometry or dynamics of the ID protein interactions.

ID pathways

Table 3(a) shows that 267 of 405 ID proteins are components of pathways. We define this subset of pathways as ID pathways (Fig. 2). Those pathways with the highest number of ID protein components are shown in Table 5. These data will be described in detail elsewhere (manuscript in preparation) and here we mention only a few features. Among the top ID pathways, three are metabolic pathways: metabolism of amino acids and derivatives has 31 ID protein components, metabolism of lipids and lipoproteins has 15, and metabolism of proteins has 14. Consistent with the broad functions of many ID proteins, two cancer pathways are represented with 24 and 14 ID components, respectively. Nine pathways in Table 5 are directly relevant to learning and memory and/or brain plasticity, among them MAP kinase signaling pathway with 21 and signaling by NGF with 17 ID protein components. In total, there are 836 ID pathways.

Table 5. ID pathways with highest number of ID protein components

Pathway (db)	# C	# ID
Metabolism of amino acids and derivatives (R)	194	31
Pathways in cancer (K)	278	24
MAP kinase signaling pathway (K)	265	21
Hemostasis (R)	466	17
Signaling by NGF (R)	220	17
Axon guidance (R)	265	15
Metabolism of lipids and lipoproteins (R)	288	15
Transmission at chemical synapses (R)	191	14
Synaptic transmission (R)	196	14
Metabolism of proteins (R)	294	14
Prostate cancer (K)	81	14
NGF signaling via TRKA (R)	135	13
Insulin receptor signaling cascade (R)	84	13
Signaling in the immune system (R)	501	13
Neurotrophin signaling (K)	123	12
Long-term potentiation (K)	70	11
Long-term depression (K)	70	11

C, total number of pathway components; # ID, number of components that are ID proteins.

DS-ID pathways

We define a DS pathway (Fig. 2) as an extended pathway that contains at least one HSA21 protein either as a component or as a PPI. Of the 161 HSA21 RefSeq proteins, 69 are present as components of one or more pathways, and 122 are present as PPI (99 as level 1 PPI), for a total non-redundant representation of 132 HSA21 proteins. To connect HSA21 genes to molecular processes underlying ID, we examined ID pathways (listed in Supplementary Table S1) for direct or indirect involvement with HSA21 proteins.

Based on their association with ID and HSA21 proteins as components and PPI, 16 classes of pathways can be discriminated (Table 6 and Fig. 2). We defined DS-ID pathways as the subset of pathways that are both DS and ID pathways. As shown in Table 6, there are six types of DS-ID pathways. A total of 710 pathways, approximately 40% of all pathways, are DS-ID. These pathways become candidates for perturbation in DS caused by increased expression of HSA21 components or interactions. Additional ID pathways that lack any HSA21 impact are assumed to be relevant to ID but not to ID in DS. Similarly, DS pathways that lack any ID protein component may be perturbed in DS and contribute to some subset of DS phenotypic features, but are assumed to be irrelevant to ID in DS.

High-impact HSA21 proteins

We first sorted HSA21 proteins based on the total number of ID pathways (i.e., DS-ID pathways) they impact as components and/or PPI. Of the 161 HSA21 proteins, 43 impact no ID pathway, either as a component or as PPI. A further 58 impact a relatively small number of pathways, ≤25, and 31 proteins impact between 25 and 100 pathways. At the high end, 29 HSA21 proteins impact more than 100 pathways, and 16 of these impact more than 150. From this set of 29 genes, we discarded two, C21orf91 and C21orf56, because they are among the ORF genes lacking functional domain annotations and the only information associated with them is from high-throughput yeast-two-hybrid interaction experiments. This left 27 HSA21 proteins which were termed high impact

Table 6. DS-ID pathways

Class	#	ID component	HSA21 component	HSA21 PPI	ID PPI	
1	178	Y	Y	Y	Y	DS-ID
2	1	Y	Y	Y	N	DS-ID
3	19	Y	Y	N	Y	DS-ID
4	3	Y	Y	N	N	DS-ID
5	504	Y	N	Y	Y	DS-ID
6	5	Y	N	Y	N	DS-ID
7	84	Y	N	N	Y	ID
8	42	Y	N	N	N	ID
9	38	N	Y	Y	Y	DS
10	0	N	Y	Y	N	DS
11	6	N	Y	N	Y	DS
12	8	N	Y	N	N	DS
13	496	N	N	Y	Y	DS
14	23	N	N	Y	N	DS
15	164	N	N	N	Y	–
16	175	N	N	N	N	–

Sixteen classes of pathways can be described based on their having components and/or PPI that are ID proteins and/or HSA21 proteins. Numbers of pathways in each class are indicated. The subset of pathway that have ID proteins as components and HSA21 proteins as components and/or PPI are designated as DS-ID pathways (see also Fig. 2).

because their increased expression in DS would be predicted to cause the greatest number of pathway perturbations.

High-impact HSA21 proteins are listed in Table 7. They include five proteins involved in regulation of transcription and four involved in PTM. Given the large number of interaction partners and targets found in general for transcription factors and kinases, it is reasonable that these proteins would impact larger numbers of pathways. The most intensively studied HSA21 proteins (based on number of literature reports and mouse models analyzed), APP, DYRK1A, RCAN1, SOD, ITSN1, and S100B, are also high-impact proteins, suggesting that the initial characteristics prompting their study for relevance to DS were soundly based and interpreted. However, many other proteins in Table 7 have received very little

Table 7. High-impact HSA21 proteins and number of direct and indirect pathway impacts

			# Pathways impacted		
Protein	Description	Functional class	Total	Comp	PPI-1
APP[a]	Amyloid precursor protein		405	39	318
IFNAR1[a]	Interferon α/β receptor 1		356	10	326
AIRE	Autoimmune regulator	TF	354	0	303
TIAM1[a]	Guanine nucleotide exchange factor for RAC		316	14	301
RCAN1[a]	Inhibitor of protein phosphatase calcineurin	PTM	264	2	262
ITSN1[a]	Multidomain protein (SH3,EH, RhoGEF, PH)		238	11	216
ETS2[a]	v-ets erythroblastosis virus E26 oncogene homolog 2	TF	221	3	215
HMGN1	High mobility group nucleosome binding domain 1		216	1	215
ITGB2	Integrin, beta 2		216	10	186
RUNX1[a]	runt-related transcription factor 1	TF	213	4	208
NRIP1[a]	Nuclear receptor interacting protein 1	TF	208	3	166
ATP5O	ATP synthase, H+transporting, mitochondrial F1 complex, O subunit		206	2	58
IFNAR2[a]	Interferon αβ receptor 2		176	11	124
DYRK1A[a]	Dual-specificity tyrosine-(Y)-phosphorylation regulated kinase 1A	PTM	174	0	161
SUMO3[a]	Small ubiquitin-like modifier protein 3	PTM	167	1	129
CCT8[a]	Chaperonin containing TCP1, subunit 8 (theta)		161	6	136
SYNJ1[a]	Synaptojanin 1		146	4	141
S100B[a]	S100 calcium binding protein B		129	24	70
RIPK4	Receptor-interacting serine–threonine kinase 4	PTM	124	0	124
CBR1[a]	Carbonyl reductase 1		122	1	98
USP16[a]	Ubiquitin-specific peptidase 16	Ubiquitin	120	0	120
TTC3[a]	Tetratricopeptide repeat domain 3	Ubiquitin E3 ligase	119	0	24
BACH1[a]	BTB and CNC homology 1, basic leucine zipper transcription factor 1	TF	118	0	117
SAMSN1	SAM domain, SH3 domain, and nuclear localization signals 1		111	0	98
SOD1[a]	Superoxide dismutase		110	8	100
UBASH3A	Ubiquitin-associated and SH3 domain containing A	Ubiquitin	110	0	110
MCM3AP[a]	Minichromosome maintenance complex component 3 associated protein		107	0	106

TF, transcription factor/regulator; PTM, protein involved in posttranslational modification. Total, number of pathways impacted as a component or as a PPI; comp, number of pathways impacted as a component; PPI-1, number of pathways impacted by level 1 PPI only. Proteins are listed in order of the number of pathways they impact.

[a]Reported as most often observed with elevated expression in trisomy in Vilardell et al. (2011).

attention overall: SAMSN1, RIPK4, ITGB2, HMGN1, CCT8, MCM3AP, and CBR1. Others, such as TIAM1, NRIP1, SUMO3, and AIRE, have an extensive literature but little of it has been related to DS. Two of the four interferon-related HSA21 proteins, IFNAR1 and IFNAR2, but not IL10RB and IFNGR1, are included as high-impact proteins. Three of the six ubiquitin-associated proteins, UBASH3A, USP16, and TTC3, are included of which TTC3 has motivated DS-related studies (Suizu et al., 2009).

Eighteen of the 27 proteins in Table 7 impact ID pathways as components, with APP, S100B, and TIAM1 impacting the greatest number, 39, 24, and 14, respectively. The IFNARs, the cell adhesion protein ITGB2, and the endocytosis multidomain protein ITSN1, each is a component of more than 10 pathways. Notably, two intensively studied proteins, DYRK1A and RCAN1, are annotated as components of 0 and 2 pathways only. Indeed, overall, HSA21 proteins impact ID pathways largely indirectly, as PPI. This is true even for APP and TIAM1 that impact 10 and 20 times more pathways as PPI than as components. This supports the roles of HSA21 proteins as modulators rather than as master regulators of cellular processes and is consistent with the (relatively) mild phenotype of DS. Last, two additional proteins, IFNGR2 and U2AF1, are components of 12 and 15 pathways, respectively; however, they are not included in Table 7 because they have relatively few PPI (45 and 8) and therefore in total impact only 57 and 23 pathways each.

Certainly, pathway annotations and PPI data are incomplete. PPI data, in particular, are likely to be seriously biased toward the subset of RefSeq proteins whose initial, perhaps fortuitous, identification and characterization revealed interesting features that prompted their selection for PPI experiments. Large-scale yeast-two-hybrid experiments will help to redress this imbalance (Venkatesan et al., 2009). However, while pathway associations of HSA21 proteins, in particular, as PPI, will increase, it seems unlikely that there will be major revisions to the relative contributions of HSA21 proteins to ID pathways overall, that is, all or almost all of the highest impact HSA21 proteins are likely included in Table 7. If overexpressed in brain in trisomy, these are worthy of further consideration for contribution to DS ID.

Prioritizing pathways

Because of the large number (710) of DS-ID pathways, some rationale for sorting is required in order to prioritize those likely to be the most informative. We considered scoring pathways based on the total number of components relative to the number of ID components or based on the total number of components plus PPI relative to the total number of HSA21 components plus HSA21 PPI. However, such a statistical analysis may obscure the biology. For example, if one pathway component has 50 protein interaction partners, only one of which is an HSA21 protein, a 50% increase in the level of that HSA21 protein would appear statistically trivial with respect to the total PPI contributions. However, PPIs will have cellular and intracellular specificity in their localization, and if the HSA21 protein was the sole interaction in a sequestered cellular location, for example, dendritic spines, then a 50% increase might have a very significant effect on the dynamics of that specific interaction. Given the limitations of current knowledge regarding pathways, the functions of ID and HSA21 proteins, and their intracellular localizations and dynamics, we elected to rank DS-ID pathways based total number of HSA21 protein components plus HSA21 level 1 PPI. Complete data are presented in Supplementary Table S1. In Table 8, we show attributes of 34 pathways selected for known and potential biological relevance to ID in DS or learning/memory in DS mouse models. When partially redundant pathways were available either in different databases or within the same database (e.g., Nerve growth factor-related pathways), the highest level pathway (greatest number of HSA21 protein

Table 8. Selected ID pathways enriched for HSA21 protein associations

	Pathway	Number of Components			Number of HSA21 PPI				Total[a]
		Total	ID	HSA21	Levels 1 and 2		Level 1		
					All	HI	All	HI	
Abnormalities in Ts65Dn									
77	**LTP (K)**	70	11	0	20	13	19	13	1567
237	**LTD (K)**	70	11	0	12	9	12	9	1185
94	**Cholinergic synapse (K)**	95	4	1	20	17	17	14	1730
4	Signaling by NGF (R)	220	17	2	41	27	35	25	4095
675	**Retrograde NGF signaling (R)**	13	1	0	5	5	4	4	308
Drug responses in Ts65Dn									
446	***NMDAR activation by Glu (R)***	36	7	0	9	6	8	6	902
574	***GABA receptor cycle (B)***	9	2	0	9	6	6	4	605
229	***Hedgehog signaling by Gli1(P)***	29	4	0	16	14	13	12	960
841	***One carbon pool by folate (K)***	18	4	2	0	0	0	0	82
Neuronal structure and synapses									
104	**NCAM signaling, neurite outgrowth (R)**	69	9	2	21	16	16	13	1857
8	Axon guidance(R)	265	15	2	34	21	28	18	3372
97	Transmission across chemical synapses (R)	191	14	3	17	12	15	11	1856
98	Synaptic transmission (R)	196	14	3	17	12	15	11	1863
153	Glutamatergic synapse (K)	110	6	1	14	10	14	10	1125
219	Glutamate binding, AMPAR activation (R)	29	1	0	13	8	13	8	808
Signaling pathways									
5	**MAPK signaling pathway (K)**	265	21	0	42	26	36	24	4192
11	TOLL receptor cascade (R)	104	5	2	28	18	27	18	2673
38	Glucocorticoid receptor network (P)	71	2	0	29	19	26	17	2532
43	CaN-NFAT signaling (P)	46	2	1	26	16	24	16	2097
44	**WNT signaling (K)**	144	1	0	24	18	24	19	2760
51	ErbB signaling pathway (K)	86	9	0	28	19	23	17	2615
54	B-catenin signaling (P)	71	2	0	27	15	17	11	2421
63	Signaling by HDAC class 1 (P)	56	2	0	21	12	21	12	1796
86	**Notch signaling pathway (P)**	44	1	0	21	14	19	12	1331
91	Calcium signaling pathway (K)	167	3	0	19	13	18	13	1484
93	JAK-STAT signaling pathway (K)	99	3	4	19	14	14	12	2004
139	mTOR signaling (P)	46	7	0	17	10	16	10	1642
162	Insulin receptor signaling cascade (R)	107	13	0	21	15	15	11	1979
745	Phosphatidylinositol signaling (K)	73	3	1	6	3	5	3	772
Other processes									
20	Apoptosis (R)	146	1	0	27	19	27	19	3037
57	Tight junction (K)	118	6	5	18	12	17	11	1938
Diseases									
81	Glioma (K)	62	10	0	24	17	19	15	2398
370	Alzheimer's disease (K)	48	1	1	11	5	8	5	1209
844	Huntington's disease (K)	19	1	0	3	1	2	1	540

Pathway: R, Reactome; K, KEGG; P, PID; B, Biocarta. Bold, pathways perturbed in DS based on molecular or electrophysiological assays in mouse models of DS. Bold italics, pathways inferred to be perturbed in DS based on rescue of learning/memory deficits in the Ts65Dn mouse model by treatment with drugs targeting these pathways. HI, number of high impact proteins. a, total number of PPI for all pathway components impacts) was selected for Table 8. Relative to the number of components, each pathway has a very large number of total PPI. This is due to pathway components that are "hub" proteins, proteins that have >20 PPI. Many hub proteins have several hundred PPI, but because hubs are typically also components of multiple pathways, not all PPI are expected to be relevant to any one specific pathway. Each pathway in Table 8 also has a significant number of HSA21 PPI, with the majority of these being high-impact HSA21 proteins (Table 7). For example, the MAP kinase pathway has 265 protein components and 42 HSA21 proteins as PPI. Of these, 26, that is, all but one of the high-impact proteins, are included.

The Ts65Dn is the DS mouse model that has been the most thoroughly studied. Five pathways in Table 8 describe functions that are abnormal in the Ts65Dn. LTP is reduced (Siarey et al., 1997; Costa and Grybko, 2005; Kleschevnikov et al., 2004), and LTD is increased (Siarey et al., 1999). Cholinergic neurons of the basal forebrain show loss of functional markers as the Ts65Dn age, starting as young as 4–6 months (Granholm et al., 2000; Holtzman et al., 1996), and abnormalities in nerve growth factor signaling, including decreased retrograde signaling, have been described (Cooper et al., 2001). In each of these reports, causal HSA21 proteins underlying the abnormalities have not been identified. Perturbations in each of these pathways may be caused both by HSA21 components (TIAM1 and ITSN1 as components of the NGF signaling pathway, and KCNJ6 as a component of cholinergic synapses) and by HSA21 interactions with components of these pathways (Table 9).

Recent studies have reported rescue by a number of drug treatments of learning/memory deficits and other abnormalities in the Ts65Dn (reviewed in Gardiner, 2010; Shichiri et al., 2011; Incerti et al., 2011; Bianchi et al., 2010; Faizi et al., 2011; Braudeau et al., 2011). In the majority of these cases, the relationship between the targets of the drugs and/or the pathways assumed to be perturbed in the Ts65Dn, and therefore rescued by the drug treatments, were neither known nor predicted. Table 8 includes three relevant pathways: activation of NMDA receptor upon glutamate binding (Reactome) which is targeted by memantine, an antagonist of NMDA receptor; GABA receptor life cycle (Biocarta) which is targeted by pentylenetetrazole, an antagonist of GABA receptor; and Hedgehog signaling events mediated by Gli proteins (PID) which is targeted by SAG1.1, a Hedgehog pathway agonist. In none

Table 9. Orthologous mouse chromosomal location of HSA21 PPIs to selected DS-ID pathways

Pathway	MMU16	MMU16	MMU17	MMU10
	Ts65Dn+	Ts65Dn−		
LTP	GABPA, TIAM1, SOD1, OLIG2, SYNJ1, IFNAR1, IFNAR2, KCNE1, ITSN1, RUNX1, RCAN1, PCP4, ETS2, HMGN1, RIPK4		RRP1B	TRPM2, PCNT, ITGB2, AIRE
LTD	TIAM1, CCT8, IFNAR1, IFNAR2, ITSN1, KCNE1, RCAN1, HMGN1, RIPK4			TRPM2, ITGB2, PCNT
NMDAR	GRIK1, TIAM1, IFNAR1, ITSN1, RCAN1, DYRK1A, PCP4, HMGN1	NRIP1		PCNT
GABAR	TIAM1, ITSN1, DYRK1A, ETS2, TFF1	HSPA13	UBASH3A	AIRE, UBE2G2
Hedgehog	APP, GABPA, BACH1, TIAM1, IFNAR1, IFNAR2, RUNX1, RCAN1, DYRK1A, CHAF1B, ETS2, HMGN1, MX1, RIPK4	NRIP1, SAMSN1		SUMO3, PCBP3, PCNT, ITGB2, DNMT3L, AIRE

MMU16, 17, 10: mouse chromosome 16, 17, 10. Ts65Dn+, trisomic in the Ts65Dn. Ts65Dn−, gene maps to mouse chromosome 16 but proximal to the Ts65Dn breakpoint and therefore is not trisomic.

of these cases are HSA21 proteins annotated as components of the pathways, but there are numerous interactions.

A fourth treatment, using folinic acid, has had moderate success in human clinical trials, but results were significant only in those individuals with DS that were also receiving thyroid hormone replacement (Blehaut et al., 2010). Reactome and KEGG contain relevant pathways called "metabolism of folate and pterines" and "one carbon pool by folate," respectively, which contain the HSA21 proteins SLC19A1, GART, and FTCD as components, but no additional HSA21 protein interactions. Integrating these pathways with thyroid signaling has so far not been possible with current database content; however, the thyroid hormone receptor, THRB, is an ID protein and has interactions with HSA21 proteins PRMT2 and NRIP1 (Table 4).

Neuronal structural abnormalities have been described in the DS brain and brains of DS mouse models, in particular, in size, density, and morphology of dendritic spines (reviewed in Benavides-Piccione et al., 2004). Pathways called axon guidance and neurite outgrowth are of potential relevance and both have HSA21 protein components, COL6A1 and COL6A2. Four pathways describe synapses and HSA21 proteins KCNJ6, KCNJ15, and GRIK1 are annotated as components of pathways titled chemical synapses and synaptic transmission.

Among signaling pathways in Table 8, perturbation of MAP kinase signaling has been reported in the Ts65Dn (Gardiner et al., 2004; Siddiqui et al., 2008). Perturbation of Calcineurin-NFAT signaling was shown in a transgenic mouse overexpressing cDNAs for the HSA21 genes *RCAN1* and *DYRK1A* (Arron et al., 2006). While this model used cDNA constructs and therefore was not ideal, the data in Table 8 support the perturbation of this pathway in DS, although with more HSA21 protein contributions than just DYRK1A and RCAN1. Perturbations in WNT and Notch, at the RNA level in neural stem cells derived from DS, were reported in Cairney et al. (2009) from microarray gene expression analysis.

The foregoing examples illustrate that pathway associations of HSA21 proteins identified in diverse non-DS-directed experiments have support from results of independent HSA21/DS-directed experiments. This suggests that additional predictions are worth investigation. Additional signaling pathways predicted to be perturbed include glucocorticoid signaling, where perturbations may affect stress responses in DS. In Fig. 3, the JAK-STAT pathway is illustrated. Four HSA21 proteins are components, the interferon-related receptors, IFNAR1 and 2, IFNGR2, and IL10RB, and numerous additional HSA21 proteins interact with components. The TOLL, ERBB, b-catenin, HDAC, calcium, mTOR, and insulin signaling pathways, each have between 17 and 28 HSA21 PPI, including >10 high-impact HSA21 proteins. Based on these data, these pathways appear as candidates for perturbation in DS that are as reasonable as NMDAR, MAP kinase, and SHH, but none so far has been investigated in DS or its mouse models.

Other processes in Table 8 include apoptosis and tight junctions. Figure 4 illustrates the apoptosis pathway and HSA21 PPI, downstream of the JAK-STAT pathway in Fig. 3. Five HSA21 proteins are components of tight junctions. Last, three disease pathways are listed. AD is strongly associated with DS with the development of an AD-like neuropathology in essentially all individuals by age 30 and an AD-like dementia in approximately 50% by age 50 (Zigman and Lott, 2007). The APP gene mapping to HSA21 is likely relevant, but eleven additional HSA21 proteins may influence the process through interactions. Huntington's disease is included because REST/NSRF levels have been shown to be perturbed by overexpression of DYRK1A and to alter regulation of expression of multiple neuronal genes (Canzonetta et al., 2008; Lepagnol-Bestel et al. 2009). In the current pathway databases, the REST protein is annotated as a component of only the Huntington's disease pathway. It does, however, also interact with 23 proteins, allowing for influence in numerous pathways. Pathway annotations will continue to expand in detail

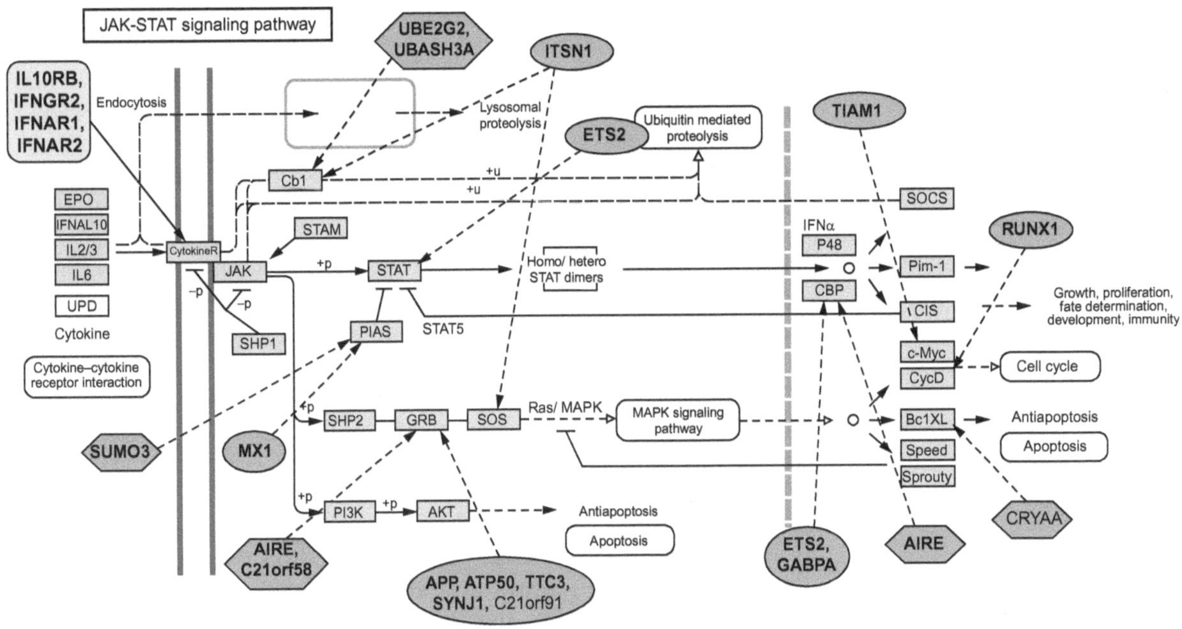

Fig. 3. JAK-STAT pathway with HSA21 protein components and interactions. Pathway diagram is from the KEGG database. Shaded shapes contain HSA21 proteins with pathway interactions indicated by dashed arrows. Ovals, orthologs of HSA21 genes map mouse chromosome 16; bold indicates trisomic within the Ts65Dn. Hexagons, orthologs of HSA21 proteins map to mouse chromosome 10 (bold) or mouse chromosome 17. Rectangle at upper left, HSA21 proteins that are components of the pathway; all are trisomic in the Ts65Dn mouse. (For color version of this figure, the reader is referred to the Web version of this chapter.)

and be extended to new cellular processes. Some relevant data are already present in the literature; with their formal annotation, the effects of HSA21 proteins on REST and REST functions may become more easily predicted. Discussion

Discussion

DS-ID pathway analysis: Advantages and limitations

We have presented a pathway analysis approach, based on integration of curated, peer-reviewed data, designed to aid in identifying critical molecular abnormalities that contribute to ID in DS. Such pathway analysis does not replace traditional investigations of single gene function and behavioral analysis of mouse models. Indeed, it depends absolutely on results of such experiments. Bioinformatics-based pathway analysis is also not an end in itself. By integrating data from disparate experiments, it can predict perturbations in DS, as in Table 8, but these must then be validated by experiments in appropriate model systems.

Pathways were first sorted for relevance to ID and then for influence by HSA21 proteins. While the number of DS-ID pathways identified by these criteria suggests unwieldy complexity, it must be recognized that such complexity was already suggested by the number and functions of HSA21 genes, by the number of HSA21 genes found to be overexpressed in trisomy, and by the functional diversity of drugs that rescue learning and memory deficits in the Ts65Dn mouse model of DS. Pathway analysis should, therefore, be considered for its advantages in complementing traditional, DS-focused experimental approaches.

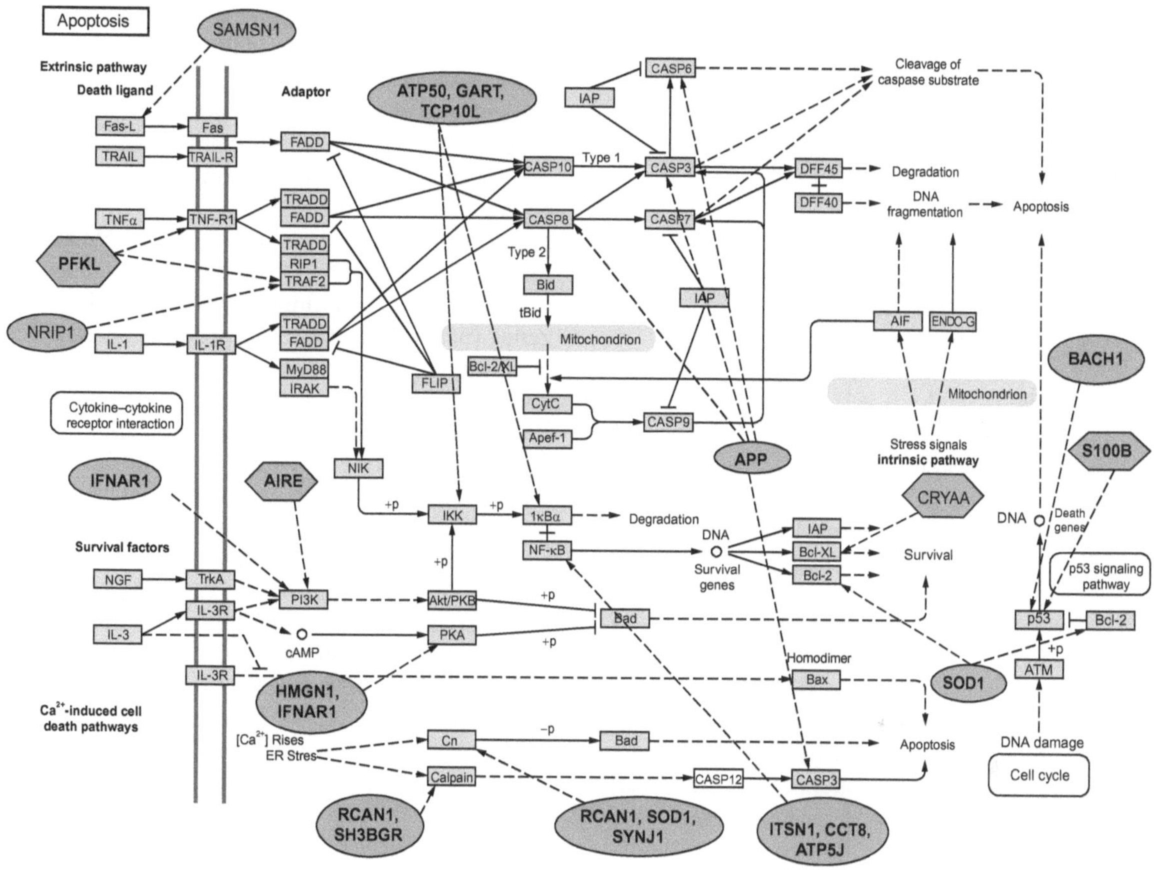

Fig. 4. Apoptosis pathway with HSA21 protein interactions. Pathway diagram is from the KEGG database. Ovals and hexagons contain HSA21 proteins with orthologs mapping to mouse chromosomes 16 or 10/17, as described in Fig. 3. (For color version of this figure, the reader is referred to the Web version of this chapter.)

First, pathway data provided a ranking of HSA21 proteins based on the number of pathways each impacts. This allows for speculation on the severity of the consequences for overexpression of each protein. A subset of 27 HSA21 proteins were found to impact a disproportionately large number of pathways. This set included the HSA21 genes most frequently studied and well established with respect to DS (*APP, DYRK1A*, etc.) as well as genes that are novel in the DS perspective. Also among this subset were 20 genes identified as among those most often found to be overexpressed in DS-related microarray experiments (Vilardell et al., 2011). Second, pathway data provide a molecular basis for many observed abnormalities, including perturbations in LTP documented in several mouse models of DS, and successes in manipulating NMDA receptor and sonic hedgehog signaling in the Ts65Dn. Given these successes, consideration should be given to similarly weighted predictions of perturbations in additional pathways, such as others listed in Table 8 and Supplementary Table S1. Third, pathway analysis identifies the specific HSA21 genes that potentially contribute to the observed abnormalities in a pathway or cellular process, information that has been largely lacking in selecting drug treatments for testing in the Ts65Dn. Using information on the chromosome location of orthologous mouse genes allows evaluation of how well-potential DS perturbations, and their responses to drug treatments, can be modeled in the partially trisomic DS mouse models.

Pathway analysis to inform development of pharmacotherapies for ID in DS

Given recent successes in rescue of learning and memory deficits in the Ts65Dn mouse model of DS, pharmacotherapies for cognitive deficits in DS seems a practical goal. However, it is unlikely that therapeutics targeting HSA21 genes/proteins will provide a practical solution. The large number of HSA21 proteins (27) impacting many pathways and commonly overexpressed in trisomy suggests that concomitant use of multiple drugs would be required, a plan not generally considered advisable, given the possibility of drug interactions. Such a plan also cannot account for effects of non-protein-coding genes, which have been largely unexplored, nor for consequences of functional interactions between HSA21 proteins where only one is targeted.

There also exist challenges in interpretation of results from successful treatment of learning and memory deficits in the Ts65Dn. Table 10 lists the successful drugs and the targets or processes that they are "known" to affect. Two facts are important. First, that drugs of such diverse mechanisms can all be successful in rescuing deficits in hippocampal tasks, in some cases in the same tasks, implies a significant gap in our understanding of what these drugs do, and/or what molecular pathways underlie the specific learning and memory tasks assayed, and/or how overexpression of trisomic genes perturbs these pathways. Second, the Ts65Dn is trisomic for only 88 of 161 protein-coding genes and only 19 of the 27 high-impact HSA21 protein-coding genes (never mind that interpreting consequences of trisomy of minimally conserved and apparent species-specific genes is beyond our current understanding). In Table 2, functional interactions among some HSA21 proteins were listed, and in Table 9, further HSA21 protein interactions with components of pathways of demonstrated relevance to the Ts65Dn phenotypic features are listed. All interactions and pathways include genes that are not trisomic in the Ts65Dn, some mapping to the proximal region of mouse chromosome 16 and others to orthologous segments of mouse chromosomes 17 and 10. Examples illustrating the complexity of pathway interactions are shown in Figs. 3 and 4. In both cases, proteins mapping throughout HSA21 interact with many different pathway components. Whether these interactions are synergistic or antagonistic is unknown, but their diversity suggests that targeting any pathway for correction will require judicious choice of target, for example, accommodating possible effects of downstream HSA21 interactions.

Table 10. Pharmacological rescue of Ts65Dn learning and memory deficits

Drug	Target/process	References (reviewed in Gardiner, 2010)
Minocycline	Anti-inflammatory	
Pentylenetetrazole	Antagonist of $GABA_A$ receptor	
Fluoxetine	Serotonin reuptake inhibitor	Bianchi et al. (2010)
Choline	DNA methylation (possibly)	Moon et al. (2010)
Memantine	Antagonist of NMDA receptor	Rueda et al. (2010), Lockrow et al. (2011)
NAPVSIPQ+SALLRSIPA	Neuroprotective peptides	Incerti et al. (2011)
Vitamin E	Antioxidant	Shichiri et al. (2011)
Xamoterol	Dopamine agonist	Faizi et al. (2011)
α5IA	Antagonist of $GABA_A$ receptor	Braudeau et al. (2011)

A conservative view for development of pharmacotherapies for ID in DS, therefore, has to be that it cannot be assumed that abnormalities observed in the Ts65Dn have an identical genetic basis, that is, that they result from overexpression of the same set of trisomic genes, as similar features observed in DS. In other words, it cannot be assumed that additional genes on HSA21 do not contribute, positively or negatively, to the DS phenotype. The extension of this conservative view is that the molecular underpinnings of drug responses in the Ts65Dn also cannot be assumed to be identical to those that will occur in people with DS. Such a concept possibly was illustrated by a recent report that antioxidant therapy failed to produce improvement or even prevent decline in cognitive function in a DS trial (Lott et al., 2011), although it was successful in the Ts65Dn (Lockrow et al., 2009; Shichiri et al., 2011). The response to antioxidants of an HSA21 protein, S100B, that is not trisomic in the Ts65Dn, may contribute to the differing response in human and mouse (Bialowas-McGoey et al., 2008).

Pursuing pathway analysis at the experimental level

We have shown that curated pathway and protein interaction data, generated in a wide range of non-DS-directed experiments, can be combined to successfully predict perturbations that have been observed in DS model systems. However, to be useful beyond prediction, pathway analysis in DS must expand to broader experimental assessments, ideally at the protein and protein modification level. Some information has been obtained from both pathway-specific, hypothesis-driven analyses and discovery-based approaches. Siddiqui et al. (2008) used quantitative Western analysis to measure levels of ~24 proteins, including phosphorylation and nonphosphorylated forms of MAP kinase pathway components and NMDA receptor subunits, and the HSA21 orthologs App, Tiam1, Itsn1, and Dyrk1a, in hippocampus and cortex of the Ts65Dn and a second DS model, the Ts1Cje, that is trisomic for a subset of genes in the Ts65Dn. The measurements demonstrated perturbations in the MAP kinase pathway, both basally and in response to the NMDA receptor antagonist, MK-801. Wang et al. (2009) used 2D gels and mass spectrometry to survey abnormalities in the embryonic stem cell line that is the parent of the Tc1 mouse model of DS. Among the observed alterations, they found elevated levels of LRP2 and LRPAP1, both of which are components of the Hedgehog pathway, suggesting that this pathway is indeed perturbed in the Tc1 and probably also in the Ts65Dn. Sun et al. (2011) also used 2D gels and mass spectrometry in the analysis of cortex from DS and control fetal brain. They reported significantly altered levels of eight proteins. Five have no annotations in the pathway databases used here, but sequence homology and GO terms suggest functions for each. The small datasets obtained in such efforts preclude real insight because altered levels of a single protein of unknown significance to pathway dynamics cannot be interpreted.

Focused efforts in two areas could potentially move the field forward in a productive manner. First, targeted measurement of the levels of HSA21 proteins, in a relevant tissue sample, would allow refinement of predicted pathway perturbations; if proteins encoded by trisomic genes are not overexpressed, by logical inference, they will not perturb associated pathways. Comparing HSA21 protein levels in a full trisomy 21 to those in the Ts65Dn could indicate how well the Ts65Dn recapitulates pathway perturbations in DS. However, such a targeted effort requires antibodies, and generating these is in itself a major undertaking. Second, identification of abnormalities in levels of PTM in trisomy could identify signaling pathway perturbations and suggest targets for more detailed analysis in mouse models exposed to learning and memory. Mass spectrometry has the potential sensitivity, and methods for detection of phosphorylation are reasonably well established. However, PTMs, in general, are labile, and the uncontrolled postmortem handling of human tissue samples likely will produce artifacts (Ahmed and Gardiner 2011), making comparison with mouse models difficult. Establishing versatile, high-throughput, and reliable methods for measurements of proteins and PTMs from tissues or cell systems that realistically reflect human DS neurobiology, in all its genetic variation, will increase the reliability of preclinical prediction and validation of drug responses, and the likely positive outcome of human clinical trials.

Acknowledgments

This work was supported by the Fondation Jerome Lejeune (X. S.), the National Institutes of Health HD056235 (K. J. G.), the Vietnamese Government (T. L.), and the Linda Crnic Institute for Down syndrome.

Abbreviations

EST	expressed sequence tag
HSA21	human chromosome 21
ID	intellectual disability
PPI	protein–protein interaction
PTM	posttranslational modification

References

Aebi, M., Fäh, J., Hurt, N., Samuel, C. E., Thomis, D., Bazzigher, L., et al. (1989). cDNA structures and regulation of two interferon-induced human Mx proteins. *Molecular and Cellular Biology*, *9*(11), 5062–5072.

Ahmed, M. M., & Gardiner, K. J. (2011). Preserving protein profiles in tissue samples: Differing outcomes with and without heat stabilization. *Journal of Neuroscience Methods*, *196*, 99–106.

Ahn, K. J., Jeong, H. K., Choi, H. S., Ryoo, S. R., Kim, Y. J., Goo, J. S., et al. (2006). DYRK1A BAC transgenic mice show altered synaptic plasticity with learning and memory defects. *Neurobiology of Disease*, *22*, 463–472.

Allore, R., O'Hanlon, D., Price, R., Neilson, K., Willard, H. F., Cox, D. R., et al. (1988). Gene encoding the beta subunit of S100 protein is on chromosome 21: Implications for Down syndrome. *Science*, *239*, 1311–1313.

Aranda, B., Achuthan, P., Alam-Faruque, Y., Armean, I., Bridge, A., Derow, C., et al. (2010). The IntAct molecular interaction database in 2010. *Nucleic Acids Research*, *38*, D525–D531.

Arron, J. R., Winslow, M. M., Polleri, A., Chang, C. P., Wu, H., Gao, X., et al. (2006). NFAT dysregulation by increased dosage of DSCR1 and DYRK1A on chromosome 21. *Nature*, *441*, 595–600.

Becker, W., Weber, Y., Wetzel, K., Eirmbter, K., Tejedor, F. J., & Joost, H. G. (1998). Sequence characteristics, subcellular localization, and substrate specificity of DYRK-related kinases, a novel family of dual specificity protein kinases. *The Journal of Biological Chemistry*, *273*, 25893–25902.

Benavides-Piccione, R., Ballesteros-Yáñez, I., de Lagrán, M. M., Elston, G., Estivill, X., Fillat, C., et al. (2004). On dendrites in Down syndrome and DS murine models: A spiny way to learn. *Progress in Neurobiology*, *74*, 111–126.

Bialowas-McGoey, L. A., Lesicka, A., & Whitaker-Azmitia, P. M. (2008). Vitamin E increases S100B-mediated microglial activation in an S100B-overexpressing mouse model of pathological aging. *GLIA*, *56*(16), 1780–1790.

Bianchi, P., Ciani, E., Guidi, S., Trazzi, S., Felice, D., Grossi, G., et al. (2010). Early pharmacotherapy restores neurogenesis and cognitive performance in the Ts65Dn mouse model for Down syndrome. *Journal of Neuroscience*, *30*, 8769–8779.

Blehaut, H., Mircher, C., Ravel, A., Conte, M., de Portzamparc, V., Poret, G., et al. (2010). Effect of leucovorin (folinic acid) on the developmental quotient of children with Down's syndrome (trisomy 21) and influence of thyroid status. *PLoS One*, *5*, e8394.

Braudeau, J., Delatour, B., Duchon, A., Lopes-Pereira, P., Dauphinot, L., de Chaumont, F., et al. (2011). Specific targeting of the GABA-A receptor {alpha}5 subtype by a selective inverse agonist restores cognitive deficits in Down syndrome mice. *Journal of Psychopharmacology*, 1–13.

Busciglio, J., & Yankner, B. A. (1995). Apoptosis and increased generation of reactive oxygen species in Down's syndrome neurons in vitro. *Nature*, *378*, 776–779.

Cairney, C. J., Sanguinetti, G., Ranghini, E., Chantry, A. D., Nostro, M. C., Bhattacharyya, A., et al. (2009). A systems biology approach to Down syndrome: Identification of Notch/Wnt dysregulation in a model of stem cells aging. *Biochimica et Biophysica Acta*, *1792*, 353–363.

Camon, E., Magrane, M., Barrell, D., Binns, D., Fleischmann, W., Kersey, P., et al. (2003). The Gene Ontology Annotation (GOA) project: Implementation of GO in SWISS-PROT, TrEMBL, and InterPro. *Genome Research*, *13*(4), 662–672.

Canzonetta, C., Mulligan, C., Deutsch, S., Ruf, S., O'Doherty, A., Lyle, R., et al. (2008). DYRK1A-dosage imbalance perturbs NRSF/REST levels, deregulating pluripotency and embryonic stem cell fate in Down syndrome. *The American Journal of Human Genetics*, *83*, 388–400.

Chen, H., Chrast, R., Rossier, C., Morris, M. A., Lalioti, M. D., & Antonarakis, S. E. (1996). Cloning of 559 potential exons of genes of human chromosome 21 by exon trapping. *Genome Research*, *6*, 747–760.

Cooper, J. D., Salehi, A., Delcroix, J. D., Howe, C. L., Belichenko, P. V., Chua-Couzens, J., et al. (2001). Failed retrograde transport of NGF in a mouse model of Down's syndrome: Reversal of cholinergic neurodegenerative phenotypes following NGF infusion. *Proceedings of the National Academy of Sciences of the United States of America*, *98*, 10439–10444.

Costa, A. C., & Grybko, M. J. (2005). Deficits in hippocampal CA1 LTP induced by TBS but not HFS in the Ts65Dn mouse: A model of Down syndrome. *Neuroscience Letters*, *1382*, 317–322.

Croft, D., O'Kelly, G., Wu, G., Haw, R., Gillespie, M., Matthews, L., et al. (2011). Reactome: A database of reactions, pathways and biological processes. *Nucleic Acids Research*, *39*(Database issue), D691–D697.

Dahmane, N., Ghezala, G. A., Gosset, P., Chamoun, Z., Dufresne-Zacharia, M. C., Lopes, C., et al. (1998). Transcriptional map of the 2.5-Mb CBR-ERG region of chromosome 21 involved in Down syndrome. *Genomics*, *48*, 12–23.

Davies, L., Karthikeyan, N., Lynch, J. T., Sial, E. A., Gkourtsa, A., Demonacos, C., et al. (2008). Cross talk of signaling pathways in the regulation of the glucocorticoid receptor function. *Molecular Endocrinology*, *22*, 1331–1344.

Dinger, M. E., Pang, K. C., Mercer, T. R., & Mattick, J. S. (2008). Differentiating protein-coding and noncoding RNA: Challenges and ambiguities. *PLoS Computational Biology*, *4*, e1000176.

Faizi, M., Bader, P. L., Tun, C., Encarnacion, A., Kleschevnikov, A., Belichenko, P., et al. (2011). Comprehensive behavioral phenotyping of Ts65Dn mouse model of Down syndrome: Activation of β(1)-adrenergic receptor by xamoterol as a potential cognitive enhancer. *Neurobiology of Disease*, *43*, 397–413.

Fritah, A., Christian, M., & Parker, M. G. (2010). The metabolic coregulator RIP140: An update. *American Journal of Physiology, Endocrinology and Metabolism*, *299*, E335–E340.

Gardiner, K. J. (2010). Molecular basis of pharmacotherapies for cognition in Down syndrome. *Trends in Pharmacological Sciences*, *31*, 66–73.

Gardiner, K., & Costa, A. C. (2006). The proteins of human chromosome 21. *American Journal of Medical Genetics. Part C, Seminars in Medical Genetics*, *142C*, 196–205.

Gardiner, K., Davisson, M. T., & Crnic, L. S. (2004). Building protein interaction maps for Down's syndrome. *Briefings in Functional Genomics & Proteomics*, *3*, 142–156.

Gardiner, K., Fortna, A., Bechtel, L., & Davisson, M. T. (2003). Mouse models of Down syndrome: How useful can they be? Comparison of the gene content of human chromosome 21 with orthologous mouse genomic regions. *Gene*, *318*, 137–147.

Gardiner, K., Horisberger, M., Kraus, J., Tantravahi, U., Korenberg, J., Rao, V., et al. (1990). Analysis of human chromosome 21: Correlation of physical and cytogenetic maps; gene and CpG island distributions. *The EMBO Journal*, *9*, 25–34.

Gardiner, K., Slavov, D., Bechtel, L., & Davisson, M. (2002). Annotation of human chromosome 21 for relevance to Down syndrome: Gene structure and expression analysis. *Genomics*, *79*, 833–843.

Gardiner, K., & Yaspo, M. L. (1998). Report of the seventh international workshop on human chromosome 21 mapping 1997. *Cytogenetics and Cell Genetics*, *82*, 1–12.

Goldgaber, D., Lerman, M. I., McBride, O. W., Saffiotti, U., & Gajdusek, D. C. (1987). Characterization and chromosomal localization of a cDNA encoding brain amyloid of Alzheimer's disease. *Science*, *235*(4791), 877–880.

Granholm, A. C., Sanders, L. A., & Crnic, L. S. (2000). Loss of cholinergic phenotype in basal forebrain coincides with cognitive decline in a mouse model of Down's syndrome. *Experimental Neurology*, *161*, 647–663.

Groner, Y., Elroy-Stein, O., Avraham, K. B., Schickler, M., Knobler, H., Minc-Golomb, D., et al. (1994). Cell damage by excess CuZnSOD and Down's syndrome. *Biomedicine and Pharmacotherapy*, *48*, 231–240.

Guipponi, M., Brunschwig, K., Chamoun, Z., Scott, H. S., Shibuya, K., Kudoh, J., et al. (2000). C21orf5, a novel human chromosome 21 gene, has a Caenorhabditis elegans ortholog (pad-1) required for embryonic patterning. *Genomics*, *68*, 30–40.

Harris-Cerruti, C., Kamsler, A., Kaplan, B., Lamb, B., Segal, M., & Groner, Y. (2004). Functional and morphological alterations in compound transgenic mice overexpressing Cu/Zn superoxide dismutase and amyloid precursor protein [correction]. *The European Journal of Neuroscience*, *19*, 1174–1190.

Hattori, M., Fujiyama, A., Taylor, T. D., Watanabe, H., Yada, T., Park, H. S., et al. (2000). Chromosome 21 mapping and sequencing consortium. The DNA sequence of human chromosome 21. *Nature*, *405*, 311–319.

Hawkins, J. W., Van Keuren, M. L., Piatigorsky, J., Law, M. L., Patterson, D., & Kao, F. T. (1987). Confirmation of assignment of the human alpha 1-crystallin gene (CRYA1) to chromosome 21 with regional localization to q22.3. *Human Genetics*, *76*(4), 375–380.

Hecquet, C. M., Ahmmed, G. U., Vogel, S. M., & Malik, A. B. (2008). Role of TRPM2 channel in mediating H2O2-induced Ca^{2+} entry and endothelial hyperpermeability. *Circulation Research, 102*, 347–355.

Holtzman, D. M., Santucci, D., Kilbridge, J., Chua-Couzens, J., Fontana, D. J., Daniels, S. E., et al. (1996). Developmental abnormalities and age-related neurodegeneration in a mouse model of Down syndrome. *Proceedings of the National Academy of Sciences of the United States of America, 93*, 13333–13338.

Incerti, M., Toso, L., Vink, J., Roberson, R., Nold, C., Abebe, D., et al. (2011). Prevention of learning deficit in a Down syndrome model. *Obstetrics and Gynecology, 117*, 354–361.

Inlow, J. K., & Restifo, L. L. (2004). Molecular and comparative genetics of mental retardation. *Genetics, 166*, 835–881.

Jones, A. R., Overly, C. C., & Sunkin, S. M. (2009). The Allen Brain Atlas: 5 years and beyond. *Nature Reviews Neuroscience, 10*(11), 821–828.

Kanehisa, M., & Goto, S. (2000). KEGG: Kyoto encyclopedia of genes and genomes. *Nucleic Acids Research, 28*(1), 27–30.

Kleschevnikov, A. M., Belichenko, P. V., Villar, A. J., Epstein, C. J., Malenka, R. C., & Mobley, W. C. (2004). Hippocampal long-term potentiation suppressed by increased inhibition in the Ts65Dn mouse, a genetic model of Down syndrome. *Journal of Neuroscience, 24*, 8153–8160.

Korbel, J. O., Tirosh-Wagner, T., Urban, A. E., Chen, X. N., Kasowski, M., Dai, L., et al. (2009). The genetic architecture of Down syndrome phenotypes revealed by high-resolution analysis of human segmental trisomies. *Proceedings of the National Academy of Sciences of the United States of America, 106*, 12031–12036.

Korenberg, J. R., Chen, X. N., Schipper, R., Sun, Z., Gonsky, R., Gerwehr, S., et al. (1994). Down syndrome phenotypes: The consequences of chromosomal imbalance. *Proceedings of the National Academy of Sciences of the United States of America, 91*, 4997–5001.

Lejeune, J., Turpin, R., & Gauthier, M. (1959). Chromosomic diagnosis of mongolism. *Archives Françaises de Pédiatrie, 16*, 962–963.

Lepagnol-Bestel, A. M., Zvara, A., Maussion, G., Quignon, F., Ngimbou, B., Ramoz, N., et al. (2009). DYRK1A interacts with the REST/NRSF-SWI/SNF chromatin remodelling complex to deregulate gene clusters involved in the neuronal phenotypic traits of Down syndrome. *Human Molecular Genetics, 18*, 1405–1414.

Lockrow, J., Boger, H., Bimonte-Nelson, H., & Granholm, A. C. (2011). Effects of long-term memantine on memory and neuropathology in Ts65Dn mice, a model for Down syndrome. *Behavioural Brain Research, 221*, 610–622.

Lockrow, J., Prakasam, A., Huang, P., Bimonte-Nelson, H., Sambamurti, K., & Granholm, A. C. (2009). Cholinergic degeneration and memory loss delayed by vitamin E in a Down syndrome mouse model. *Experimental Neurology, 216*(2), 278–289.

Lott, I. T. (1982). Down's syndrome, aging, and Alzheimer's disease: A clinical review. *Annals of the New York Academy of Sciences, 396*, 15–27.

Lott, I. T., Doran, E., Nguyen, V. Q., Tournay, A., Head, E., & Gillen, D. L. (2011). Down syndrome and dementia: A randomized, controlled trial of antioxidant supplementation. *American Journal of Medical Genetics. Part A, 155A*, 1939–1948.

Maglott, D., Ostell, J., Pruitt, K. D., & Tatusova, T. (2011). Entrez Gene: Gene-centered information at NCBI. *Nucleic Acids Research, 39*(Database issue), D52–D57.

Maier, T., Güell, M., & Serrano, L. (2009). Correlation of mRNA and protein in complex biological samples. *FEBS Letters, 583*, 3966–3973.

Michiels, F., Habets, G. G., Stam, J. C., van der Kammen, R. A., & Collard, J. G. (1995). A role for Rac in Tiam1-induced membrane ruffling and invasion. *Nature, 375*, 338–340.

Moon, J., Chen, M., Gandhy, S. U., Strawderman, M., Levitsky, D. A., Maclean, K. N., et al. (2010). Perinatal choline supplementation improves cognitive functioning and emotion regulation in the Ts65Dn mouse model of Down syndrome. *Behavioral Neuroscience, 124*(3), 346–361.

Mouton-Liger, F., Thomas, S., Rattenbach, R., Magnol, L., Larigaldie, V., Ledru, A., et al. (2011). PCP4 (PEP19) overexpression induces premature neuronal differentiation associated with Ca(2+) /calmodulin-dependent kinase II delta activation in mouse models of down syndrome. *The Journal of Comparative Neurology, 13*, doi:10.1002/cne.22651.

O'Doherty, A., Ruf, S., Mulligan, C., Hildreth, V., Errington, M. L., Cooke, S., et al. (2005). An aneuploid mouse strain carrying human chromosome 21 with Down syndrome phenotypes. *Science, 309*(5743), 2033–2037.

Olson, L. E., Roper, R. J., Sengstaken, C. L., Peterson, E. A., Aquino, V., Galdzicki, Z., et al. (2007). Trisomy for the Down syndrome 'critical region' is necessary but not sufficient for brain phenotypes of trisomic mice. *Human Molecular Genetics, 16*, 774–782.

Paoloni-Giacobino, A., Chen, H., & Antonarakis, S. E. (1997). Cloning of a novel human neural cell adhesion molecule gene (NCAM2) that maps to chromosome region 21q21 and is potentially involved in Down syndrome. *Genomics, 43*, 43–51.

Peri, S., Navarro, J. D., Kristiansen, T. Z., Amanchy, R., Surendranath, V., Muthusamy, B., et al. (2004). Human protein reference database as a discovery resource for proteomics. *Nucleic Acids Research, 32*(Database issue), D497–D501.

Perluigi, M., di Domenico, F., Fiorini, A., Cocciolo, A., Giorgi, A., Foppoli, C., et al. (2011). Oxidative stress occurs early in Down syndrome pregnancy: A redox proteomics analysis of amniotic fluid. *Proteomics. Clinical Applications*, *5*, 167–178.

Pontius, J. U., Wagner, L., & Schuler, G. D. (2003). UniGene: A unified view of the transcriptome. In *The NCBI handbook*. Bethesda, MD: National Center for Biotechnology Information.

Rahmani, Z., Blouin, J. L., Creau-Goldberg, N., Watkins, P. C., Mattei, J. F., Poissonnier, M., et al. (1989). Critical role of the D21S55 region on chromosome 21 in the pathogenesis of Down syndrome. *Proceedings of the National Academy of Sciences of the United States of America*, *86*, 5958–5962.

Reymond, A., Camargo, A. A., Deutsch, S., Stevenson, B. J., Parmigiani, R. B., Ucla, C., et al. (2002). Nineteen additional unpredicted transcripts from human chromosome 21. *Genomics*, *79*, 824–832.

Reymond, A., Friedli, M., Henrichsen, C. N., Chapot, F., Deutsch, S., Ucla, C., et al. (2001). From PREDs and open reading frames to cDNA isolation: Revisiting the human chromosome 21 transcription map. *Genomics*, *78*, 46–54.

Robakis, N. K., Ramakrishna, N., Wolfe, G., & Wisniewski, H. M. (1987). Molecular cloning and characterization of a cDNA encoding the cerebrovascular and the neuritic plaque amyloid peptides. *Proceedings of the National Academy of Sciences of the United States of America*, *84*(12), 4190–4194.

Roubertoux, P. L., & Carlier, M. (2010). Mouse models of cognitive disabilities in trisomy 21 (Down syndrome). *American Journal of Medical Genetics. Part C, Seminars in Medical Genetics*, *154C*, 400–416.

Rueda, N., Llorens-Martín, M., Flórez, J., Valdizán, E., Banerjee, P., Trejo, J. L., et al. (2010). Memantine normalizes several phenotypic features in the Ts65Dn mouse model of Down syndrome. *Journal of Alzheimer's Disease*, *21*(1), 277–290.

Sananbenesi, F., Fischer, A., Schrick, C., Spiess, J., & Radulovic, J. (2002). Phosphorylation of hippocampal Erk-1/2, Elk-1, and p90-Rsk-1 during contextual fear conditioning: Interactions between Erk-1/2 and Elk-1. *Molecular and Cellular Neuroscience*, *21*(3), 463–476.

Schaefer, C. F., Anthony, K., Krupa, S., Buchoff, J., Day, M., Hannay, T., et al. (2009). PID: The Pathway Interaction Database. *Nucleic Acids Research*, *37*(Database issue), D674–D679.

Schwanhäusser, B., Busse, D., Li, N., Dittmar, G., Schuchhardt, J., Wolf, J., et al. (2011). Global quantification of mammalian gene expression control. *Nature*, *473*, 337–342.

Shichiri, M., Yoshida, Y., Ishida, N., Hagihara, Y., Iwahashi, H., Tamai, H., et al. (2011). α-Tocopherol suppresses lipid peroxidation and behavioral and cognitive impairments in the Ts65Dn mouse model of Down syndrome. *Free Radical Biology & Medicine*, *50*, 1801–1811.

Shirazi Fard, S., Kele, J., Vilar, M., Paratcha, G., & Ledda, F. (2010). Tiam1 as a signaling mediator of nerve growth factor-dependent neurite outgrowth. *PLoS One*, *5*, e9647.

Siarey, R. J., Carlson, E. J., Epstein, C. J., Balbo, A., Rapoport, S. I., & Galdzicki, Z. (1999). Increased synaptic depression in the Ts65Dn mouse, a model for mental retardation in Down syndrome. *Neuropharmacology*, *38*, 1917–1920.

Siarey, R. J., Stoll, J., Rapoport, S. I., & Galdzicki, Z. (1997). Altered long-term potentiation in the young and old Ts65Dn mouse, a model for Down Syndrome. *Neuropharmacology*, *36*, 1549–1554.

Sichitiu, S., Sinet, P. M., Lejeune, J., & Frézal, J. (1974). Overactivity of the indophenol oxidase (form dimeric) in the trisomy 21, secondary to triplicated gene (author's transl). *Humangenetik*, *23*, 65–72.

Siddiqui, A., Lacroix, T., Stasko, M. R., Scott-McKean, J. J., Costa, A. C., & Gardiner, K. J. (2008). Molecular responses of the Ts65Dn and Ts1Cje mouse models of Down syndrome to MK-801. *Genes, Brain, and Behavior*, *7*, 810–820.

Slemmon, J. R., Feng, B., & Erhardt, J. A. (2000). Small proteins that modulate calmodulin-dependent signal transduction: Effects of PEP-19, neuromodulin, and neurogranin on enzyme activation and cellular homeostasis. *Molecular Neurobiology*, *22*, 99–113.

St George-Hyslop, P. H., Tanzi, R. E., Polinsky, R. J., Haines, J. L., Nee, L., Watkins, P. C., et al. (1987). The genetic defect causing familial Alzheimer's disease maps on chromosome 21. *Science*, *235*(4791), 885–890.

Stark, C., Breitkreutz, B. J., Reguly, T., Boucher, L., Breitkreutz, A., & Tyers, M. (2006). BioGRID: A general repository for interaction datasets. *Nucleic Acids Research*, *34*(Database issue), D535–D539.

Sturgeon, X., & Gardiner, K. J. (2011). Transcript catalogs of human chromosome 21 and orthologous chimpanzee and mouse regions. *Mammalian Genome*, *22*, 261–271.

Suizu, F., Hiramuki, Y., Okumura, F., Matsuda, M., Okumura, A. J., Hirata, N., et al. (2009). The E3 ligase TTC3 facilitates ubiquitination and degradation of phosphorylated Akt. *Developmental Cell*, *17*(6), 800–810.

Sun, Y., Dierssen, M., Toran, N., Pollak, D. D., Chen, W. Q., & Lubec, G. (2011). A gel-based proteomic method reveals several protein pathway abnormalities in fetal Down syndrome brain. *Journal of Proteomics*, *74*, 547–557.

Tan, Y. H., Tischfield, J., & Ruddle, F. H. (1973). The linkage of genes for the human interferon-induced antiviral protein and indophenol oxidase-B traits to chromosome G-21. *The Journal of Experimental Medicine*, *137*, 317–330.

Tassone, F., Xu, H., Burkin, H., Weissman, S., & Gardiner, K. (1995). cDNA selection from 10 Mb of chromosome 21 DNA: Efficiency in transcriptional mapping and reflections of genome organization. *Human Molecular Genetics*, *4*, 1509–1518.

Toiber, D., Azkona, G., Ben-Ari, S., Torán, N., Soreq, H., & Dierssen, M. (2010). Engineering DYRK1A overdosage yields Down syndrome-characteristic cortical splicing aberrations. *Neurobiology of Disease*, *40*, 348–359.

Venkatesan, K., Rual, J. F., Vazquez, A., Stelzl, U., Lemmens, I., Hirozane-Kishikawa, T., et al. (2009). An empirical framework for binary interactome mapping. *Nature Methods*, *6*(1), 83–90.

Vilardell, M., Rasche, A., Thormann, A., Maschke-Dutz, E., Pérez-Jurado, L. A., Lehrach, H., et al. (2011). Meta-analysis of heterogeneous Down syndrome data reveals consistent genome-wide dosage effects related to neurological processes. *BMC Genomics*, *12*, 229.

Villard, L., Tassone, F., Haymowicz, M., Welborn, R., & Gardiner, K. (1997). Map location, genomic organization and expression patterns of the human RED1 RNA editase. *Somatic Cell and Molecular Genetics*, *23*, 135–145.

Wang, Y., Mulligan, C., Denyer, G., Delom, F., Dagna-Bricarelli, F., Tybulewicz, V. L., et al. (2009). Quantitative proteomics characterization of a mouse embryonic stem cell model of Down syndrome. *Molecular & Cellular Proteomics*, *8*(4), 585–595.

White, R., Morganstein, D., Christian, M., Seth, A., Herzog, B., & Parker, M. G. (2008). Role of RIP140 in metabolic tissues: Connections to disease. *FEBS Letters*, *582*, 39–45.

Wilming, L. G., Gilbert, J. G., Howe, K., Trevanion, S., Hubbard, T., & Harrow, J. L. (2008). The vertebrate genome annotation (Vega) database. *Nucleic Acids Research*, *36*(Database issue), D753–D760.

Wu, G., Feng, X., & Stein, L. (2010). A human functional protein interaction network and its application to cancer data analysis. *Genome Biology*, *11*, R53.

Xu, K., Rajagopal, S., Klebba, I., Dong, S., Ji, Y., Liu, J., et al. (2010). The role of fibroblast Tiam1 in tumor cell invasion and metastasis. *Oncogene*, *29*, 6533–6542.

Yang, S. H., Jaffray, E., Hay, R. T., & Sharrocks, A. D. (2003). Dynamic interplay of the SUMO and ERK pathways in regulating Elk-1 transcriptional activity. *Molecular Cell*, *12*, 63–74.

Yu, T., Li, Z., Jia, Z., Clapcote, S. J., Liu, C., Li, S., et al. (2010). A mouse model of Down syndrome trisomic for all human chromosome 21 syntenic regions. *Human Molecular Genetics*, *19*, 2780–2791.

Zigman, W. B., & Lott, I. T. (2007). Alzheimer's disease in Down syndrome: Neurobiology and risk. *Mental Retardation and Developmental Disabilities Research Reviews*, *13*, 237–246.

Zuo, P., & Maniatis, T. (1996). The splicing factor U2AF35 mediates critical protein-protein interactions in constitutive and enhancer-dependent splicing. *Genes & Development*, *10*, 1356–1368.

M. Dierssen and R. de la Torre (Eds.)
Progress in Brain Research, Vol. 197
ISSN: 0079-6123

CHAPTER 6

Neurological phenotypes for Down syndrome across the life span

Ira T. Lott*

Department of Pediatrics and Neurology, School of Medicine, University of California Irvine (UCI), Orange, CA, USA

Abstract: This chapter reviews the neurological phenotype of Down syndrome (DS) in early development, childhood, and aging. Neuroanatomic abnormalities in DS are manifested as aberrations in gross brain structure as well as characteristic microdysgenetic changes. As the result of these morphological abnormalities, brain circuitry is impaired. While an intellectual disability is ubiquitous in DS, there is a wide range of variation in cognitive performance and a growing understanding between aberrant brain circuitry and the cognitive phenotype. Hypotonia is most marked at birth, affecting gait and ligamentous laxity. Seizures are bimodal in presentation with infantile spasms common in infancy and generalized seizures associated with cognitive decline observed in later years. While all individuals have the characteristic neuropathology of Alzheimer's disease (AD) by age 40years, the prevalence of dementia is not universal. The tendency to develop AD is related, in part, to several genes on chromosome 21 that are overexpressed in DS. Intraneuronal accumulation of β-amyloid appears to trigger a cascade of neurodegeneration resulting in the neuropathological and clinical manifestations of dementia. Functional brain imaging has elucidated the temporal sequence of amyloid deposition and glucose metabolic rate in the development of dementia in DS. Mitochondrial abnormalities contribute to oxidative stress which is part of AD pathogenesis in DS as well as AD in the general population. A variety of medical comorbidities threaten cognitive performance including sleep apnea, abnormalities in thyroid metabolism, and behavioral disturbances. Mouse models for DS are providing a platform for the formulation of clinical trials with intervention targeted to synaptic plasticity, brain biochemistry, and morphological brain alterations.

Keywords: Down syndrome; brain development; seizures; hypotonia; dementia; clinical trials.

*Corresponding author.
Tel.: (714) 456-5333; Fax: (714) 456-8466
E-mail: itlott@uci.edu

DOI: 10.1016/B978-0-444-54299-1.00006-6

Neurological phenotypes in Down syndrome

The neurological phenotype in Down syndrome (DS) is the product of genetic expression and environmental influences. Like the other forms of genetically determined developmental disability, the neurological phenotype in DS changes across the life span. Changes in gene expression can determine differentiation of tissue involved in development and in functional decline associated with aging. Put differently, from the moment of conception we begin to age, a process involving decay in cellular structures, gene regulation, and DNA sequencing. Biological findings indicate that individuals with DS provide a link between development and aging. This review will examine the neurological phenotype at different age epochs in DS.

Neuroanatomic abnormalities and cognitive implications

The morphology of the brain in DS is a characteristic of the disorder and includes reduced brain weight with diminished proportions in the volumes of the frontal and temporal lobes. The brains of adults with DS are about 20% smaller than typically developing brains even when the measure is corrected for reduced body size (Kemper, 1991). This reduction in brain size appears in 4–5-month fetuses and progresses during the last 3months of gestation (Engidawork and Lubec, 2003; Guihard-Costa et al., 2006). Magnetic resonance imaging (MRI) studies show an approximate 17% decrease in children with DS between 10 and 20years (Pinter et al., 2001a). MRI studies confirm a selective decrease in the volumes of the hippocampus and the temporal lobes. The brain is brachycephalic with a small cerebellum, simplified gyral appearance, and a narrow superior temporal gyrus (Coyle et al., 1986; Wisniewski, 1990). These anatomic findings have also been seen by voxel-based morphometric MRI studies of the brain in children and adults with DS (Pinter et al., 2001a,b; Teipel et al., 2003, 2004; White and Alkire, 2003). The MRI studies have shown a remarkable preservation of subcortical gray matter structures in the face of a generally diminished brain volume, suggesting that there may be a temporal disassociation in development between the cortical versus subcortical areas. This may be an example of the disruption of developmental timing and homeostasis that occurs in DS and possibly other examples of aneuploidy (Pritchard and Kola, 1999). Unexplainably, the parahippocampal gyrus may be larger than normal in DS (Kesslak et al., 1994).

These gross anatomic abnormalities suggest that aberrations in fetal and early postnatal development may underlie the neuroanatomic abnormalities in DS. On the histological level, morphometric studies of the cortex show a reduced number of neurons, decreased neuronal densities, and abnormal neuronal distribution most marked in cortical layers II and IV. Ultrastructural studies show that synaptic density, synaptic length, and the contact zones between neurons are abnormal (Raz et al., 1995) (Fig. 1). Individuals with DS show a 20–50% reduction of neurons from birth to age 60months, suggesting that a deceleration in neurogenesis begins after 22weeks gestation (Schmidt-Sidor et al., 1990). However, neurogenetic cell differentiation is also impaired in fetuses with DS starting at 17weeks gestation as demonstrated by reduced numbers of dividing cells in the dentate gyrus and periventricular white matter (Contestabile et al., 2007). Factors cited to explain the decreased generation of neurons and their subsequent atrophy in DS have included a deleterious effect of the extra chromosome 21 on the timing of the mitotic cell cycle (Mittwoch and Wilkie, 1971). An analysis of cell proteins expressed in various stages of the cell cycle has shown a prolongation of the G_2 phase of the cell cycle, the last step in interphase prior to mitotic division (Contestabile et al., 2007). In fetuses with DS, an analysis of cell phenotypes shows less reduction in astrocytes as compared to neurons (Guidi et al., 2008). In this study, there was an increased incidence of apoptotic cell death, suggesting that both neurogenesis and death of neurons are accounting

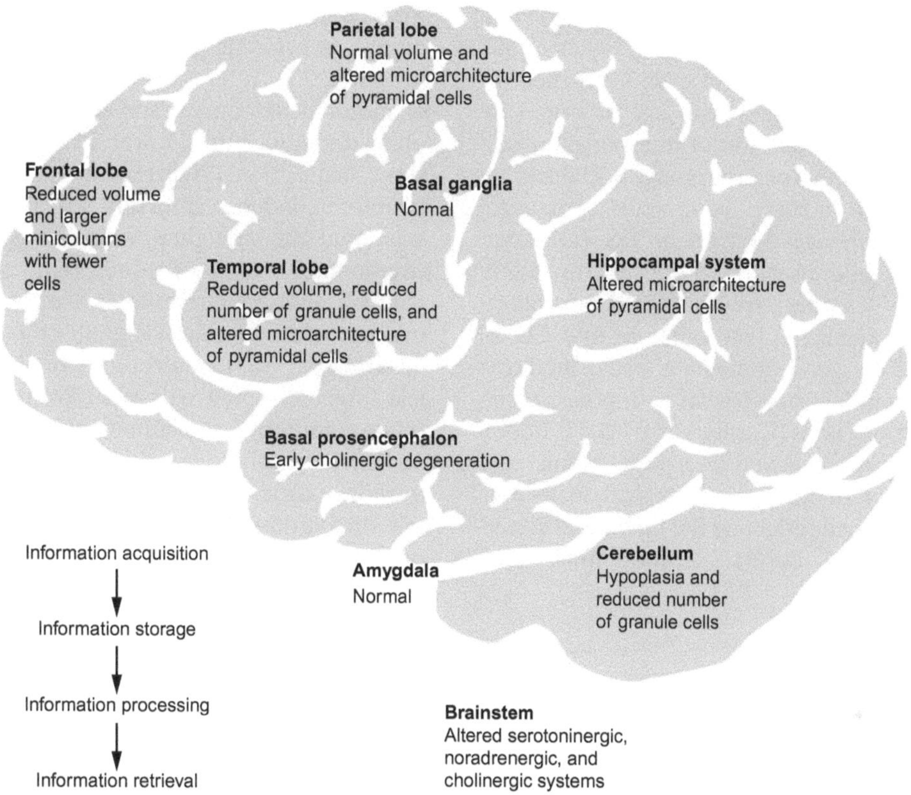

Fig. 1. The learning circuit in DS. Structures involved in information acquisition, processing, or storage are affected in DS. The figure depicts some of the described changes that may lead to cognitive dysfunction in individuals with DS. (Reprinted from Lancet Neurology, Lott IT and Dierssen M, Cognitive deficits and associated neurological complications in individuals with Down's syndrome, 2010, June 9(6):623–33 with permission from Elsevier). (For color version of this figure, the reader is referred to the Web version of this chapter.)

for some of the microscopic alterations in the brain. Children with DS actually display greater dendritic branching and total dendritic length than controls up to age 6months, but then show a decrease in dendritic arborization during childhood and adult life (Becker et al., 1991). The functional consequence of these morphogenetic changes results in abnormal neuronal connectivity and limited processing of information. Smaller neurons and shorter dendritic lengths may result in shorter connections in which networks tend to be local rather than interhemispheral in scope. As recently reviewed (Lott and Dierssen, 2010), the cerebellum may be implicated in both cognition and affect in individuals with DS. The cerebellum is small. It seems plausible that neuropsychological abnormalities in DS such as impaired attention, executive control, language learning, working memory, and emotional responses may reflect dysfunction in the cerebellar–cortical–limbic circuitry (Strick et al., 2009).

Cognitive development in DS has been described elsewhere in this volume (Edgin et al., 2010a). The precise effect of the above-described cytoarchitectonic abnormalities on cognitive development in DS is not clear, in part due to individual variation in performance measures with age (Couzens et al., 2011; Tsao and Kindelberger, 2009).

Weakness in language abilities of children with DS has been noted (Dykens et al., 2006; Fidler et al., 2005; Rondal et al., 2003). But even here, there is performance improvement with age as noted in lexical store, comprehension of interpersonal relations, and visual motor processing. Socialization and competence in daily living skills appear to improve through age 30years in DS (Dressler et al., 2010). Children with DS appear to present memory profiles that are distinct from other Williams and fragile-X syndromes in that DS is characterized by good immediate visual memory and rapid phonological retrieval with poor verbal working memory skills (Conners et al., 2011; Edgin et al., 2010b). As reviewed elsewhere in this volume, a battery of neuropsychological measures has been developed reflecting the functional status of the prefrontal cortex, hippocampus, and cerebellum.

Hypotonia

Hypotonia is ubiquitous in infants with DS and is defined as decreased resistance to passive muscle stretch (Fig. 2). The ligamentous laxity resulting from hypotonia is associated with a delay in motor development (Carr, 1970; Melyn and White, 1973). Infants with DS show a sequence of motor development similar to typically developing toddlers but a lower rate of motor milestone acquisition (Agiovlasitis et al., 2009). The hypotonia induces difficulty in postural control such that an individual with DS is focused on overcoming a lack of equilibrium compared to controls and this necessitates a different strategy in motor development (Rigoldi et al., 2011). The instability in trunk control may be linked to atypical finger grasping patterns (Jover et al., 2010). In a quantitative movement protocol measuring functional mobility, individuals with DS appeared to show longer durations of execution across all of the standardized tasks (Galli et al., 2010). As a consequence of hypotonia, individuals with DS have inherent joint laxity resulting in reduced gait stability and increased energetic costs for physical exertion (Agiovlasitis et al., 2009). These atypical gait patterns include longer stance time, decreased hip extension, and increased hip abduction during the swing phase. Hypotonia in DS is often associated with low levels of physical activity with the result of decreased bone mass accrual and predisposition to fractures (Hawli et al., 2009). The low levels of physical activity in DS result in a higher body mass index, lower levels of lean mass, and reduced bone mass-related parameters, which, in turn, may affect cardiovascular strength capacities (Gonzalez-Aguero et al., 2010).

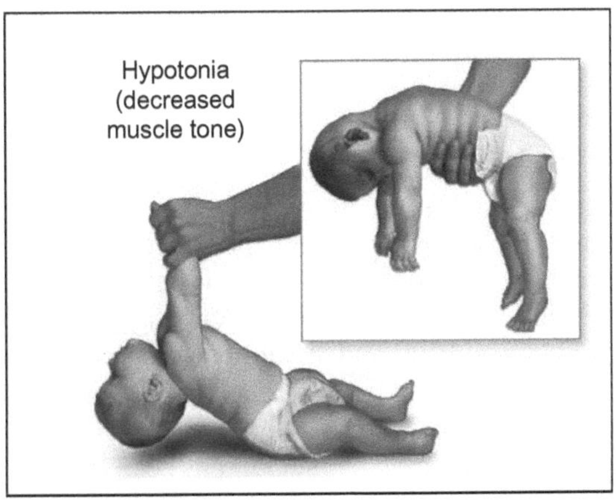

Fig. 2. Hypotonia in Down syndrome. Note the head lag upon pull to sitting and the inability to support posture in ventral suspension. (For color version of this figure, the reader is referred to the Web version of this chapter.)

Ligamentous laxity in DS provokes instability of vertebral movement at the atlanto-axial junction. The occiput, the atlas (C1), and the axis (C2) normally form a functional unit which assures a high degree of mobility of the upper cervical spine providing that strong ligaments keep these structures in place. In individuals with DS, the excessive laxity of the posterior transverse ligament which attaches the odontoid bone to C1 appears to be the most common cause of atlanto-axial subluxation although there is no universal agreement on this point (Merrick et al., 2000). In addition, C-1 hypoplasia and occult spinal canal stenosis have been linked to the increased risk of compressive cervical myelopathy in individuals with DS.

The anatomic locus for hypotonia in DS appears to be the cerebellum, a structure which shows hypocellularity due to impairment in the granular and periventricular zones during fetal development (Guidi et al., 2011). Additionally, the cerebellum plays a major role in the regulation of proprioceptive-motor control and motor learning. Beyond its effect on motor functioning, disorganized cerebellar output may impair higher order functioning such as emotion and cognition (Schmahmann, 2004; Teipel et al., 2004).

Seizures

It is estimated that 5–13% of children with DS have seizures (Arya et al., 2011; Lujic et al., 2011). The occurrence is bimodal with 40% having seizures before 1 year of age—generally infantile spasms—and with 40% developing seizures after the third decade, generally tonic–clonic or myoclonic in manifestation (Pueschel et al., 1991). Infantile spasms are associated with electroencephalographic (EEG) characteristics of idiopathic rather than symptomatic epilepsy. Children with DS have better seizure control following intervention with antiepileptic drugs than typically developing children. It appears that children with DS have better seizure control compared to other patients with symptomatic infantile spasms, and early initiation of appropriate treatment may contribute to the prevention of late seizure development and better developmental outcome (Arya et al., 2011).

When Lennox–Gastaut syndrome occurs in DS, the onset is usually later than seen in the typical population and may be associated with reflex seizures (Ferlazzo et al., 2009). There is a high rate of EEG abnormalities in DS without the recognition of clinical seizures (40%); therefore, EEG findings do not correlate with outcome (Smigielska-Kuzia et al., 2009). Electroclinical characteristics of seizures in adults with DS have included progressive slowing of background activity with frontal sharp and slow waves while those with myoclonic epilepsy typically display generalized spike and slow wave discharges (Vignoli et al., 2011).

Individuals with DS over age 45 years are more likely to develop dementia than those with seizures onset at younger ages (Menendez, 2005). Up to 84% of demented individuals with DS develop seizures. Senile myoclonic epilepsy appears to be a common manifestation of dementia in DS (De Simone et al., 2010).

The cellular basis of seizures may relate to the areas of brain dysgenesis that are commonly part of the neuroanatomic presentation in DS. Areas of focal cortical dysplasia may be associated with epilepsy in the general population related to the location and size of the lesions (Andrade, 2009; Leventer et al., 2008; Sisodiya et al., 2009). These areas of dysgenesis often indicate a pattern of molecular disruption that contributes to structural disorganization of the cortex and subsequent susceptibility to epilepsy. Audiogenic seizures have been observed in the rodent model (Ts65Dn) of DS and can be attenuated with antagonists to the metabotropic glutamate receptor 5 (mGluR5). These findings suggest that the amyloid precursor protein (APP), which is overexpressed in DS, may mediate this seizure susceptibility through the glutamate receptor (Westmark et al., 2009, 2010). Seizures in the temporal lobe may damage hippocampal circuitry, a situation which is worsened by the toxic accumulation of amyloid-β peptides seen in Alzheimer's disease (AD) (Noebels, 2011). There appears to be an association between the mGluR5 and APP in that activation of the glutamate receptor elevates APP in the dendrites of primary neuronal culture as well as synaptosomes (Westmark et al., 2009). These findings suggest that APP increases seizure incidence in AD. Treatment with glutamate receptor antagonists may help repress APP and serve as a therapeutic target for AD (Sokol et al., 2011).

Seizures appear to be associated with rapid cognitive decline in demented individuals with DS (Lott et al., 2012). In a retrospective study of individuals with DS and dementia, those with seizures became untestable by standard cognitive techniques employed for individuals with DS much sooner than the group without seizures. Adjustments were made for potentially confounding covariates of age, gender, *APOE4* status, baseline cognitive impairment, years since dementia onset at baseline, and treatment assignment. Like those individuals with AD in the general population who develop seizures, cognitive decline is worsened in the group with DS and dementia. Prospective studies of this association are indicated.

Dementia

The characteristic neuropathology of AD is present in the brains of individuals with DS by age 40 years (Mann and Esiri, 1989; Wisniewski et al., 1985). The findings include the accumulation of senile plaques (amyloid-β-protein) and neurofibrillary tangles (hyperphosphorylated tau protein). In the most common form of trisomy 21 in DS, there is an overexpression of the gene for APP from which the amyloid-β-protein is derived (Rumble et al., 1989). Intraneuronal accumulation of β-amyloid appears to trigger a cascade of neurodegeneration. Abnormal subcellular processing of β-amyloid leads to increased amyloid secretion and the generation of free radicals, thereby increasing the burden of oxidative stress (Zigman and Lott, 2007). Immune dysfunction in DS has been associated with the occurrence of AD (Kusters et al., 2009). The immune dysfunction begins in children with DS as reflected in abnormalities seen on the analysis of peripheral T-cell subsets, natural killer cells, and serum cytokines (Cetiner et al., 2010). DNA damage, repair, and abnormal dynamic flux have been linked to oxidative stress and AD-type neurodegeneration (Martin, 2008; Vasudevaraju et al., 2008). These observations have led to neurobiological studies of APP and the temporal events in AD pathogenesis in DS. Other genes on chromosome 21 that have been implicated in the pathogenesis of AD include superoxide dismutase, *Ets-2* transcription factors, DS critical region stress-inducible factor, β-site APP cleaving enzyme, and *S100β* and dual-specificity tyrosine phosphorylation-regulation kinase 1A (*DYRK1A*) (Head and Lott, 2004).

Another consequence of the gene dosage increase in APP in DS is the early deposition of β-amyloid in brain (Fig. 3). The parahippocampal and inferior temporal gyri demonstrate β-amyloid staining as young as 8years of age with subsequent accumulation in the CA-1/subiculum, dentate molecular layer, and the remainder of the hippocampal area (Leverenz and Raskind, 1998). Molecular dating techniques have shown that β-amyloid is deposited in the superficial layers of the frontal cortex in DS with subsequent transition to deeper cortical areas with age (Azizeh et al., 2000). In a report examining the distribution of intraneuronal and extracellular brain–protein aggregates associated with AD in the general population, only the individual with DS showed extracellular β-amyloid and neuritic plaques (Braak and Del Tredici, 2011). Functional neuroimaging and pathological studies have been directed toward the underlying processes of aging and dementia in DS. Positron emission imaging studies of glucose metabolic rate (GMR) in the brains of individuals with DS who are concentrating on a task show increased GMR in the medial temporal lobe in the same areas that are reduced in individuals with AD in the general population (Fig. 4a). This finding suggests that there may be a compensatory reaction occurring in the entorhinal cortex and hippocampus in young adults with DS prior to the onset of dementia. Neurite sprouting has been seen in approximately the same areas by postmortem examination (Haier et al., 2003; Head et al., 2007). Dynamic [11C] PiB PET (11C-labeled Pittsburgh Compound B that specifically binds

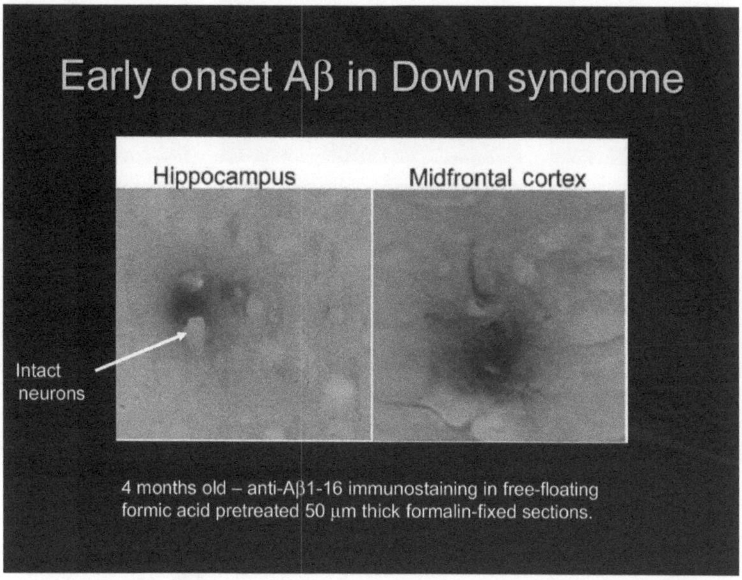

Fig. 3. Aβ deposition in brain of an infant with Down syndrome. (Image provided by Dr. Elizabeth Head, Sanders-Brown Center on Aging, University of Kentucky, Lexington, KY). (For color version of this figure, the reader is referred to the Web version of this chapter.)

fibrillar β-amyloid plaques) imaging has been carried out in a small sample of adults with DS and preliminary findings suggest a correlation between amyloid deposition and age (Fig. 4b) (Landt et al., 2011). In a study of [18F] FDDNP PET (FDDNP, molecule that binds to plaques and tangles *in vivo*) functional brain imaging, there is high binding levels of the ligand in DS comparable to those seen in AD (Fig. 4c). Measures of behavioral dysfunction in the same study are consistent with age-related progression of AD neuropathology in this population (Nelson et al., 2011). Future functional imaging studies in DS are likely to elucidate the dynamics of amyloid plaque formation and the occurrence of neurofibrillary tangles as well as their relationship to neuropsychological measures.

However, β-amyloid immunoreactivity has been reported in neuronal structures that have no AD pathology, suggesting that brain amyloid may not have an invariable role in the formation. An alternative explanation may relate to the *DYRK1A* which is located on chromosome 21 and regulates multiple genes that may contribute to deregulation of neural pathways responsible for dementia (Ji et al., 2010). Specifically, *DYRK1A* overexpression in the brains of individuals with DS likely contributes to early onset neurofibrillary degeneration, in part, through the hyperphosphorylation of tau. The several-fold increase in the number of DYRK1A and tau-positive neurofibrillary tangles in DS supports this hypothesis. DYRK1A also appears to enhance phosphorylation of APP, thereby elevating $A\beta_{40}$ and $A\beta_{42}$ levels. Thus, DYRK1A-mediated hyperphosphorylation of tau may provide a functional link between DS and AD (Ryoo et al., 2007).

The role of plasma levels of β-amyloid as a biomarker for AD in DS has been studied extensively. Among adults with DS, decreasing levels of plasma $A\beta_{42}$, a decline in the $A\beta_{42}/A\beta_{40}$ ratio, or increasing levels of $A\beta_{40}$ may be sensitive

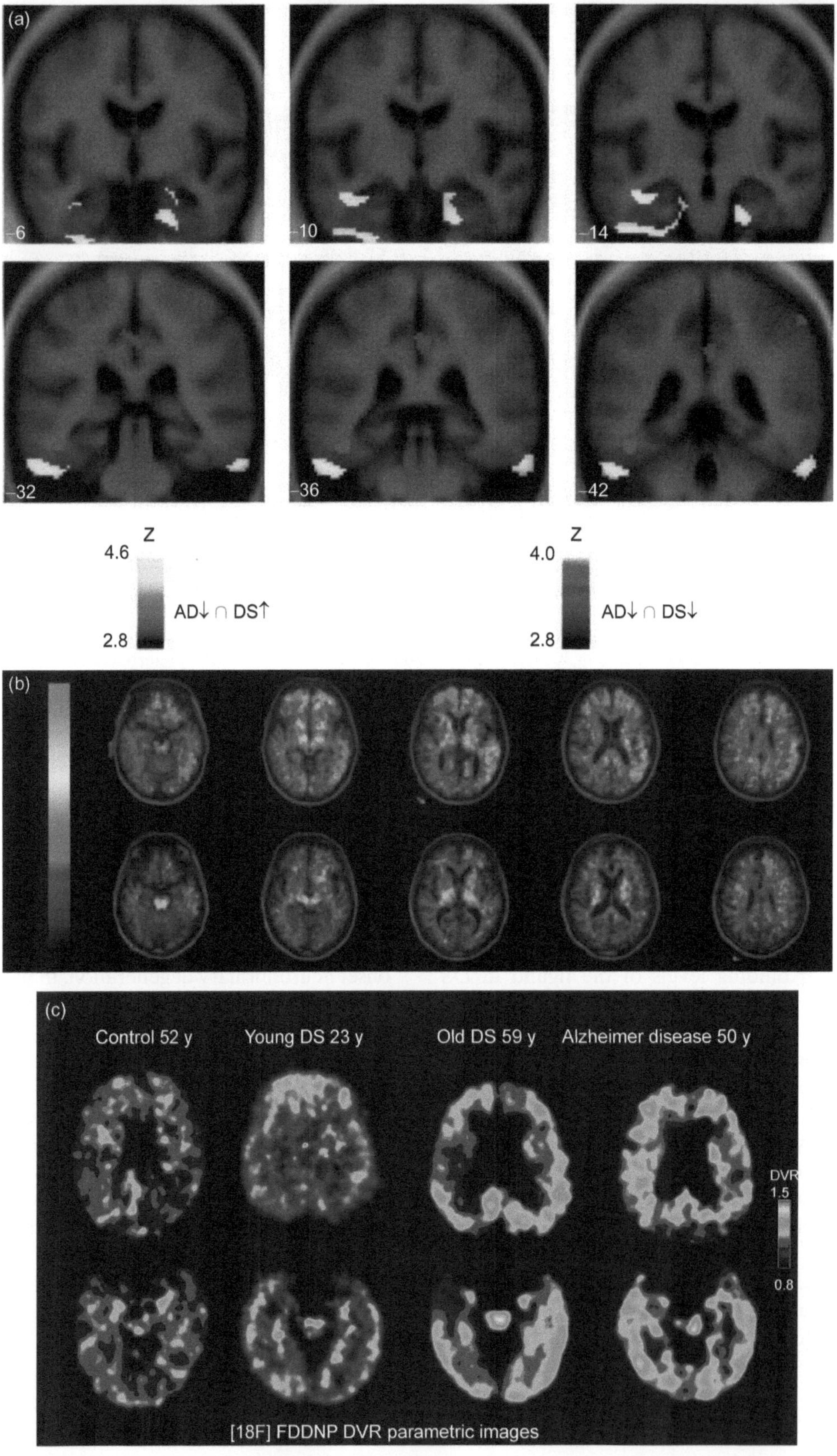
(a)
−6
−10
−14
−32
−36
−42
Z
4.6
AD↓ ∩ DS↑
2.8
Z
4.0
AD↓ ∩ DS↓
2.8
(b)
(c)
Control 52 y
Young DS 23 y
Old DS 59 y
Alzheimer disease 50 y
DVR
1.5
0.8
[18F] FDDNP DVR parametric images

indicators of conversion to AD, possibly reflecting compartmentalization of β-amyloid peptides in the brain (Schupf et al., 2010). Plasma β-amyloid levels alone did not dissociate DS adults with and without dementia. However, in demented adults with DS, *APOE4* allele was associated with higher $A\beta_{40}$ but not $A\beta_{42}$. After controlling for level of intellectual disability (mild, moderate, severe) and the presence or absence of dementia, there was an improved prediction of neuropsychological scores by plasma levels of β-amyloid (Head et al., 2011). There is an association of several biomarkers for AD for in the general population that are also seen in individuals with DS including *APOE*, *SORL1*, *RUNX1*, *BACE1*, and *ALDH18a1* (Patel et al., 2011). However, other studies have not found a relationship between the *APOE* genotype and plasma levels of $A\beta_{40}$ or $A\beta_{42}$ (Jones et al., 2009).

Individuals with DS have an increased tendency toward oxidative stress. Lack of balance in the metabolism of free radicals may have a direct role in the development of neuropathological changes of AD in DS (Zigman and Lott, 2007). Urinary markers of oxidative stress are increased in adolescents and adults with DS (Campos et al., 2010). Children with DS have an imbalance in the plasma and urinary levels of melatonin and kynurenine, thereby enhancing susceptibility toward oxidative stress (Uberos et al., 2010). DNA damage that occurs during the process of oxidative stress has been implicated in both DS and AD, but a pathogenetic precedent has not yet been established (Martin, 2008).

Mitochondrial control region mutations are seen in the brains of individuals with DS and dementia as well as AD (Coskun et al., 2010) (Fig. 5). These mitochondrial mutations are also seen in peripheral blood and DNA from lymphoblastoid cell lines. A complex I deficit has been described in mitochondria from skin fibroblasts in DS that produces free radicals associated with oxidative stress (Valenti et al., 2011). Methylation, which is a necessary event in mitochondrial functioning, is impaired in DS (Infantino et al., 2011). Dysregulated chromosome 21 genes in cortical neural progenitor cells from individuals with DS lead to elevated levels of S100β-induced radical oxygen species along with loss of the water channel aquaporin 4, causing an increase in programmed cell death (Lu et al., 2011). Caspase activation, another precursor of programmed cell death, occurs in association with amyloid plaques and neurofibrillary tangles in DS (Head et al., 2002). Fibroblasts in DS have a higher than normal level of apoptosis and are more susceptible to the proapoptotic effects of okadaic acid, a toxin known to induce malformations and inhibit differentiation in cell

Fig. 4. (a) FDG imaging in Down syndrome (DS) (Reprinted from Neurology, Haier RJ, et al., Temporal cortex hypermetabolism in Down syndrome prior to the onset of dementia, 2003, Dec 23;61(12):1673–1679, with permission from Wolter Kluwer Health). Results of SPM'99 conjunction analyses. Blue areas show where glucose metabolic rate (GMR) is lower in subjects with Alzheimer's disease (AD) compared with matched controls and where GMR is lower in nondemented DS subjects compared with matched controls. Yellow areas show the conjunction of where GMR is higher in the nondemented DS group compared with matched controls and where GMR is lower in the AD subjects compared with their matched controls. *Note:* The yellow areas are in the inferior temporal/entorhinal cortex; blue areas are in the posterior cingulate and left fusiform gyrus. Statistical results are shown on MRI templates for six coronal slices (-6-42), which show a large section of inferiotemporal/entorhinal cortex. Talairach and Tournoux atlas coordinates of all findings ($p<0.001$). (b) PiB amyloid staining for dementia in DS (Reprinted from Archives of Neurology, Landt J et al., Using positron emission tomography and carbon 11-labeled Pittsburgh compound B to image brain fibrillar β-amyloid in adults with Down syndrome: Safety, acceptability, and feasibility, 2011, July; 68 (7):890–896 with permission from American Medical Association). Fused carbon 11-labeled Pittsburgh Compound B nondisplaceable binding potential and magnetic resonance images for a subject with DS and AD (top) and a control without DS (bottom). (c) FDDNP imaging in DS (Reprinted from Archives of Neurology, Nelson LD et al., Positron emission tomogrpahy of brain β-amyloid and tau levels in adults with Down syndrome, 2011, June; 68(6):768–774 with permission from American Medical Association). 18F-FDDNP PET imaging in control, young DS, old DS, and AD showing binding characteristics. (For interpretation of the references to color in this figure legend, the reader is referred to the Web version of this chapter.)

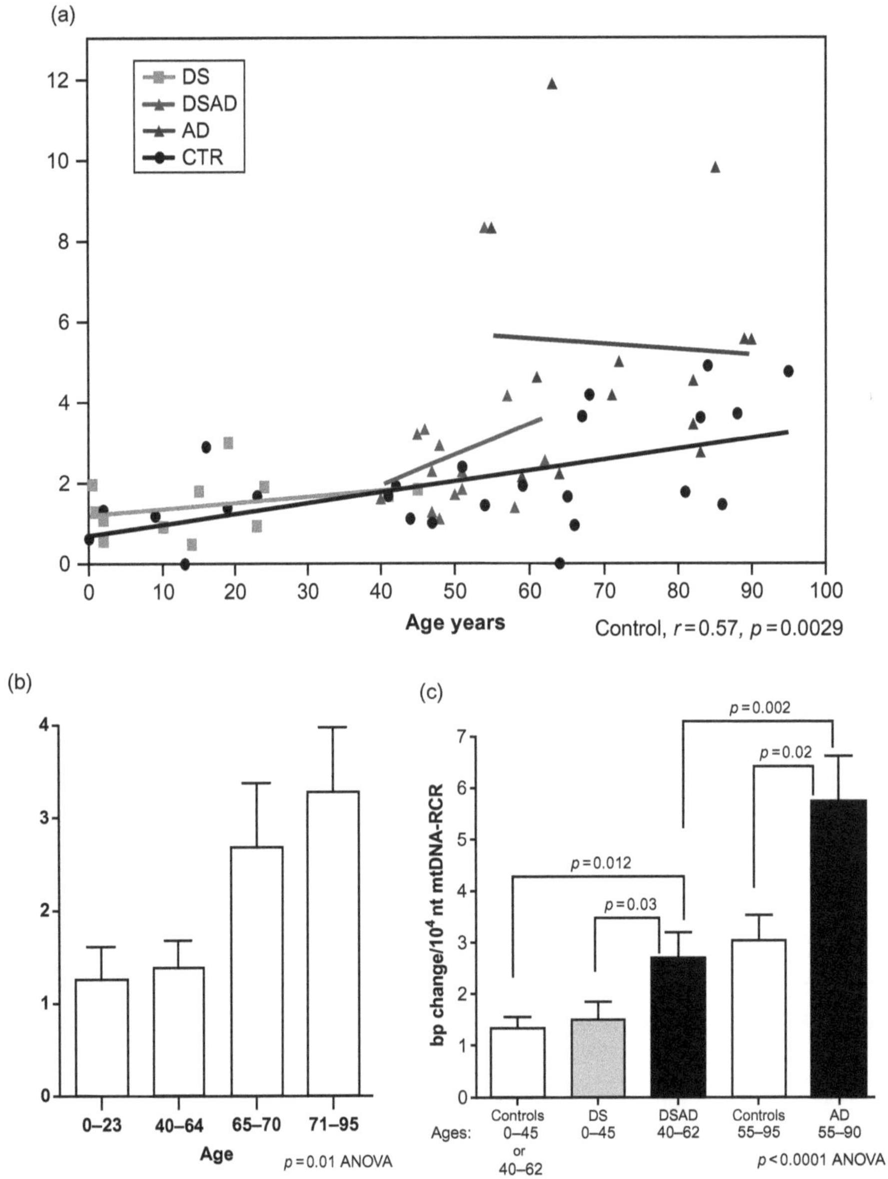

Fig. 5. mtDNA Regulatory Control Region (RCR) mutation frequency in the frontal cortices of control, DS, DSAD, and AD patients. (a) The correlation of mtDNA RCR mutation frequency with aging in all four groups. (b) The comparison of mtDNA RCR mutation frequency in four different age groups: 0–23, 40–64, 65–70, and 71–95years. (c) The comparison of mtDNA RCR mutation frequencies of DSAD and AD brains to age-matched controls and DS brains. Because the control groups for DS (ages 0–40) and DSAD (ages 40–62) were not significantly different (see panel b), DS and DSAD control groups were grouped together. (Reprinted from Journal of Alzheimer's Disease, Coskun P et al., Mitochondrial dysfunction and the etiology of Alzheimer's disease and Down syndrome dementia, 2010:20, S293–S310, with permission from IOS press). (For color version of this figure, the reader is referred to the Web version of this chapter.)

lines (Dogliotti et al., 2010). Multiple genes that are upregulated in trisomy 21 may contribute to transcriptional disruption and increase the liability to oxidative stress (Lockstone et al., 2007; Strydom et al., 2009). Oxidative stress may be a common pathogenetic mechanism across multiple genetic disorders including ataxia-telangiectasia, Fanconi anemia, Werner syndrome, and DS (Lloret et al., 2011). Clinical trials to treat oxidative stress in DS are reported elsewhere in this volume.

Comorbidities affecting the neurological phenotype

Obstructive sleep apnea occurs in over half of children with DS aged 2–4years and is related to otolaryngological problems associated with the disorder (Shott et al., 2006). There have been few studies of the cognitive consequences of obstructive sleep apnea in DS, but there is known to be an association of sleep apnea with lower IQ testing and behavioral difficulties (Bass et al., 2004; Melendres et al., 2004). In a small study of children with DS, sleep apnea was associated with difficulties in visual-perceptual skills involving the right hemisphere (Andreou et al., 2002). The consequence for cognition may be higher for adults with DS based upon studies of aging in the general population. Findings here show impaired working memory, attention, vigilance, and the ability to carry out simulated activities in the face of obstructive sleep apnea (Jackson et al., 2011). Volume reduction in the hippocampus has been observed in adults with chronic sleep apnea (Torelli et al., 2011). In patients with mild cognitive impairment in the general population, sleep apnea was associated with an increase in language dysfunction (Kim et al., 2011). Further studies on the neurological effect of sleep disorders in individuals with DS are indicated.

Hearing and visual disturbances can complicate the neurological evaluation of all patients with intellectual disability. Individuals with DS are prone to develop hyperopia, strabismus, blepharitis, cataracts, and glaucoma (Esbensen, 2010; van Allen et al., 1999). Hearing loss in adults in the general population may be independently associated with lower test scores of memory and executive functioning (Gates et al., 2010; Lin et al., 2011). Therefore, hearing and visual evaluation should be part of any diagnosis of cognitive dysfunction in DS.

Evaluation of thyroid functioning has been a part of health care guidelines for both children and adults with DS (Virji-Babul et al., 2007). In subclinical and overt hypothyroidism in the general population, deficits are noted in spatial, associative, and verbal memory with some amelioration by thyroid replacement therapy (Kramer et al., 2009). The impairments resulting from hypothyroidism are primarily mnemonic rather than a reflection of general cognitive slowing and suggest a disruption of normal hippocampal function or connectivity (Correia et al., 2009). However, other studies of subclinical hypothyroidism in the general population have failed to show significant neuropsychiatric problems (Baek et al., 2009). Hashimoto's encephalopathy, an autoimmune complication of hypothyroidism, which presents with myoclonus, neuropsychiatric symptoms, cognitive decline, and stroke-like episodes, has been described in DS (Brodtmann, 2009; Popova et al., 2008). The course is fluctuating in DS without a female preponderance. Histopathological brain lesions include gliosis, demyelination, spongiform transformation of neurons, and vasculitis involving venules (Vincent and Crino, 2011). There are variable abnormalities on MRI brain imaging with up to half of the studies in the normal range. There is a response in many patients to steroids or intravenous immunoglobulin.

Children with DS may be at a lower risk for significant behavioral comorbidities in that they show a lower profile of maladaptive behaviors compared to children with other intellectual disabilities (Dykens and Kasari, 1997; Stores et al., 1998). However, in comparison to typically developing age-matched peers, children with DS show higher rates of inattention, oppositional behaviors, and impulsivity (Dykens, 2007). Cleland (Cleland et al., 2010) has challenged the stereotypical perception that individuals with DS are highly social and have excellent "people" skills. Differences are cited in developing task-oriented strategies for problem solving and difficulties with goal-directed behaviors (Glenn et al., 2001; Jahromi et al., 2008; Kasari and Freeman, 2001; Wishart, 1993; Wishart and Manning, 1996). These studies provide some evidence for interpersonal difficulties experienced by individuals with DS throughout life despite general perceptions to the contrary. The curious association of DS with autistic disorder may reflect an underlying problem in social interactions in some children with DS. Recent use of screening tools and standardized diagnostic measures in 123 children with DS has yielded a prevalence of 18% for autistic spectrum disorders and 7% for the diagnosis of autism (DiGuiseppi et al., 2010). This is over a 10-fold increase compared to the general population (Lowenthal et al., 2007). Commonalities between the two conditions have been sought on several levels of research. Difficulties in word recognition, oral language, nonverbal ability, and working memory may underline reading comprehension problems in both DS and autistic spectrum disorders (Ricketts, 2011). Cortical-inhibition circuitries regulating neural plasticity has been postulated as related to the pathogenesis of brain functioning in both DS and autistic disorder (Baroncelli et al., 2011). Morphological brain studies in autism have shown dysgenetic abnormalities in the frontal lobes, amygdala, and cerebellum—all areas that have been described as dysgenetic in DS. Recent reviews have suggested that the time course of brain development may be an important determining factor (Amaral et al., 2008). Put slightly differently, deviant growth trajectories in autism may lead to developmental disruption of established patterns of functional connectivity (Lewis and Elman, 2008). Moreover, careful morphological studies of postmortem brain of individuals with autism have shown subgroup variability in neuronal density, providing a cytoarchitectonic foundation for brain heterogeneity (Simms et al., 2009). Future research is likely to highlight differences in the neuroanatomic and functional bases of the comorbidity between DS and autistic spectrum disorders. Other behavioral disorders in DS have centered on depression. While it has been thought that individuals with DS are more prone to depression than those with other forms of intellectual disability, a recent meta-review does not support the putative increased prevalence of depression in trisomy 21 (Walker et al., 2011). However, there are a number of risk factors in DS that may connote a vulnerability to depression including smaller hippocampal volumes, changes in neurotransmitter systems, deficits in attachment behaviors, and frequently occurring somatic disorders.

Altering the neurological phenotype through treatment

Understanding of the brain region affected by trisomy in mouse models has been helpful in the development of pharmacological approaches that may improve cognitive function in individuals with DS (Das and Reeves, 2011). The authors summarize the mechanisms by which pharmacological intervention appears to alter synaptic plasticity, brain biochemistry, and/or brain morphology in the mouse models for DS. The advantages of using mouse models for understanding various abnormalities in DS are numerous and include the ability to study a sufficient number of animals in a short time, genetic engineering for individual genes, control for genetic background, low cost of screening, ability to study both developing and mature subjects, modeling therapeutic interventions, and the lack of postmortem delays (Patterson, 2009; Salehi et al., 2007). Cognitive impairment may differ in various mouse models for DS and through this variation may provide a platform for mechanism-based drug design. There are many genes in the mouse that are orthologous to genes on human chromosome 21 (Das and Reeves, 2011). Studies in the mouse allow mapping of similar brain regions in the human for which there may be potential pharmacogenetic targets. However, the relationship between specific mouse targets and those in human trisomy 21 is complex. Individuals with DS show a broad spectrum of cognitive abilities and disabilities, suggesting that a single pharmacological agent may not be effective throughout the intellectual continuum in the disorder. Additionally, many of the mouse agents focus on the hippocampus while cognitive function in humans is dependent on many brain regions that interact with the hippocampus such as the entorhinal cortex, striatum, cerebellum, and cerebral cortex. The future studies of these areas in the mouse model promise to be an exciting area in the development of neurocognitive therapies for DS.

There have been a number of clinical trials in DS that are based upon trying to improve dysfunctional neuronal circuitry in the disorder. The most prominent of these has been focused on the cholinergic system in brain. Memory loss in old age is strongly correlated with a decline in the levels of acetylcholine (Mori et al., 2002; White and Ruske, 2002). The areas of the brain rich in cholinergic neurons such as the basal forebrain nuclei show a massive degeneration in cholinergic neurons. Administration of drugs that inhibit the breakdown of acetylcholine (donepezil) has reversed the memory deficits associated with aging and AD in the general population (Rusted et al., 1994; van Reekum et al., 1997).

As previously noted, individuals with DS have the characteristic neuropathology of AD by age 40years. The Ts65Dn mouse model of DS appears to parallel the cholinergic degeneration found in adults with DS and dementia which is correlated with deficient release of acetylcholine from the hippocampus (Holtzman et al., 1996). An encouraging aspect of the mouse study was the reversal of deficits in the 4-month-old Ts65Dn mouse when treated with an acetylcholinesterase inhibitor (Chang and Gold, 2008). In part, these results in the animal model led to trials of several different acetylcholinesterase inhibitors in adults with DS and dementia. However, in DS, there appears to be no consistent or compelling evidence for efficacy of acetylcholinesterase inhibitors including donepezil, galantamine, and rivastigmine in preventing the cognitive decline associated with dementia (Mohan et al., 2009a–d). Likewise, a trial of donepezil in children and adolescents with DS failed to demonstrate cognitive benefit (Kishnani et al., 2010). These results underlie the complexity of interpreting findings in animal models for human pharmacological intervention.

Basal forebrain cholinergic neuronal degeneration has been reversed by intracerebroventricular injections of nerve growth factor (Cooper et al., 2001). In evaluation of the genes within the targeted chromosomal segment in Ts65Dn mice, it became apparent that an extra copy of the gene for APP contributed to decreased transport of nerve growth factor (Salehi et al., 2006). This raises the prospect that treatment to reduce the level of APP gene expression may lessen the severity of cognitive decline in the dementia associated with DS (Megarbane et al., 2009).

Memantine is an antagonist of *N*-methyl D-aspartate receptor and appears to exert a positive effect on cognitive function by reducing oxidative damage and upregulating brain-derived neurotrophic factor (Marvanova et al., 2001). The compound is also a noncompetitive inhibitor of the serotonin receptor 5HT3 (Rammes et al., 2001). In the general population, the drug appears to alleviate some symptoms of AD (Ferris, 2003). In the Ts65Dn mouse, oral treatment with memantine improved spatial learning in the Morris water maze task (Rueda et al., 2008). These findings have been translated into a clinical trial of memantine for improving cognitive functioning in young adults with DS which is currently underway (Costa, 2011).

Individuals with DS and mice with partial trisomy contain an extra copy of the *DYRK1A* gene and this appears to be associated with changes in brain morphogenesis (Dierssen and de Lagran, 2006; Toiber et al., 2010). Learning problems similar to those seen in individuals with DS have been reported in transgenic mice using a human YAC construct that contains five genes including

DYRK1A. Increased gene dosage of *DYRK1A* results in brain morphogenesis defects, low brain-derived neurotrophic factor levels, and mnemonic deficits in these trisomy 21 mice. Overexpression of *DYRK1A* may exert its effect on apoptosis and cell survival (Tejedor and Hammerle, 2011). Epigallocatechin gallate—a member of a natural polyphenols family, found in great amount in green tea leaves—is a specific and safe DYRK1A inhibitor. In a study utilizing a polyphenol diet, the major features of the transgenic phenotype were rescued (Guedj et al., 2009).

An imbalance between inhibitory and excitatory neurotransmission may contribute to altered brain function in DS. A GABA-A antagonist to the major central nervous system inhibitory transmitter appears to restore cognitive function in Ts65Dn mice but has convulsant properties. Development of a GABA-A benzodiazepine receptor inverse agonist that is selective for the α5-subtype receptor appears to eliminate the convulsant propensity while maintaining cognitive restorative properties (Braudeau et al., 2011a,b). This compound appears to have potential significance for clinical trials.

The challenge of therapeutic trials in DS is similar to those reviewed for AD in the general population for which pleiotropic interventions have been recommended (Frautschy and Cole, 2010). In this context, there is a need not only for stage-specific intervention in AD pathogenesis but also for correcting abnormal pathways that will not be affected by targeting prodromal mechanisms. In DS, these abnormal pathways include loss of synapses, aberrant compensatory responses, defects in energy metabolism, and loss of neuroprotective mechanisms. In a useful review of potential drug candidates to improve cognitive functioning in DS, agents are categorized into those that influence synaptic plasticity, brain biochemistry, or brain pathology, respectively (Das and Reeves, 2011). An ongoing conundrum for the use of agents to improve cognitive functioning is the interaction of the drug with existing medication, an issue that has been explored in the neuropsychiatric literature and which can be addressed partially through preclinical studies (Wallace et al., 2011).

Nonpharmacological interventions such as cognitive therapy and exercise have shown some improvements in individuals with AD within the general population (Buschert et al., 2010; Littbrand et al., 2011). While cognitive improvement as the result of exercise has not been proven in DS, there are improvements in work performance variables for individuals in the exercise group. Although evidence exists to support improvements in physiological and psychological aspects from strategies using mixed physical activity programs, well-conducted research examining long-term physical outcomes, adverse effects, psychosocial outcomes, and costs is required before informed practice decisions can be made (Andriolo et al., 2010; Shields and Taylor, 2010).

Medical comorbidities, some of which have been reviewed above, can effect neurological performance in individuals with DS and negatively alter the cognitive phenotype. In a study of non-psychiatric health factors leading to psychiatric decompensation in adults with DS, 40% were found to have a significant comorbid medical condition considered causal in the behavioral decline (Charlot et al., 2011). Premature onset of medical comorbidities that may affect cognitive functioning in adults with DS include visual and hearing impairments, adult onset seizures, thyroid dysfunction, sleep apnea, and musculoskeletal problems (Esbensen, 2010). The quality of life in adults with DS who are at risk for AD as well as children whose intellectual development may be impacted by comorbid medical conditions can be improved considerably by systematic health care screening and adherence to health care guidelines (van Allen et al., 1999).

Concluding remarks

Many of the neurological characteristics of individuals with DS may be considered to reflect aberrations in the timing and sequence of development. Examples include systematic alterations in neural circuitry, age-specific onset of epilepsy, the evolution of amyloid deposition in brain, and interactions of genes on chromosome 21. Mouse models have been enormously helpful in providing a foundation for targets of pharmacological intervention. However, there are challenges in translating the results from mice studies to human clinical trials. It is encouraging that more trials are being developed upon specific brain mechanisms based upon detailed studies in the mouse. This approach is welcomed in an area that has been underserved in clinical trials. Understanding the basis for cognitive morbidity in DS and its potential treatment holds the additional promise of elucidating important aspects of development and aging within the general population.

Acknowledgments

This research was supported by NIH grants HD-065160, AG-21912, HD-37427. I thank Nina Movsesyan, PhD, for her assistance in the preparation of this manuscript.

References

Agiovlasitis, S., McCubbin, J. A., Yun, J., Pavol, M. J., & Widrick, J. J. (2009). Economy and preferred speed of walking in adults with and without Down syndrome. *Adapted Physical Activity Quarterly*, *26*, 118–130.

Amaral, D. G., Schumann, C. M., & Nordahl, C. W. (2008). Neuroanatomy of autism. *Trends in Neurosciences*, *31*, 137–145.

Andrade, D. M. (2009). Genetic basis in epilepsies caused by malformations of cortical development and in those with structurally normal brain. *Human Genetics*, *126*, 173–193.

Andreou, G., Galanopoulou, C., Gourgoulianis, K., Karapetsas, A., & Molyvdas, P. (2002). Cognitive status in Down syndrome individuals with sleep disordered breathing deficits (SDB). *Brain and Cognition*, *50*, 145–149.

Andriolo, R. B., El Dib, R. P., Ramos, L., Atallah, A. N., & da Silva, E. M. (2010). Aerobic exercise training programmes for improving physical and psychosocial health in adults with Down syndrome. *Cochrane Database of Systematic Reviews*, CD005176.

Arya, R., Kabra, M., & Gulati, S. (2011). Epilepsy in children with Down syndrome. *Epileptic Disorders*, *13*, 1–7.

Azizeh, B. Y., Head, E., Ibrahim, M. A., Torp, R., Tenner, A. J., Kim, R. C., et al. (2000). Molecular dating of senile plaques in the brains of individuals with Down syndrome and in aged dogs. *Experimental Neurology*, *163*, 111–122.

Baek, K. H., Zaslavsky, A., Lynch, R. C., Britt, C., Okada, Y., Siarey, R. J., et al. (2009). Down's syndrome suppression of tumour growth and the role of the calcineurin inhibitor DSCR1. *Nature*, *459*, 1126–1130.

Baroncelli, L., Braschi, C., Spolidoro, M., Begenisic, T., Maffei, L., & Sale, A. (2011). Brain plasticity and disease: A matter of inhibition. *Neural Plasticity*, *2011*, 286073.

Bass, J. L., Corwin, M., Gozal, D., Moore, C., Nishida, H., Parker, S., et al. (2004). The effect of chronic or intermittent hypoxia on cognition in childhood: A review of the evidence. *Pediatrics*, *114*, 805–816.

Becker, L., Mito, T., Takashima, S., & Onodera, K. (1991). Growth and development of the brain in Down syndrome. *Progress in Clinical and Biological Research*, *373*, 133–152.

Braak, H., & del Tredici, K. (2011). The pathological process underlying Alzheimer's disease in individuals under thirty. *Acta Neuropathologica*, *121*, 171–181.

Braudeau, J., Dauphinot, L., Duchon, A., Loistron, A., Dodd, R. H., Herault, Y., et al. (2011a). Chronic treatment with a promnesiant GABA-A alpha5-selective inverse agonist increases immediate early genes expression during memory processing in mice and rectifies their expression levels in a Down syndrome mouse model. *Advances in Pharmacological Sciences*, *2011*, 153218.

Braudeau, J., Delatour, B., Duchon, A., Pereira, P. L., Dauphinot, L., de Chaumont, F., et al. (2011b). Specific targeting of the GABA-A receptor alpha5 subtype by a selective inverse agonist restores cognitive deficits in Down syndrome mice. *Journal of Psychopharmacology*, *25*, 1030–1042.

Brodtmann, A. (2009). Hashimoto encephalopathy and Down syndrome. *Archives of Neurology*, *66*, 663–666.

Buschert, V., Bokde, A. L., & Hampel, H. (2010). Cognitive intervention in Alzheimer disease. *Nature Reviews Neurology*, *6*, 508–517.

Campos, C., Guzman, R., Lopez-Fernandez, E., & Casado, A. (2010). Urinary uric acid and antioxidant capacity in children and adults with Down syndrome. *Clinical Biochemistry*, *43*, 228–233.

Carr, J. (1970). Mental and motor development in young Mongol children. *Journal of Mental Deficiency Research*, *14*, 205–220.

Cetiner, S., Demirhan, O., Inal, T. C., Tastemir, D., & Sertdemir, Y. (2010). Analysis of peripheral blood T-cell subsets, natural killer cells and serum levels of cytokines in children with Down syndrome. *International Journal of Immunogenetics*, *37*, 233–237.

Chang, Q., & Gold, P. E. (2008). Age-related changes in memory and in acetylcholine functions in the hippocampus in the Ts65Dn mouse, a model of Down syndrome. *Neurobiology of Learning and Memory*, *89*, 167–177.

Charlot, L., Abend, S., Ravin, P., Mastis, K., Hunt, A., & Deutsch, C. (2011). Non-psychiatric health problems among psychiatric inpatients with intellectual disabilities. *Journal of Intellectual Disability Research*, *55*, 199–209.

Cleland, J., Wood, S., Hardcastle, W., Wishart, J., & Timmins, C. (2010). Relationship between speech, oromotor, language and cognitive abilities in children with Down's syndrome. *International Journal of Language & Communication Disorders*, *45*, 83–95.

Conners, F. A., Moore, M. S., Loveall, S. J., & Merrill, E. C. (2011). Memory profiles of Down, Williams, and fragile X syndromes: Implications for reading development. *Journal of Developmental and Behavioral Pediatrics*, *32*, 405–417.

Contestabile, A., Fila, T., Ceccarelli, C., Bonasoni, P., Bonapace, L., Santini, D., et al. (2007). Cell cycle alteration and decreased cell proliferation in the hippocampal dentate gyrus and in the neocortical germinal matrix of fetuses with Down syndrome and in Ts65Dn mice. *Hippocampus*, *17*, 665–678.

Cooper, J. D., Salehi, A., Delcroix, J. D., Howe, C. L., Belichenko, P. V., Chua-Couzens, J., et al. (2001). Failed retrograde transport of NGF in a mouse model of Down's syndrome: reversal of cholinergic neurodegenerative phenotypes following NGF infusion. *Proceedings of the National Academy of Sciences of the United States of America*, *98*, 10439–10444.

Correia, N., Mullally, S., Cooke, G., Tun, T. K., Phelan, N., Feeney, J., et al. (2009). Evidence for a specific defect in hippocampal memory in overt and subclinical hypothyroidism. *The Journal of Clinical Endocrinology and Metabolism*, *94*, 3789–3797.

Coskun, P. E., Wyrembak, J., Derbereva, O., Melkonian, G., Doran, E., Lott, I. T., et al. (2010). Systemic mitochondrial dysfunction and the etiology of Alzheimer's disease and Down syndrome dementia. *Journal of Alzheimer's Disease*, *20*(Suppl. 2), S293–S310.

Costa, A. C. (2011). On the promise of pharmacotherapies targeted at cognitive and neurodegenerative components of Down syndrome. *Developmental Neuroscience*, *33*, 414–427.

Couzens, D., Cuskelly, M., & Haynes, M. (2011). Cognitive development and Down syndrome: Age-related change on the Stanford-Binet test (fourth edition). *American Journal on Intellectual and Developmental Disabilities*, *116*, 181–204.

Coyle, J. T., Oster-Granite, M. L., & Gearhart, J. D. (1986). The neurobiologic consequences of Down syndrome. *Brain Research Bulletin*, *16*, 773–787.

Das, I., & Reeves, R. H. (2011). The use of mouse models to understand and improve cognitive deficits in Down syndrome. *Disease Models & Mechanisms*, *4*, 596–606.

de Simone, R., Puig, X. S., Gélisse, P., Crespel, A., & Genton, P. (2010). Senile myoclonic epilepsy: Delineation of a common condition associated with Alzheimer's disease in Down syndrome. *Seizure*, *19*, 383–389.

Dierssen, M., & de Lagran, M. M. (2006). DYRK1A (dual-specificity tyrosine-phosphorylated and -regulated kinase 1A): A gene with dosage effect during development and neurogenesis. *TheScientificWorldJOURNAL*, *6*, 1911–1922.

Diguiseppi, C., Hepburn, S., Davis, J. M., Fidler, D. J., Hartway, S., Lee, N. R., et al. (2010). Screening for autism spectrum disorders in children with Down syndrome: Population prevalence and screening test characteristics. *Journal of Developmental and Behavioral Pediatrics*, *31*, 181–191.

Dogliotti, G., Galliera, E., Dozio, E., Vianello, E., Villa, R. E., Licastro, F., et al. (2010). Okadaic acid induces apoptosis in Down syndrome fibroblasts. *Toxicology In Vitro*, *24*, 815–821.

Dressler, A., Perelli, V., Feucht, M., & Bargagna, S. (2010). Adaptive behaviour in Down syndrome: A cross-sectional study from childhood to adulthood. *Wiener Klinische Wochenschrift*, *122*, 673–680.

Dykens, E. M. (2007). Psychiatric and behavioral disorders in persons with Down syndrome. *Mental Retardation and Developmental Disabilities Research Reviews*, *13*, 272–278.

Dykens, E. M., Hodapp, R. M., & Evans, D. W. (2006). Profiles and development of adaptive behavior in children with Down syndrome. *Down's Syndrome, Research and Practice*, *9*, 45–50.

Dykens, E. M., & Kasari, C. (1997). Maladaptive behavior in children with Prader-Willi syndrome, Down syndrome, and nonspecific mental retardation. *American Journal of Mental Retardation*, *102*, 228–237.

Edgin, J. O., Mason, G. M., Allman, M. J., Capone, G. T., Deleon, I., Maslen, C., et al. (2010a). Development and validation of the Arizona Cognitive Test Battery for Down syndrome. *Journal of Neurodevelopmental Disorders*, *2*, 149–164.

Edgin, J. O., Pennington, B. F., & Mervis, C. B. (2010b). Neuropsychological components of intellectual disability: The contributions of immediate, working, and associative memory. *Journal of Intellectual Disability Research*, *54*, 406–417.

Engidawork, E., & Lubec, G. (2003). Molecular changes in fetal Down syndrome brain. *Journal of Neurochemistry*, *84*, 895–904.

Esbensen, A. J. (2010). Health conditions associated with aging and end of life of adults with Down syndrome. *International Review of Research in Mental Retardation, 39*, 107–126.

Ferlazzo, E., Adjien, C. K., Guerrini, R., Calarese, T., Crespel, A., Elia, M., et al. (2009). Lennox-Gastaut syndrome with late-onset and prominent reflex seizures in trisomy 21 patients. *Epilepsia, 50*, 1587–1595.

Ferris, S. H. (2003). Evaluation of memantine for the treatment of Alzheimer's disease. *Expert Opinion on Pharmacotherapy, 4*, 2305–2313.

Fidler, D. J., Most, D. E., & Guiberson, M. M. (2005). Neuropsychological correlates of word identification in Down syndrome. *Research in Developmental Disabilities, 26*, 487–501.

Frautschy, S. A., & Cole, G. M. (2010). Why pleiotropic interventions are needed for Alzheimer's disease. *Molecular Neurobiology, 41*, 392–409.

Galli, M., Cimolin, V., Patti, P., Ferrario, D., Heaney, G., Albertini, G., et al. (2010). Quantifying established clinical assessment measures using 3D-movement analysis in individuals with Down syndrome. *Disability and Rehabilitation, 32*, 1768–1774.

Gates, G. A., Gibbons, L. E., McCurry, S. M., Crane, P. K., Feeney, M. P., & Larson, E. B. (2010). Executive dysfunction and presbycusis in older persons with and without memory loss and dementia. *Cognitive and Behavioral Neurology, 23*, 218–223.

Glenn, S., Dayus, B., Cunningham, C., & Horgan, M. (2001). Mastery motivation in children with Down syndrome. *Down's Syndrome, Research and Practice, 7*, 52–59.

Gonzalez-Aguero, A., Vicente-Rodriguez, G., Moreno, L. A., Guerra-Balic, M., Ara, I., & Casajus, J. A. (2010). Health-related physical fitness in children and adolescents with Down syndrome and response to training. *Scandinavian Journal of Medicine & Science in Sports, 20*, 716–724.

Guedj, F., Sebrie, C., Rivals, I., Ledru, A., Paly, E., Bizot, J. C., et al. (2009). Green tea polyphenols rescue of brain defects induced by overexpression of DYRK1A. *PLoS One, 4*, e4606.

Guidi, S., Bonasoni, P., Ceccarelli, C., Santini, D., Gualtieri, F., Ciani, E., et al. (2008). Neurogenesis impairment and increased cell death reduce total neuron number in the hippocampal region of fetuses with Down syndrome. *Brain Pathology, 18*, 180–197.

Guidi, S., Ciani, E., Bonasoni, P., Santini, D., & Bartesaghi, R. (2011). Widespread proliferation impairment and hypocellularity in the cerebellum of fetuses with Down syndrome. *Brain Pathology, 21*, 361–373.

Guihard-Costa, A. M., Khung, S., Delbecque, K., Menez, F., & Delezoide, A. L. (2006). Biometry of face and brain in fetuses with trisomy 21. *Pediatric Research, 59*, 33–38.

Haier, R. J., Alkire, M. T., White, N. S., Uncapher, M. R., Head, E., Lott, I. T., et al. (2003). Temporal cortex hypermetabolism in Down syndrome prior to the onset of dementia. *Neurology, 61*, 1673–1679.

Hawli, Y., Nasrallah, M., & El-Hajj Fuleihan, G. (2009). Endocrine and musculoskeletal abnormalities in patients with Down syndrome. *Nature Reviews Endocrinology, 5*, 327–334.

Head, E., Doran, E., Nistor, M., Hill, M., Schmitt, F. A., Haier, R. J., et al. (2011). Plasma amyloid-beta as a function of age, level of intellectual disability, and presence of dementia in Down syndrome. *Journal of Alzheimer's Disease, 23*, 399–409.

Head, E., & Lott, I. T. (2004). Down syndrome and beta-amyloid deposition. *Current Opinion in Neurology, 17*, 95–100.

Head, E., Lott, I. T., Cribbs, D. H., Cotman, C. W., & Rohn, T. T. (2002). Beta-amyloid deposition and neurofibrillary tangle association with caspase activation in Down syndrome. *Neuroscience Letters, 330*, 99–103.

Head, E., Lott, I. T., Patterson, D., Doran, E., & Haier, R. J. (2007). Possible compensatory events in adult Down syndrome brain prior to the development of Alzheimer disease neuropathology: Targets for nonpharmacological intervention. *Journal of Alzheimer's Disease, 11*, 61–76.

Holtzman, D. M., Santucci, D., Kilbridge, J., Chua-Couzens, J., Fontana, D. J., Daniels, S. E., et al. (1996). Developmental abnormalities and age-related neurodegeneration in a mouse model of Down syndrome. *Proceedings of the National Academy of Sciences of the United States of America, 93*, 13333–13338.

Infantino, V., Castegna, A., Iacobazzi, F., Spera, I., Scala, I., Andria, G., et al. (2011). Impairment of methyl cycle affects mitochondrial methyl availability and glutathione level in Down's syndrome. *Molecular Genetics and Metabolism, 102*, 378–382.

Jackson, M. L., Howard, M. E., & Barnes, M. (2011). Cognition and daytime functioning in sleep-related breathing disorders. *Progress in Brain Research, 190*, 53–68.

Jahromi, L. B., Gulsrud, A., & Kasari, C. (2008). Emotional competence in children with Down syndrome: Negativity and regulation. *American Journal of Mental Retardation, 113*, 32–43.

Ji, L., Chauhan, V., Mehta, P., Wegiel, J., Mehta, S., & Chauhan, A. (2010). Relationship between proteolytically cleaved gelsolin and levels of amyloid-beta protein in the brains of Down syndrome subjects. *Journal of Alzheimer's Disease, 22*, 609–617.

Jones, E. L., Hanney, M., Francis, P. T., & Ballard, C. G. (2009). Amyloid beta concentrations in older people with Down syndrome and dementia. *Neuroscience Letters, 451*, 162–164.

Jover, M., Ayoun, C., Berton, C., & Carlier, M. (2010). Specific grasp characteristics of children with trisomy 21. *Developmental Psychobiology, 52*, 782–793.

Kasari, C., & Freeman, S. F. (2001). Task-related social behavior in children with Down syndrome. *American Journal of Mental Retardation*, *106*, 253–264.

Kemper, T. L. (1991). Down syndrome. In A. Peters & E. J. Jones (Eds.), *Cerebral cortex*. Plenum Press Kluwer Academic.

Kesslak, J. P., Nagata, S. F., Lott, I., & Nalcioglu, O. (1994). Magnetic resonance imaging analysis of age-related changes in the brains of individuals with Down's syndrome. *Neurology*, *44*, 1039–1045.

Kim, S. J., Lee, J. H., Lee, D. Y., Jhoo, J. H., & Woo, J. I. (2011). Neurocognitive dysfunction associated with sleep quality and sleep apnea in patients with mild cognitive impairment. *The American Journal of Geriatric Psychiatry*, *19*, 374–381.

Kishnani, P. S., Heller, J. H., Spiridigliozzi, G. A., Lott, I., Escobar, L., Richardson, S., et al. (2010). Donepezil for treatment of cognitive dysfunction in children with Down syndrome aged 10–17. *American Journal of Medical Genetics. Part A*, *152A*, 3028–3035.

Kramer, C. K., von Muhlen, D., Kritz-Silverstein, D., & Barrett-Connor, E. (2009). Treated hypothyroidism, cognitive function, and depressed mood in old age: The Rancho Bernardo study. *European Journal of Endocrinology*, *161*, 917–921.

Kusters, M. A., Verstegen, R. H., Gemen, E. F., & de Vries, E. (2009). Intrinsic defect of the immune system in children with Down syndrome: A review. *Clinical and Experimental Immunology*, *156*, 189–193.

Landt, J., D'Abrera, J. C., Holland, A. J., Aigbirhio, F. I., Fryer, T. D., Canales, R., et al. (2011). Using positron emission tomography and Carbon 11-labeled Pittsburgh Compound B to image Brain Fibrillar beta-amyloid in adults with Down syndrome: Safety, acceptability, and feasibility. *Archives of Neurology*, *68*, 890–896.

Leventer, R. J., Guerrini, R., & Dobyns, W. B. (2008). Malformations of cortical development and epilepsy. *Dialogues in Clinical Neuroscience*, *10*, 47–62.

Leverenz, J. B., & Raskind, M. A. (1998). Early amyloid deposition in the medial temporal lobe of young Down syndrome patients: A regional quantitative analysis. *Experimental Neurology*, *150*, 296–304.

Lewis, J. D., & Elman, J. L. (2008). Growth-related neural reorganization and the autism phenotype: A test of the hypothesis that altered brain growth leads to altered connectivity. *Developmental Science*, *11*, 135–155.

Lin, F. R., Ferrucci, L., Metter, E. J., An, Y., Zonderman, A. B., & Resnick, S. M. (2011). Hearing loss and cognition in the Baltimore Longitudinal Study of Aging. *Neuropsychology*, *25*, 763–770.

Littbrand, H., Stenvall, M., & Rosendahl, E. (2011). Applicability and effects of physical exercise on physical and cognitive functions and activities of daily living among people with dementia: A systematic review. *American Journal of Physical Medicine & Rehabilitation*, *90*, 495–518.

Lloret, A., Badia, M. C., Giraldo, E., Ermak, G., Alonso, M. D., Pallardo, F. V., et al. (2011). Amyloid-beta toxicity and tau hyperphosphorylation are linked via RCAN1 in Alzheimer's disease. *Journal of Alzheimer's Disease*, *27*, 701–709.

Lockstone, H. E., Harris, L. W., Swatton, J. E., Wayland, M. T., Holland, A. J., & Bahn, S. (2007). Gene expression profiling in the adult Down syndrome brain. *Genomics*, *90*, 647–660.

Lott, I. T., & Dierssen, M. (2010). Cognitive deficits and associated neurological complications in individuals with Down's syndrome. *Lancet Neurology*, *9*, 623–633.

Lott, I., Doran, E., Nguyen, V. Q., Tournay, A., Movsesyan, N., & Gillen, D. L. (2012). Down syndrome and dementia: Seizures and cognitive decline. *Journal of Alzheimer's Disease*, [Epub ahead of print].

Lowenthal, R., Paula, C. S., Schwartzman, J. S., Brunoni, D., & Mercadante, M. T. (2007). Prevalence of pervasive developmental disorder in Down's syndrome. *Journal of Autism and Developmental Disorders*, *37*, 1394–1395.

Lu, J., Esposito, G., Scuderi, C., Steardo, L., Delli-Bovi, L. C., Hecht, J. L., et al. (2011). S100B and APP promote a gliocentric shift and impaired neurogenesis in Down syndrome neural progenitors. *PLoS One*, *6*, e22126.

Lujic, L., Bosnjak, V. M., Delin, S., Duranovic, V., & Krakar, G. (2011). Infantile spasms in children with Down syndrome. *Collegium Antropologicum*, *35*(Suppl. 1), 213–218.

Mann, D. M., & Esiri, M. M. (1989). The pattern of acquisition of plaques and tangles in the brains of patients under 50 years of age with Down's syndrome. *Journal of Neurological Sciences*, *89*, 169–179.

Martin, L. J. (2008). DNA damage and repair: Relevance to mechanisms of neurodegeneration. *Journal of Neuropathology and Experimental Neurology*, *67*, 377–387.

Marvanova, M., Lakso, M., Pirhonen, J., Nawa, H., Wong, G., & Castren, E. (2001). The neuroprotective agent memantine induces brain-derived neurotrophic factor and trkB receptor expression in rat brain. *Molecular and Cellular Neurosciences*, *18*, 247–258.

Megarbane, A., Ravel, A., Mircher, C., Sturtz, F., Grattau, Y., Rethore, M. O., et al. (2009). The 50th anniversary of the discovery of trisomy 21: The past, present, and future of research and treatment of Down syndrome. *Genetics in Medicine*, *11*, 611–616.

Melendres, M. C., Lutz, J. M., Rubin, E. D., & Marcus, C. L. (2004). Daytime sleepiness and hyperactivity in children with suspected sleep-disordered breathing. *Pediatrics*, *114*, 768–775.

Melyn, M. A., & White, D. T. (1973). Mental and developmental milestones of noninstitutionalized Down's syndrome children. *Pediatrics*, *52*, 542–545.

Menendez, M. (2005). Down syndrome, Alzheimer's disease and seizures. *Brain & Development*, *27*, 246–252.

Merrick, J., Ezra, E., Josef, B., Hendel, D., Steinberg, D. M., & Wientroub, S. (2000). Musculoskeletal problems in Down Syndrome European Paediatric Orthopaedic Society Survey: The Israeli sample. *Journal of Pediatric Orthopaedics. Part B*, *9*, 185–192.

Mittwoch, U., & Wilkie, D. (1971). The effect of chlorimipramine on DNA synthesis and mitosis in cultured human cells. *British Journal of Experimental Pathology*, *52*, 186–191.

Mohan, M., Bennett, C., & Carpenter, P. K. (2009a). Galantamine for dementia in people with Down syndrome. *Cochrane Database of Systematic Reviews*, CD007656.

Mohan, M., Bennett, C., & Carpenter, P. K. (2009b). Memantine for dementia in people with Down syndrome. *Cochrane Database of Systematic Reviews*, CD007657.

Mohan, M., Bennett, C., & Carpenter, P. K. (2009c). Rivastigmine for dementia in people with Down syndrome. *Cochrane Database of Systematic Reviews*, CD007658.

Mohan, M., Carpenter, P. K., & Bennett, C. (2009d). Donepezil for dementia in people with Down syndrome. *Cochrane Database of Systematic Reviews*, CD007178.

Mori, C., Spooner, E. T., Wisniewsk, K. E., Wisniewski, T. M., Yamaguch, H., Saido, T. C., et al. (2002). Intraneuronal Abeta42 accumulation in Down syndrome brain. *Amyloid*, *9*, 88–102.

Nelson, L. D., Siddarth, P., Kepe, V., Scheibel, K. E., Huang, S. C., Barrio, J. R., et al. (2011). Positron emission tomography of brain beta-amyloid and tau levels in adults with Down syndrome. *Archives of Neurology*, *68*, 768–774.

Noebels, J. (2011). A perfect storm: Converging paths of epilepsy and Alzheimer's dementia intersect in the hippocampal formation. *Epilepsia*, *52*, 39–46.

Patel, A., Rees, S. D., Kelly, M. A., Bain, S. C., Barnett, A. H., Thalitaya, D., et al. (2011). Association of variants within APOE, SORL1, RUNX1, BACE1 and ALDH18A1 with dementia in Alzheimer's disease in subjects with Down syndrome. *Neuroscience Letters*, *487*, 144–148.

Patterson, D. (2009). Molecular genetic analysis of Down syndrome. *Human Genetics*, *126*, 195–214.

Pinter, J. D., Brown, W. E., Eliez, S., Schmitt, J. E., Capone, G. T., & Reiss, A. L. (2001a). Amygdala and hippocampal volumes in children with Down syndrome: A high-resolution MRI study. *Neurology*, *56*, 972–974.

Pinter, J. D., Eliez, S., Schmitt, J. E., Capone, G. T., & Reiss, A. L. (2001b). Neuroanatomy of Down's syndrome: A high-resolution MRI study. *The American Journal of Psychiatry*, *158*, 1659–1665.

Popova, G., Paterson, W. F., Brown, A., & Donaldson, M. D. (2008). Hashimoto's thyroiditis in Down's syndrome: Clinical presentation and evolution. *Hormone Research*, *70*, 278–284.

Pritchard, M. A., & Kola, I. (1999). The "gene dosage effect" hypothesis versus the "amplified developmental instability" hypothesis in Down syndrome. *Journal of Neural Transmission. Supplementum*, *57*, 293–303.

Pueschel, S. M., Louis, S., & McKnight, P. (1991). Seizure disorders in Down syndrome. *Archives of Neurology*, *48*, 318–320.

Rammes, G., Rupprecht, R., Ferrari, U., Zieglgansberger, W., & Parsons, C. G. (2001). The N-methyl-D-aspartate receptor channel blockers memantine, MRZ 2/579 and other amino-alkyl-cyclohexanes antagonise 5-HT(3) receptor currents in cultured HEK-293 and N1E-115 cell systems in a non-competitive manner. *Neuroscience Letters*, *306*, 81–84.

Raz, N., Torres, I. J., Briggs, S. D., Spencer, W. D., Thornton, A. E., Loken, W. J., et al. (1995). Selective neuroanatomic abnormalities in Down's syndrome and their cognitive correlates: Evidence from MRI morphometry. *Neurology*, *45*, 356–366.

Ricketts, J. (2011). Research review: Reading comprehension in developmental disorders of language and communication. *Journal of Child Psychology and Psychiatry*, *52*, 1111–1123.

Rigoldi, C., Galli, M., Mainardi, L., Crivellini, M., & Albertini, G. (2011). Postural control in children, teenagers and adults with Down syndrome. *Research in Developmental Disabilities*, *32*, 170–175.

Rondal, J. A., Elbouz, M., Ylieff, M., & Docquier, L. (2003). Francoise, a fifteen-year follow up. *Down's Syndrome, Research and Practice*, *8*, 89–99.

Rueda, N., Florez, J., & Martinez-Cue, C. (2008). Effects of chronic administration of SGS-111 during adulthood and during the pre- and post-natal periods on the cognitive deficits of Ts65Dn mice, a model of Down syndrome. *Behavioural Brain Research*, *188*, 355–367.

Rumble, B., Retallack, R., Hilbich, C., Simms, G., Multhaup, G., Martins, R., et al. (1989). Amyloid A4 protein and its precursor in Down's syndrome and Alzheimer's disease. *The New England Journal of Medicine*, *320*, 1446–1452.

Rusted, J., Graupner, L., O'Connell, N., & Nicholls, C. (1994). Does nicotine improve cognitive function? *Psychopharmacology*, *115*, 547–549.

Ryoo, S. R., Jeong, H. K., Radnaabazar, C., Yoo, J. J., Cho, H. J., Lee, H. W., et al. (2007). DYRK1A-mediated hyperphosphorylation of Tau. A functional link between Down syndrome and Alzheimer disease. *The Journal of Biological Chemistry*, *282*, 34850–34857.

Salehi, A., Delcroix, J. D., Belichenko, P. V., Zhan, K., Wu, C., Valletta, J. S., et al. (2006). Increased App expression in a mouse model of Down's syndrome disrupts NGF transport and causes cholinergic neuron degeneration. *Neuron*, *51*, 29–42.

Salehi, A., Faizi, M., Belichenko, P. V., & Mobley, W. C. (2007). Using mouse models to explore genotype-phenotype relationship in Down syndrome. *Mental Retardation and Developmental Disabilities Research Reviews*, *13*, 207–214.

Schmahmann, J. D. (2004). Disorders of the cerebellum: Ataxia, dysmetria of thought, and the cerebellar cognitive affective syndrome. *The Journal of Neuropsychiatry and Clinical Neurosciences*, *16*, 367–378.

Schmidt-Sidor, B., Wisniewski, K. E., Shepard, T. H., & Sersen, E. A. (1990). Brain growth in Down syndrome subjects 15 to 22 weeks of gestational age and birth to 60 months. *Clinical Neuropathology*, *9*, 181–190.

Schupf, N., Zigman, W. B., Tang, M. X., Pang, D., Mayeux, R., Mehta, P., et al. (2010). Change in plasma Aß peptides and onset of dementia in adults with Down syndrome. *Neurology, 75*, 1639–1644.

Shields, N., & Taylor, N. F. (2010). A student-led progressive resistance training program increases lower limb muscle strength in adolescents with Down syndrome: A randomised controlled trial. *Journal of Physiotherapy, 56*, 187–193.

Shott, S. R., Amin, R., Chini, B., Heubi, C., Hotze, S., & Akers, R. (2006). Obstructive sleep apnea: Should all children with Down syndrome be tested? *Archives of Otolaryngology—Head & Neck Surgery, 132*, 432–436.

Simms, M. L., Kemper, T. L., Timbie, C. M., Bauman, M. L., & Blatt, G. J. (2009). The anterior cingulate cortex in autism: Heterogeneity of qualitative and quantitative cytoarchitectonic features suggests possible subgroups. *Acta Neuropathologica, 118*, 673–684.

Sisodiya, S. M., Fauser, S., Cross, J. H., & Thom, M. (2009). Focal cortical dysplasia type II: Biological features and clinical perspectives. *Lancet Neurology, 8*, 830–843.

Smigielska-Kuzia, J., Sobaniec, W., Kulak, W., & Bockowski, L. (2009). Clinical and EEG features of epilepsy in children and adolescents in Down syndrome. *Journal of Child Neurology, 24*, 416–420.

Sokol, D. K., Maloney, B., Long, J. M., Ray, B., & Lahiri, D. K. (2011). Autism, Alzheimer disease, and fragile X: APP, FMRP, and mGluR5 are molecular links. *Neurology, 76*, 1344–1352.

Stores, R., Stores, G., Fellows, B., & Buckley, S. (1998). Daytime behaviour problems and maternal stress in children with Down's syndrome, their siblings, and non-intellectually disabled and other intellectually disabled peers. *Journal of Intellectual Disability Research, 42* (Pt. 3), 228–237.

Strick, P. L., Dum, R. P., & Fiez, J. A. (2009). Cerebellum and nonmotor function. *Annual Review of Neuroscience, 32*, 413–434.

Strydom, A., Dickinson, M. J., Shende, S., Pratico, D., & Walker, Z. (2009). Oxidative stress and cognitive ability in adults with Down syndrome. *Progress in Neuro-Psychopharmacology & Biological Psychiatry, 33*, 76–80.

Teipel, S. J., Alexander, G. E., Schapiro, M. B., Moller, H. J., Rapoport, S. I., & Hampel, H. (2004). Age-related cortical grey matter reductions in non-demented Down's syndrome adults determined by MRI with voxel-based morphometry. *Brain, 127*, 811–824.

Teipel, S. J., Schapiro, M. B., Alexander, G. E., Krasuski, J. S., Horwitz, B., Hoehne, C., et al. (2003). Relation of corpus callosum and hippocampal size to age in nondemented adults with Down's syndrome. *The American Journal of Psychiatry, 160*, 1870–1878.

Tejedor, F. H., & Hammerle, B. (2011). MNB/DRYK1A as a multiple regulator of neuronal development. *FEBS, 278*(2), 225–235.

Toiber, D., Azkona, G., Ben-Ari, S., Toran, N., Soreq, H., & Dierssen, M. (2010). Engineering DYRK1A overdosage yields Down syndrome-characteristic cortical splicing aberrations. *Neurobiology of Disease, 40*, 348–359.

Torelli, F., Moscufo, N., Garreffa, G., Placidi, F., Romigi, A., Zannino, S., et al. (2011). Cognitive profile and brain morphological changes in obstructive sleep apnea. *NeuroImage, 54*, 787–793.

Tsao, R., & Kindelberger, C. (2009). Variability of cognitive development in children with Down syndrome: Relevance of good reasons for using the cluster procedure. *Research in Developmental Disabilities, 30*, 426–432.

Uberos, J., Romero, J., Molina-Carballo, A., & Munoz-Hoyos, A. (2010). Melatonin and elimination of kynurenines in children with Down's syndrome. *Journal of Pediatric Endocrinology & Metabolism, 23*, 277–282.

Valenti, D., Manente, G. A., Moro, L., Marra, E., & Vacca, R. A. (2011). Deficit of complex I activity in human skin fibroblasts with chromosome 21 trisomy and overproduction of reactive oxygen species by mitochondria: Involvement of the cAMP/PKA signalling pathway. *The Biochemical Journal, 435*, 679–688.

van Allen, M. I., Fung, J., & Jurenka, S. B. (1999). Health care concerns and guidelines for adults with Down syndrome. *American Journal of Medical Genetics, 89*, 100–110.

van Reekum, R., Black, S. E., Conn, D., & Clarke, D. (1997). Cognition-enhancing drugs in dementia: A guide to the near future. *Canadian Journal of Psychiatry, 42*(Suppl. 1), 35S–50S.

Vasudevaraju, P., Bharathi, , Garruto, R. M., Sambamurti, K., & Rao, K. S. (2008). Role of DNA dynamics in Alzheimer's disease. *Brain Research Reviews, 58*, 136–148.

Vignoli, A., Zambrelli, E., Chiesa, V., Savini, M., la Briola, F., Gardella, E., et al. (2011). Epilepsy in adult patients with Down syndrome: A clinical-video EEG study. *Epileptic Disorders, 13*, 125–132.

Vincent, A., & Crino, P. B. (2011). Systemic and neurologic autoimmune disorders associated with seizures or epilepsy. *Epilepsia, 52*(Suppl. 3), 12–17.

Virji-Babul, N., Eichmann, A., Kisly, D., Down, J., & Haslam, R. H. (2007). Use of health care guidelines in patients with Down syndrome by family physicians across Canada. *Paediatrics & Child Health, 12*, 179–183.

Walker, J. C., Dosen, A., Buitelaar, J. K., & Janzing, J. G. (2011). Depression in Down syndrome: A review of the literature. *Research in Developmental Disabilities, 32*, 1432–1440.

Wallace, T. L., Ballard, T. M., Pouzet, B., Riedel, W. J., & Wettstein, J. G. (2011). Drug targets for cognitive enhancement in neuropsychiatric disorders. *Pharmacology, Biochemistry, and Behavior*, *99*, 130–145.

Westmark, C. J., Westmark, P. R., & Malter, J. S. (2009). MPEP reduces seizure severity in Fmr-1 KO mice over expressing human Abeta. *International Journal of Clinical and Experimental Pathology*, *3*, 56–68.

Westmark, C. J., Westmark, P. R., & Malter, J. S. (2010). Alzheimer's disease and Down syndrome rodent models exhibit audiogenic seizures. *Journal of Alzheimer's Disease*, *20*, 1009–1013.

White, N. S., & Alkire, M. T. (2003). Impaired thalamocortical connectivity in humans during general-anesthetic-induced unconsciousness. *NeuroImage*, *19*, 402–411.

White, K. G., & Ruske, A. C. (2002). Memory deficits in Alzheimer's disease: The encoding hypothesis and cholinergic function. *Psychonomic Bulletin and Review*, *9*, 426–437.

Wishart, J. G. (1993). The development of learning difficulties in children with Down's syndrome. *Journal of Intellectual Disability Research*, *37*(Pt. 4), 389–403.

Wishart, J. G., & Manning, G. (1996). Trainee teachers' attitudes to inclusive education for children with Down's syndrome. *Journal of Intellectual Disability Research*, *40*(Pt. 1), 56–65.

Wisniewski, K. E. (1990). Down syndrome children often have brain with maturation delay, retardation of growth, and cortical dysgenesis. *American Journal of Medical Genetics. Supplement*, *7*, 274–281.

Wisniewski, K. E., Wisniewski, H. M., & Wen, G. Y. (1985). Occurrence of neuropathological changes and dementia of Alzheimer's disease in Down's syndrome. *Annals of Neurology*, *17*, 278–282.

Zigman, W. B., & Lott, I. T. (2007). Alzheimer's disease in Down syndrome: Neurobiology and risk. *Mental Retardation and Developmental Disabilities Research Reviews*, *13*, 237–246.

M. Dierssen and R. de la Torre (Eds.)
Progress in Brain Research, Vol. 197
ISSN: 0079-6123

CHAPTER 7

Human and mouse model cognitive phenotypes in Down syndrome: implications for assessment

Jamie O. Edgin*, Gina M. Mason, Goffredina Spanò, Andrea Fernández and Lynn Nadel

University of Arizona, Down Syndrome Research Group, Tucson, AZ, USA

Abstract: The study of cognitive function in Down syndrome (DS) has advanced rapidly in the past decade. Mouse models have generated data regarding the neurological basis for the specific cognitive profile of DS (i.e., deficits in aspects of hippocampal, prefrontal, and cerebellar function) and have uncovered pharmacological treatments with the potential to affect this phenotype. Given this progress, the field is at a juncture in which we require assessments that may effectively translate the findings acquired in mouse models to humans with DS. In this chapter, we describe the cognitive profile of humans with DS and associated mouse models, discussing the ways in which we may merge these findings so as to more fully understand cognitive strengths and weaknesses in this population. New directions for approaches to cognitive assessment in mice and humans are discussed.

Keywords: cognitive assessment; mouse phenotype; pharmacological intervention; Down syndrome; mouse models; medial temporal lobe function; prefrontal function; cerebellum.

Introduction

The study of cognitive function in Down syndrome (DS), the most common genetic cause of intellectual disability (ID), has advanced rapidly in the past decade. We now know that DS results in a specific cognitive phenotype in humans (Fidler and Nadel, 2007; Pennington et al., 2003; Vicari, 2006) and that this cognitive profile can be reliably measured with validated and well-replicated cognitive assessments (Edgin et al., 2010a). In general, individuals with DS present variable deficits in learning, memory, and language with global levels of cognitive function in the range of ID (i.e., IQ < 70) (Lott and Dierssen, 2010).

The study of cognitive deficits in DS has benefited substantially from the use of mouse models, which have provided data highlighting the neural mechanisms that may underscore cognitive

*Corresponding author.
Tel.: +1-520-626-0244; Fax: +1-520-621-9306
E-mail: jamie.edgin@gmail.com

DOI: 10.1016/B978-0-444-54299-1.00007-8

difficulties. These advances have generated several targets for interventions to ameliorate the cognitive deficits. Recent findings using such models have shown that aspects of the associated cognitive deficits can be modified through pharmacological agents and environmental enrichment (Fernandez et al., 2007; Guedj et al., 2009; Martinez-Cue et al., 2002; Roper et al., 2006; Salehi et al., 2009).

The mouse has proven to be an especially useful model organism, not only because of the general mammalian and genetic similarities with humans (Crnic and Pennington, 2000), but also more specifically because many of the protein-coding genes currently identified on human chromosome 21 (HSA21) are conserved on mouse chromosomes (MMU) 10, 16, and 17 (Pletcher et al., 2001). While most of the research to date has been carried out with the Ts65Dn mouse (Davisson et al., 1990), models that more closely approximate the genetic basis of DS in humans have recently been developed, involving triplication of different chromosomal regions. These include a "transchromosomic" (trans-species aneuploid) model containing nearly the entire HSA21, "Tc1" (O'Doherty et al., 2005) and a more recent model syntenic for all regions of HSA21 orthologs on mouse 16, 17, and 10, "Dp (16)1Yey/+, Dp(17)1Yey/+, Dp(10)1Yey/+" (Yu et al., 2010). Other models have been developed to test the importance of segments of genes included in the triplication, such as the DS "critical" region (e.g., Ts1rhr; Olson et al., 2004a). While the newer models have been developed to contain more complete segments of HSA21 (or mouse orthologs), there are still no "perfect" models of the human case. For instance, Tc1, which could have replicated more closely the genetic alteration, results in mosaicism. Table 1 displays the current models in use and the extent of their genetic alteration (i.e., both in terms of mouse and HSA structure). The Ts65Dn model has been the most widely used in studies of behavioral outcomes as well as tests of pharmacological interventions; we review the data for this model in detail throughout the chapter, adding the findings from newer models that include a more complete representation of the genes on HSA21 where relevant. Beyond offering testable mechanisms of drug treatment in this population, mouse models that prove sensitive to the same treatment as humans may also help us understand

Table 1. Mouse models of Down syndrome and descriptions of their genetic alteration

Mouse model	Mouse chromosomes affected	HSA21 syntenic regions
Ts65Dn (Davisson et al., 1990)	MMU16, subcentromeric region of MMU17	Region spanning from *Mrpl39* to *Zfp295*; MMU 17 trisomic region not syntenic to any HSA21 region
Ts1Cje (Sago et al., 1998)	MMU16	Region spanning from 21q22.1 to 22.3
Ts1Rhr (Olson et al., 2004a)	MMU16	"Down Syndrome Critical Region" (DSCR); spanning from *cbr3* gene to *mx2* gene on chr21 (~21q22.2–22.3)
Ts1Yah (Pereira et al., 2009)	MMU17	*Abcg1-U2af1* telomeric region
Tc1 (O'Doherty et al., 2005)	N/A	Most of HSA21 inserted through microcell-mediated chromosome transfer; regions inserted bound by markers CXADR and D21S1922, and IFNAR1 and RUNX1
Dp(16)1Yu (also see Ts1Yu, Dp(16)1Yey) (Li et al., 2007)	MMU16	Large region spanning from 21q11.2 to 22.3
Dp(16)1Yey/+, Dp(17)1Yey/+, Dp(10)1Yey/+ (Yu et al., 2010)	MMU16, MMU17, MMU10	Distal portion of 21q22.3 (MMU10); proximal portion of 21q22.3 (MMU17); region spanning from 21q11.2 to 22.3 (MMU16)

the reasons for treatment success, the projected time course of treatment effects, and in the identification of biomarkers to establish treatment efficacy.

Given that hypothesis-based drug treatment in humans with DS is ongoing, the field urgently requires cognitive measures that bridge findings between mouse models and humans with DS. The success of animal-tested interventions in humans with DS rests on two central assumptions, (1) that particular drugs may affect mice and humans in essentially the same way, by targeting the same neural systems, and (2) that the neural systems so targeted play similar functional roles in mice and humans, such that the drugs will have a substantial impact on cognitive function in the human. Beyond these assumptions, it is also critical for the measures used in both species to have adequate reliability, validity, and sensitivity to detect effects. While it is often difficult to have measures in both species that tap similar constructs (i.e., demonstrate construct validity), maximizing this characteristic in measures may help to support translation of the findings. Therefore, in order to maximize success of interventions based on animal model work, the tests administered in humans and mice should be as directly comparable as possible. Our primary goal in this chapter is to facilitate the translation of findings from the mouse model to humans with DS. To do so, we (1) detail the specific profile of cognitive and neurological deficits in humans with DS, (2) detail findings regarding brain and behavioral deficits in the range of mouse models, and (3) discuss ways to merge these findings so as to more fully understand cognitive function in DS and measure the outcome of clinical trials in the future.

While theories regarding ID syndromes have sometimes posited that the various IDs include similar core cognitive deficits (e.g., a common impairment in working memory; Cornoldi and Vecchi, 2003; Crnic and Pennington, 2000), the bulk of the evidence supports the view that each syndrome has its own unique profile of cognitive strengths and weaknesses (Edgin et al., 2010b; Nadel, 1999; Vicari et al., 2005). Basic neuroscience work has also shown that multiple learning and memory systems exist within the brain (e.g., Nadel, 1994), with dissociations in cognitive processes apparent even within single brain systems, such as the hippocampus (e.g., CA3/dentate gyrus vs. CA1/subiculum functions). We assume that taking such dissociations into account will prove essential in detailing the unique profile of spared and impaired cognitive abilities in ID syndromes and that this will have critical implications for behavioral and pharmacological intervention. Therefore, our discussion focuses on neurocognitive dissociations that are apparent in humans with DS and animal models.

The importance of a developmental approach

While our understanding of the DS cognitive phenotype has advanced rapidly in the past 10 years, we still are lacking an understanding of the developmental trajectory of these cognitive deficits. The large majority of studies conducted to date have measured cognitive outcomes in adult mice and older children or adult humans. Interventions, with a few exceptions (Moon et al., 2010; Roper et al., 2006), have also been conducted with adult mice.

There are a number of problems with assessing the cognitive and behavioral phenotype in individuals with DS, or any other developmental disability, at one "snapshot" in time, and then comparing that snapshot to data generated in adults without any developmental disability or experimental animals with lesions. First, we have substantial evidence that the cognitive profile in individuals with DS evolves across the lifespan, demonstrating periods of slowing in development or decline. Data collected at a single time point will not reflect the complexity of these transitions. Similarly, even when patterns of behavior appear exactly the same, the underlying brain bases of these behaviors may differ across developmental periods or in different populations (e.g., DS vs.

typically developing controls). A recent example of this point comes from a study showing that pediatric brain injuries may result in what appear to be "prefrontal" impairments, no matter the site of the lesion (Jacobs et al., 2011). This study suggests that end-state impairments that appear specific to a region (e.g., frontal lobe) may actually result from injury or dysfunction in any of several regions that are part of a network including the prefrontal cortex. In this case, improper construction of networks of connectivity may cause the deficit rather than dysfunction restricted to one region alone (e.g., in this case, the prefrontal cortex). With this kind of information about networks in hand, one would suggest a rather different treatment than if the outcome was examined at one point in development (i.e., a treatment supporting connectivity vs. a treatment to enhance PFC function). Keeping this complexity of the developmental process in mind, we also review the available findings from early development in humans with DS and in mouse models.

Brain development in DS

Brain development is a dynamic process: in DS this seems to involve a sequence of relative declines across childhood and into adulthood. Shortly after birth, the brains of individuals with DS appear to be within the normal range in terms of major structural indices (i.e., brain size, shape, lobular proportions, and neurotransmitter development; Bar-Peled et al., 1991; Brooksbank et al., 1989; Flórez et al., 1990; Pazos et al., 1994; Schmidt-Sidor et al., 1990; Wisniewski and Schmidt-Sidor, 1989). By 6 months of age differences in brain structure are clearly apparent (Engidawork and Lubec, 2003; Golden and Hyman, 1994; Schmidt-Sidor et al., 1990; Wisniewski and Kida, 1994).

Neuropathological evidence demonstrates foreshortening of the frontal lobes, narrowing of the superior temporal gyrus, and diminished size of the cerebellum and brainstem in infants with DS (Benda, 1971; Blackwood and Corsellis, 1976; Crome et al., 1966). Additionally, structural differences in the medial temporal lobes (MTL), prefrontal cortex, and cerebellum have also been observed in studies using magnetic resonance imaging later in development and into adulthood (Menghini et al., 2011; Nadel, 2003). This pattern of neurological development corresponds well with patterns of cognitive deficits found in humans and mice with DS.

Nadel (1986) suggested that the pattern of findings of brain dysfunction in DS described above is suggestive of compromised development in late-developing systems. The prefrontal cortex, hippocampus, and cerebellum are regions with relatively protracted neural development, including postnatal generation of neurons and synapses, and myelination of the tracts connecting these regions and the rest of the brain persisting into later childhood. Recent evidence suggests that there may be dissociations in the developmental trajectory of functional subregions within these structures as well, and that the later developing components are again at greatest risk. It remains unclear why having an extra copy of HSA21 differentially affects late-developing structures, but the pattern of differences in brain structure and function observed in these regions seems relatively well established in both humans and mouse models.

Given this background, we focus on dysfunction in the MTL, cerebellum, and prefrontal cortex as well as on language impairments. It is, of course, difficult to build a bridge between findings of language deficits in the human and behavior in the mouse. However, ability of the mice to hear and learn in comparison to their healthy littermates is interesting to discriminative association learning. Given the extent of language impairments in DS, this set of skills demands attention when testing the effects of interventions, even though direct parallels cannot be made with animal models. Our discussion of data addressing cognitive and neurological differences in DS will predominantly focus on findings in children and young adults, given that declines associated with

Alzheimer's disease (AD) are prevalent in older adults and can complicate the interpretation of the cognitive profile (Lott and Dierssen, 2010).

MTL functions

The cognitive functions of the MTL have been extensively studied in humans and animal models of DS (Crnic and Pennington, 2000; Pennington et al., 2003; Uecker et al., 1993), and this brain region has been identified as the target of several potential drug agents in mouse (Fernandez et al., 2007; Guedj et al., 2009; Salehi et al., 2009). The MTL contains several distinct areas, each of which may serve a specialized role in mnemonic processing and, as evidence has recently suggested, aspects of perception (Bussey and Saksida, 2005; Nadel and Hardt, 2011). Figure 1 shows the various MTL regions and their interactions. Figure 2 displays the functional regions of the hippocampal circuit itself.

The hippocampus is a site of convergence of information from multiple sources, including the dorsal and the ventral visual streams (Ungerleider and Mishkin, 1982). Within these processing streams, representations of visual information become increasingly more complex moving from the periphery to the center, with perhaps the most complex representations being processed in the hippocampus (e.g., objects in specific contexts). Evidence suggests that one role of the hippocampus is to orthogonalize these representations of episodic contexts, using uniquely varying details of events to do so.

The majority of evidence suggests that the different regions in the hippocampus and surrounding MTL are specialized to process differing types of input (Kesner and Goodrich-Hunsaker, 2010). One specialization comes from the segregation of object and spatial information routed into the hippocampus via the entorhinal cortex. Object information is processed in the ventral visual stream,

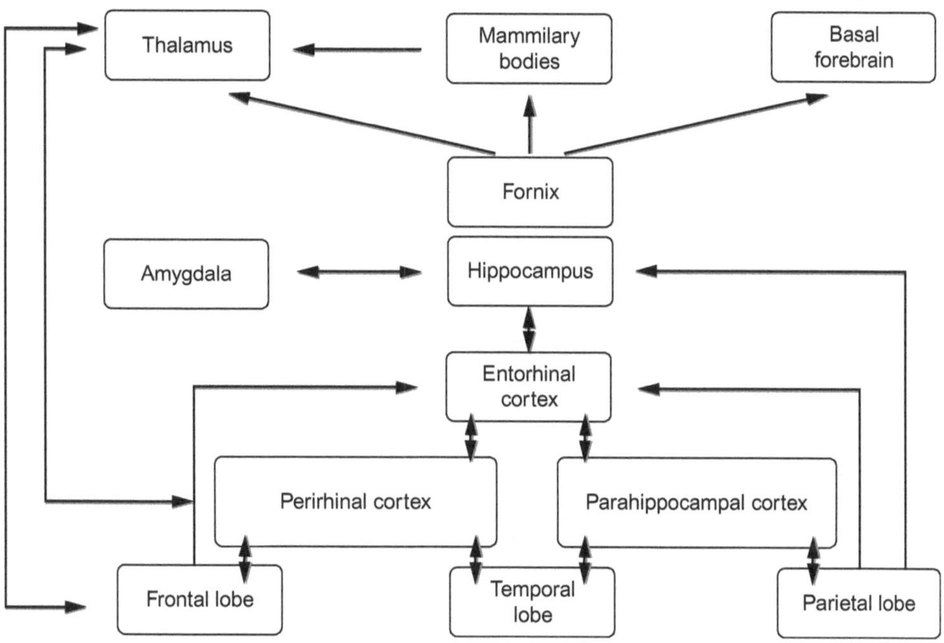

Fig. 1. Schematic diagram of the connectivity between regions of the medial temporal lobe system. Adapted from Nadel and Hardt (2011).

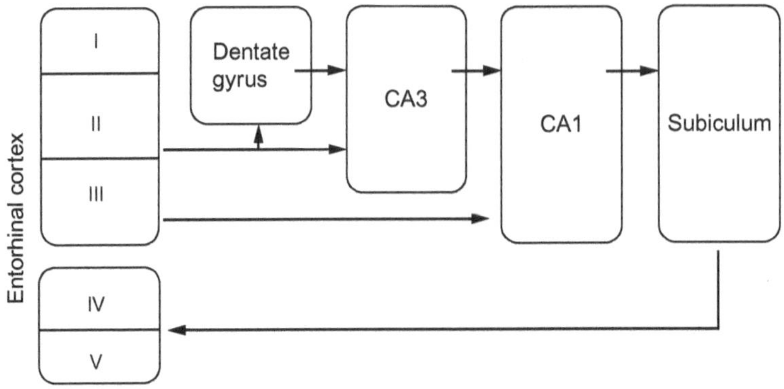

Fig. 2. Schematic diagram of the hippocampus proper and circuits originating from the entorhinal cortex. Inputs from the entorhinal cortex (EC) to hippocampus travel through the dentate gyrus and CA3 (the preforant path) or directly to CA1.

including the perirhinal cortex, which projects predominantly to the lateral entorhinal cortex, and hence to the hippocampus. The perirhinal cortex is specifically involved in the processing of objects with highly overlapping features (i.e., in resolving feature ambiguity). Monkeys with perirhinal lesions are impaired at discriminating morphed pairs of stimuli, particularly when featural overlap between the pairs is high (Bussey et al., 2003). In contrast, hippocampally lesioned monkeys are unimpaired on this task (Saksida et al., 2006). The role of the perirhinal cortex differs from the object processing undertaken by the fusiform gyrus in the temporal lobe, which allows for view independent recognition of single distinct objects and faces. Forming a set of inputs into the hippocampus that are primarily spatial in nature, the parahippocampal cortex encodes spatial locations or visual scenes, and receives direct projections from the dorsal visual stream (Sommer et al., 2005). Within the hippocampus, the dentate gyrus and CA3 are involved in memory tasks that require spatial pattern separation (Bakker et al., 2008) and the CA1 and the subiculum have been shown to be involved in novelty detection and spatial navigation (Kesner and Goodrich-Hunsaker, 2010). Other functional dissociations have been noted in the hippocampus, including variation in the number and size of place-sensitive fields along the septotemporal axis (i.e., dissociations in dorsal and ventral hippocampus; Jung et al., 1994).

The evidence linking these areas to distinct information processing functions derives from studies of adult patients with lesions, intact adults studied with neuroimaging methods, or animal work. Developmental disorders such as DS do not reveal such sharply defined distinctions. Deficient processing in the MTL regions projecting to the hippocampus likely leads to a degradation of the representations processed by these regions. Therefore, deficits in the mnemonic functions of the MTL could arise from dysfunction specific to the hippocampus proper *or* from impairments at any point in this pathway (e.g., Zola-Morgan et al., 1989). Given this possibility, our review of MTL dysfunction in humans with DS and mouse models considers not only findings relating to the function of the hippocampus itself but also tasks that may tap MTL inputs into the hippocampus. We focus on nonverbal tasks as the findings from measures of verbal memory are difficult to interpret, given deficits in auditory short-term memory (STM) and language function (Pennington et al., 2003).

Early dorsal and ventral visual stream processing

Dendritic abnormalities (in branching, length, spine density, and arborization) have been observed in various brain areas within DS (see Dierssen and Ramakers, 2006 for a review). In the cortex, studies of dendritic branching have suggested the possibility of deficits early in the visual stream in DS. An analysis of dendritic arborization in visual cortex showed a greater number of branches than typical in early infancy in humans with DS (i.e., less than 6 months), with decreases to lower than normal levels after 2 years (Becker et al., 1986). The finding of greater branching of dendrites has been replicated in other areas of cortex and may reflect altered connectivity patterns in the associated circuits, which ultimately results in reductions in dendritic arborization.

Despite these findings, the few behavioral studies examining early visual stream function have failed to observe defects. For instance, individuals with DS have been shown to perform at the level of mental age (MA) matched controls on tasks requiring immediate memory for spatial locations across several studies (i.e., CANTAB Spatial Span or CORSI blocks; Pennington et al., 2003; Visu-Petra et al., 2007; Edgin et al., 2010a), suggesting spared function of the dorsal visual stream, at least in relation to MA controls. In another study, Fidler et al. (2006) found that 2–3-year-old children with DS showed no deficits in relation to MA-matched infants on the visual reception scale of the Mullen Scales of Early Learning, which included tasks of basic visuomotor and perceptual ability (e.g., visual tracking and scanning of objects and simple visual discriminations). In an eye-tracking study, Brown et al. (2003) showed that 2–3year-olds with DS were unimpaired at visual tracking and integration when compared to children with William's syndrome, and to MA, and chronological age (CA) matched controls.

However, in early studies of the visual–spatial processing of global and local elements of a visual display, Wang et al. (1995) presented data suggesting individuals with DS had an abnormal tendency toward globally oriented visual perception. When asked to reconstruct elements of a hierarchical visual display, with one letter—say an "X"—being constructed out of many smaller "Os," they would focus on the global features of the display (the X) with less attention paid to the local features (the Os).

In a subsequent study examining the perception of global and local details in DS, Edgin, J. O. and Pennington, B. F. (unpublished data) examined global and local perception using two tasks: (1) the Navon figures and (2) a task described by Banks and Prinzmetal (1976) in which children have to detect a letter (either a T or an F) that is embedded within a visual display with varying global and local features (see Fig. 3). Twenty-four children with DS (mean age 14.9 years) were compared to 24 MA-matched controls. Results showed that children with DS were equally focused on the global and local levels when reproducing the drawings and showed better reproduction of detail at the local level than MA controls. Results on the Banks and Prinzmetal task were also in line with the usual pattern of data in the typical population, with measured biases at both the local and global levels (see Fig. 3).

Taken together, these data suggest integrity in the early stages of both dorsal and ventral visual stream processing. While more work could be conducted in this area, these findings suggest that the inputs into the medial temporal lobe from these processing streams may be intact. Therefore, memory dysfunction is likely to originate not from dysfunction in primary visual function processes but from differences in the hippocampus or surrounding MTL.

MTL in humans with DS

Magnetic resonance imaging studies have shown consistent evidence for reductions in gray matter density in the hippocampus in children and young adults with DS (Menghini et al., 2011; Pinter et al., 2001). Alterations in MTL microstructure are also apparent, with levels of dendritic branching in the temporal cortex, CA1, CA2, and CA3 in the hippocampus particularly affected in patients with DS (Ferrer et al., 1990; Takashima et al., 1989). White et al. (2003) also

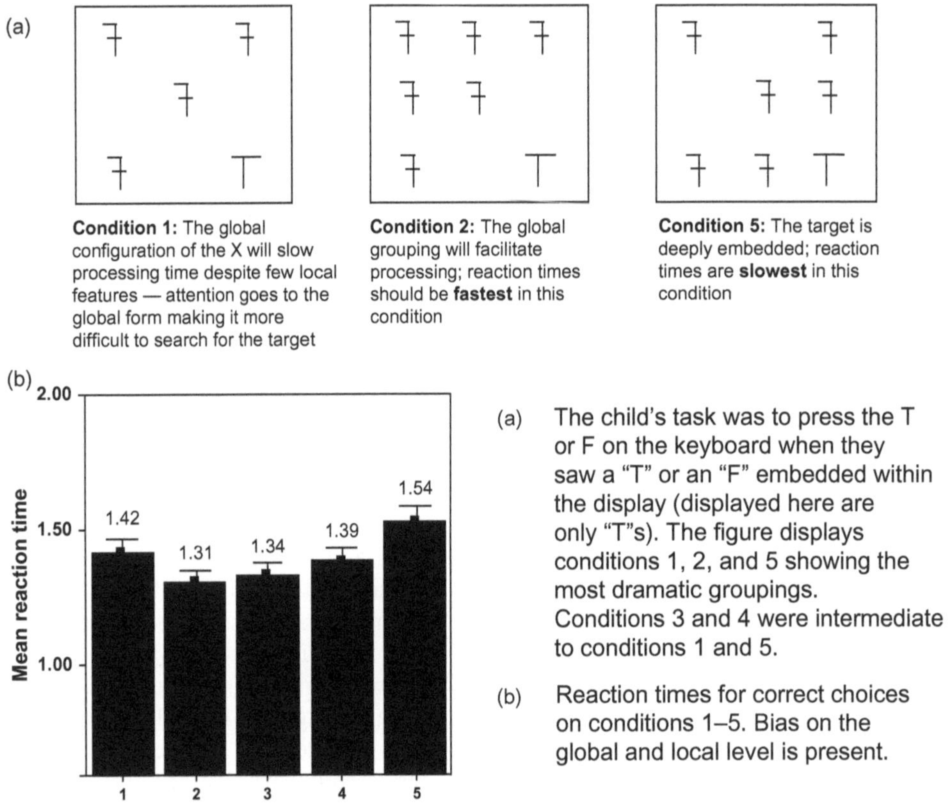

Fig. 3. Banks and Prinzmetal Task (1976) and results in 24 individuals with DS.

found reductions in gray matter density in non-demented adults with DS, including specific reductions in CA2 and CA3. A recent as yet unpublished study (Abraham, H. et al., personal communication) demonstrated not only reduced hippocampal volume in DS, but also an indication of some regional differences, with the dentate gyrus being more affected than other areas. There was a hint in this work that specific reductions in neurogenesis might be responsible for this impact on the dentate gyrus and on hippocampal volume overall. In addition, the onset of myelination was delayed in DS relative to controls, particularly so in the dentate gyrus. It is of interest to note that in both controls and DS, myelination occurs later in dentate gyrus than in other parts of the hippocampus (Arnold and Trojanowski, 1996; Contestabile et al., 2007).

Data on MTL-related neuropsychological measures in humans with DS are presented in Table 2. Uecker et al. (1993) provide a comprehensive review of early research suggesting behavioral deficits on MTL tasks in those with DS. Described in detail in this review, Mangan (1992) carried out one of the first investigations of the integrity of hippocampal functions. CA-matched control infants and infants with DS were tested on three related spatial tasks, one of which, a place-learning task, assessed the state of function of the hippocampal system. Two other spatial tasks included in the study could be solved without engagement of the hippocampus, involving

Table 2. Studies of MTL tasks in humans with DS

Study	Findings	MTL region implicated
Mangan (1992)	Toddler with DS<CA on place–learning task DS=CA on response learning (make a specific body turn), cue learning (approach a specific cue)	CA1
Carlesimo et al. (1997)	DS<ID<MA on word list learning, prose recall, reproduction of a complex figure	general MTL function
Pennington et al. (2003)	DS<MA on CANTAB PRM	perirhinal
	DS<MA on CANTAB PAL	CA3/dentate gyrus
	DS<MA on virtual maze	CA1
general MTL	DS<MA on everyday episodic memory	General
Vicari et al. (2005)	DS<MA on visual-object patterns DS=MA on learning of visual-spatial material	Perirhinal cortex, CA3/dentate gyrus
Visu-Petra et al. (2007)	DS<MA on CANTAB PAL, PRM, SRM[a]	CA3/dentate gyrus, perirhinal
	DS=MA on CANTAB SSP and SWM	Parahippocampal spared
Edgin et al. (2010a)	DS<MA on CANTAB PAL	CA3/dentate gyrus, perirhinal
	DS=MA on virtual maze	CA1 spared

[a]Authors note the presence of significant floor effects; MA, mental age match; CA, chronological age match; ID, intellectual disability match.

either response learning (make a specific body turn) or cue learning (approach a specific cue). In all three tasks, the child searched for a toy hidden under a pie-plate, using a place, response, or cue strategy. After viewing the hiding of the toy, the infants were removed from the apparatus for a delay interval and then were given a "memory" test. On the critical memory probes, children with DS performed similarly to typical children on the response and cue tasks but were severely impaired on the place task. These findings point to deficits in functions of the hippocampus. However, MA-matched controls were not used in this study, making the findings somewhat difficult to interpret. See Table 3 for a comparison of controls used in studies of animals and humans with DS.

In another study examining dissociations in mnemonic processing in DS versus other IDs, Carlesimo et al. (1997) found impairments in explicit memory in the context of spared implicit memory (e.g., intact word stem and visual priming). Explicit memory deficits were apparent on word list learning and prose recall as well as in the reproduction of a complex figure over a delay. However, it is difficult to further define the specific functions of the MTL tapped by these tasks, given their language demands as well as the possibility that deficits on the complex figure

Table 3. Studies of prefrontal tasks in humans with DS

Experimental type	Control type	Comparison value
Human	Mental age (MA)-matched control (TD children 3–6 years; IQ-matched same-age individuals with other ID)	*Benefits*: controls for effects of IQ on cognitive performance. *Limitations*: developmental or phenotype-specific confounds may be present
	Chronological age (CA)-matched control (TD same-age individuals; IQ-matched same-age children with other ID)	*Benefits*: controls for effects of age and experience on cognitive differences observed. *Limitations*: IQ or phenotype-specific confounds may be present
Mouse model	Wild type (euploid)	*Benefits*: controls for developmental period of the mouse. *Limitations:* Not able to determine whether or not deficits result from global or specific dysfunction

TD: typically developing.

reproduction could reflect dysfunction in a number of brain regions.

Pennington et al. (2003) reported deficits in relation to MA controls on several tasks that may map onto the function of the MTL, including immediate memory for patterns (CANTAB Pattern Recognition Memory, PRM), binding between object and location (CANTAB PAL), spatial memory through navigation (a virtual Morris water maze, MWM; Morris, 1984), and list learning across four trials. The group with DS also showed impairment on one ecological episodic memory measure (e.g., remembering what they did in the last testing session). Many findings of Pennington et al. (2003) were replicated by Visu-Petra et al. (2007) in a study that also showed deficits on the CANTAB PAL and PRM tests in relation to MA-matched controls. Together, these findings suggest deficits on tasks dependent on MTL structures adjacent to the hippocampus (Heuer and Bachevalier, 2011), such as visual recognition memory, as well as on tasks dependent on the hippocampus proper (i.e., spatial navigation relying on CA1/subiculum).

However, in one of the largest samples tested on neuropsychological measures to date (Edgin et al., 2010a), we did not replicate one of the deficits suggesting involvement of the CA1/subiculum circuitry. This study measured MTL, prefrontal, and cerebellar function in 74 children and young adults. Individuals with DS showed clear deficits on an object in place memory task in comparison to MA controls (i.e., CANTAB PAL), but performance on a test of spatial memory with navigation (the c–g arena) was not impaired.

Similarly, other studies seem to suggest greater deficits in some MTL regions than others (Ellis et al., 1989). For instance, Vicari et al. (2005) found that a group of 15 individuals with DS were more impaired on a task tapping visual-object long-term memory than on a task tapping memory for spatial locations, which was completed at the level of MA-matched controls. However, this finding is difficult to interpret in the context of the MTL model presented here, because the tasks assessing spatial memory did not necessarily involve allocentric spatial representations or spatial navigation, as did the virtual arena task utilized in Pennington et al. (2003) and Edgin et al. (2010a). Further, not all studies have shown differences in performance on object and spatial memory tests (CANTAB PRM vs. CANTAB SRM (Spatial Recognition Memory); Visu-Petra et al., 2007). The latter note that the CANTAB SRM test showed substantial floor effects, which could have obscured any differences.

In recent work in our laboratory, 11 children with DS and a matched sample of 11 MA controls were tested on their recognition of single objects displayed on a white background or in varying contexts after a 5-min delay (i.e., objects in congruent and incongruent contexts, see Fig. 4, e.g., stimuli and results). This task had been linked to the parahippocampal cortex in adults in a recent imaging study (Hayes et al., 2007). Individuals with DS displayed striking impairments across all conditions, including the recognition of single objects and objects embedded in a context. Again, these findings implicate multiple regions of the MTL, in addition to the hippocampus itself.

Given that a large number of treatment studies target the function of the hippocampus and MTL for neurocognitive intervention in DS, it is important to establish the impact of these deficits on the broader profile of cognitive and behavioral development. Pennington et al. (2003) reported an association between hippocampal measures (a composite score) and both adaptive behavior and language, including expressive and receptive syntax. Edgin et al. (2010b) also replicated the association between the CANTAB PAL and adaptive behavior. The composite used by Pennington et al. (2003) summed the scores of tests likely tapping the function of both the hippocampus proper (e.g., c–g arena and CANTAB PAL) and the surrounding MTL (e.g., CANTAB

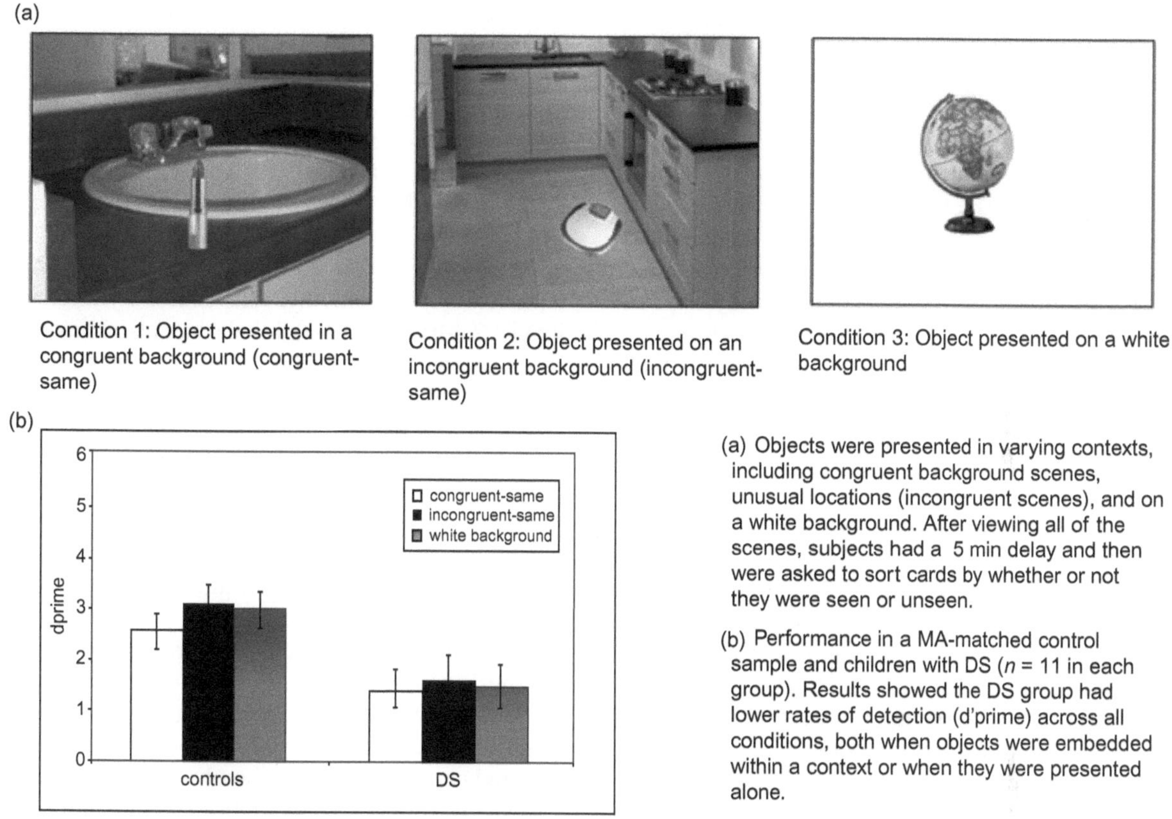

Fig. 4. Memory for objects in context and on a white background in individuals with DS and controls.

PRM). In a recent reanalysis of these data (Fernández, A., Nadel, L., & Edgin, J. O. unpublished), we examined how each specific memory measure from Pennington et al. (2003), including auditory STM, may predict language outcomes. In this analysis, auditory and pattern STM (i.e., CANTAB PRM) were related to both expressive and receptive syntax scores. However, hippocampal tests did not relate beyond the influence of STM impairments. This pattern was different from MA controls, for whom auditory and spatial STM related to expressive and receptive syntax and one hippocampal test related to expressive syntax. Future studies should aim to understand the brain basis of STM impairments in DS in order to better target treatments.

While more research is clearly needed to clarify this body of work on MTL deficits in DS, these findings do suggest that individuals with DS are impaired on tasks dependent upon functions of MTL processing streams. The data point to deficits in recognition memory, which may depend upon areas adjacent to the hippocampus proper, as well as deficits on tasks tapping functions of the hippocampus itself. However, little research has been conducted with paradigms capable of determining which specific MTL structures are implicated. Many tasks that have been utilized in past work, such as the reproduction of a complex figure, are too diverse in their cognitive demands to determine the neural basis of any observed deficit. Recent work has identified tasks that tap the functions of specific MTL subregions (Bakker et al., 2008; Bussey et al., 2003), which could provide a basis for more targeted batteries of assessments in humans and animal models. Given that successful pharmacological modification of MTL deficits may depend on targeting specific regions (e.g., perirhinal vs. CA1), this direction could be an important next step in neuropsychological work with humans with DS.

MTL evidence from mouse models

Mouse models have added invaluable information to our understanding of the mechanisms of MTL dysfunction as well as potential treatments targeted at modifying the function of this region. For instance, models have revealed neuropathology in the MTL, including abnormalities of hippocampal long-term potentiation (LTP), long-term depression (LTD), and excessive inhibition in the dentate gyrus (Kleschevnikov et al., 2004; Siarey et al., 1997 and see Chapter 10). Data from Ts65Dn (see Chapter 9) show robust evidence for involvement of the dentate gyrus/CA3 (preforant) pathway, including extensive loss of granule cells in the dentate gyrus and disruptions of synaptic function in CA3 (Hanson et al., 2007; Insausti et al., 1998; Lorenzi and Reeves, 2006). In addition to evidence regarding involvement of dentate gyrus and CA3, a recent study evaluating gene expression in the Ts65dn (Ginsberg et al., 2012) found differences in the gene expression profiles of CA1 neurons and across regions. It is important to note that gene expression among the two types of samples taken from each group (i.e., CA1 neurons vs. regional dissections) differed within the groups as well as within the postmortem tissue samples taken from humans with AD and human controls.

Recent targeted attempts to remedy these specific hippocampal deficiencies have been successful in reversing MTL-dependent memory deficits with pharmacological and environmental interventions in mouse models (Fernandez et al., 2007; Guedj et al., 2009; Martinez-Cue et al., 2002; Salehi et al., 2009). For instance, Braudeau et al. (2011) demonstrated that Ts65Dn mice treated with an α5-selective GABA inverse agonist (3-(5-methylisoxazol-3-yl)-6-[(1-methyl-1,2,3-triazol-4-yl)methyloxy]-1,2,4-triazolo[3,4-*a*]phthalazine, also known as α51A) showed improved acquisition of the platform location and decreased thigmotaxis on the MWM task, as well as improvements in the ability to discriminate between novel and familiar objects in the novel object recognition (NOR) task. Regarding environmental enrichment, Chakrabarti et al. (2011) recently showed that both short- and long-term exposure to enriched conditions (including a larger colony cage with a running wheel and various "toys") restored hippocampal cell proliferation and neurogenesis within the dentate gyrus of Ts65Dn mice to levels comparable to those observed in control (unenriched euploid) mice. Increased cell proliferation and neurogenesis was also observed in the forebrain subventricular zone, though the results were a bit more selective: these effects were only seen with long-term (4 weeks) enrichment, with more robust increases found in female mice than in males. Ultimately, treatment approaches for DS will be multimethod, and therefore, moving forward, approaches that test the joint influence of environmental stimulation and drug therapy could be particularly revealing.

Seregaza et al. (2006), Roubertoux and Carlier (2010), and Das and Reeves (2011) have reviewed the cognitive phenotype across the various DS models. Similar to humans, Ts65Dn mice that have been screened for blindness do not display visual or sensory motor deficits, with spared performance on simple cued learning tasks (Crnic and Pennington, 2000). Reeves et al. (1995) reported hippocampal-dependent deficits in the Ts65Dn, showing that affected mice took longer to reach the hidden platform than wild-type mice on the MWM. Ts65Dn have also shown impairments on a broad range of MTL dependent tasks, including context fear conditioning (CFC), spontaneous alternation in a T-maze, NOR, and nesting behavior (Das and Reeves, 2011; Gardiner, 2010; Shamloo et al., 2010). While many of these tasks might tap CA1/subiculum due to their navigational components, the deficit in NOR provides the closest mapping onto the deficits in the human on tasks such as CANTAB PRM or PAL which assess recognition memory or the pattern separation functions relating to the dentate gyrus/CA3 pathway.

However, in one study utilizing immediate measures of spatial or object novelty, there was no noticeable difference between older Ts65Dn and wild-type mice (Hyde and Crnic, 2002), suggesting immediate recognition memory is not impaired in the Ts65Dn, in contrast to the PRM deficits seen in humans (Pennington et al., 2003). Das and Reeves (2011) also note that object in place tasks have not been found to be impaired across two models (Ts1Cje and Ts65Dn), which is in sharp contrast to the well-replicated deficits on object in place measures in humans (i.e., CANTAB PAL).

While the Ts65Dn findings suggest MTL impairments broadly, MTL findings have not been consistent across models (reviewed in detail in Das and Reeves, 2011). For instance, the Ts1Rhr and Ts1Yah did not show MWM deficits, and the Tc1 only displayed MWM and NOR deficits with a short delay. Also, in Tc1, LTP was disrupted at 1h, but not disrupted in the long term. Therefore, the Tc1 shows less evidence for task impairment in the long term, with some evidence of deficits in recognition memory (NOR) and STM (Morice et al., 2008; O'Doherty et al., 2005). The Dp(16)1Yey/+, Dp(17)1Yey/+, Dp(10)1Yey/+ model that includes all mouse orthologs of Has21 showed MWM deficits, but in the context of slower swimming speed (Yu et al., 2010). Dp(16)1Yey/+, Dp(17)1Yey/+, Dp(10)1Yey/+ mice did, however, exhibit deficits on contextual fear conditioning and CA1 LTP was altered. Given the inconsistency of findings across models and various studies, it seems important that both human and rodent studies also include a consistent set of measures, including measures targeted toward specific aspects of MTL function.

Merging mouse and human MTL findings

Research in both species using more consistently administered, comprehensive batteries of tests that dissociate MTL function is clearly needed. Given the consistent findings of object in place deficits in humans with DS, more work should be done using similar measures in mice. Two studies suggest that mouse models do not show deficits on these tasks (Fernandez and Garner, 2007; Fernandez and Garner, 2008), but given the consistent presence of this deficit in humans, more work in the mouse is warranted. Further, to our knowledge, no study has explored tasks specifically related to perirhinal cortex in DS mouse models, including immediate tests of visual discrimination in conditions of high feature ambiguity. Given that the cognitive profile in the human is suggestive of the involvement of MTL structures adjacent to the hippocampus, it is necessary to explore the function of these regions in more detail.

While MTL functions are impaired in humans and seem relevant to developmental gains in other areas (i.e., language and adaptive behavior), these findings were examined mostly later in development. More developmental work is needed in both species to determine the pattern of deficits in early development and the impact of such deficits on learning across time. This developmental work is essential if we are to target those systems that would have the maximum impact in supporting cognitive development across childhood in DS.

Prefrontal functions

The prefrontal cortex, and particularly, the working memory functions associated with this region, has been shown to be highly important for cognitive function, academic achievement, and behavior in typical and atypical cognitive development (Alloway, 2009; Robinson et al., 2003). The prefrontal cortex mediates a variety of functions, including our ability to juggle incoming information, plan everyday tasks, and successfully adapt to the changing demands of the external environment. As with the MTL, the prefrontal cortex subserves a number of arguably distinct functions. These functions (often grouped under the label "executive functions, EF") include (1) the ability to hold information in mind and manipulate it (i.e., working memory), (2) the ability to inhibit actions for which a response tendency has been established (i.e., inhibitory control), and (3) the ability to flexibly switch between response sets (set-shifting).

In evidence suggesting distinct cognitive processes in typical adults, Miyake et al. (2000) utilized a latent variable approach to show that the three EF factors—working memory, inhibitory control, and set-shifting—are intercorrelated but separate. Friedman et al. (2008) also showed that these factors shared common variance but were differentially heritable in a behavioral genetics study. Given these results, Miyake has argued that there are distinct components of prefrontal function that may benefit from an overarching process, such as executive attention (Baddeley, 1986; Engle et al., 1999; Norman and Shallice, 1986).

The executive factors found by Miyake and colleagues are consistent with the cognitive profile of adult patients with differing prefrontal lesions, which suggests dissociations in the behavioral profile based on the location of the lesion within the frontal lobe. While frontal lesions can result in a diverse set of outcomes, often differing substantially based on the background of the patient, there appear to be two main subtypes. Mesulum (2002) describes them as follows: (1) a frontal abulic syndrome involving a loss of initiative, creativity, and concentration and (2) a frontal disinhibition syndrome which, in sharp contrast, involves behavioral excess, impulsivity, and lack of forethought. Patients with frontal abulic syndrome often have lesions to anterior frontal cortex. In contrast, patients with frontal disinhibition syndrome have damage to the orbital and medial frontal cortex including areas connecting to the striatum, which has been shown to be involved in inhibitory control across a number of studies (Durston et al., 2002). Working memory, on the other hand, has been shown to relate to dorsolateral prefrontal cortex, with the anterior cingulate also recruited on tasks requiring the resolution of conflicting stimuli (i.e., set-shifting tasks) (D'Esposito et al., 1995). The two prefrontal syndrome subtypes roughly correspond to the dissociated EF from Miyake et al. (2000), relating to reduced working memory and attentional capacity versus deficits in inhibitory control.

There is, however, some debate in the literature regarding the separability of these component processes of prefrontal function, especially early on in development. For instance, recent findings have suggested that in very young children executive processes may be unitary and that only during the course of development do they become dependent on somewhat separate neural mechanisms (Wiebe et al., 2010). Given the altered maturation of the PFC in DS, one might imagine global delays in most PFC-driven cognitive functions. However, given that these functions come to rely on separate neural mechanisms over development, individuals with DS may show a mixed profile of spared and impaired prefrontal functions.

Prefrontal function in humans with DS

Table 4 displays the findings from studies of prefrontal function in DS. Impairments in prefrontal function have been found in some, but not all,

Table 4. Studies of prefrontal tasks in humans with DS

Study	Findings
Inhibition	
Kopp et al. (1983)	DS<MA on reward delay latency
Pennington et al. (2003)	DS=MA on stopping task
Lanfranchi et al. (2010)	DS<MA on stroop type task–day/night version
Edgin et al. (2010a)	DS=MA on inhibition phase of modified dots task
Edgin et al. (2011)	DS normal range on BRIEF inhibition, emotional control
Raitano Lee et al. (2011)	Toddlers with DS normal range on BRIEF inhibition, emotional control
Set-shifting	
Zelazo et al. (1996)	DS<MA on DCCS
Edgin (2003)	DS<WS on DCCS
Rowe et al. (2006)	DS<ID on weigl color-form sort test
Lanfranchi et al. (2010)	DS<MA on rule shift card and modified card sorting test
Working memory	
Pennington et al. (2003)	DS=MA on CANTAB spatial working memory and counting span task
Lanfranchi et al. (2004)	DS<MA on verbal and visuospatial tasks
Lanfranchi et al. (2010)	DS<MA on verbal and visuospatial dual tasks
Visu-Petra et al. (2007)	DS<MA on CANTAB Spatial WM
Edgin et al. (2010a)	DS<MA on working memory phase of modified dots task

MA, mental age match; ID, intellectual disability match; DCCS, dimensional change card sorting task.

studies; our early work involving the administration of a large battery of executive function tasks yielded one of the negative results (Pennington et al., 2003). However, most recent research has found deficits, with the majority of studies showing specific deficits in working memory and attention (Brown et al., 2003; Edgin et al., 2010a; Lanfranchi et al., 2004; Lanfranchi et al., 2010; Visu-Petra et al., 2007) and set-shifting (Edgin, 2003; Lanfranchi et al., 2010; Rowe et al., 2006; Zelazo et al., 1996), with less consistent impairments found on tasks requiring inhibitory control. These findings are consistent with structural imaging evidence showing reductions in gray matter volumes in the frontal cortex and cingulate gyrus (White et al., 2003).

In Rowe et al. (2006), young adults with DS were impaired on a range of prefrontal measures. However, after control for differences in verbal memory and psychomotor speed, only measures of attention and set-shifting were significant. In a large sample, Edgin et al. (2010a) also confirmed deficits in working memory and set-shifting in comparison to MA-matched controls, with impairments found on the CANTAB intra-dimensional/extra-dimensional set-shifting task as well as the working memory phase of a directional Stroop task (the Modified "Dots" task; Davidson et al., 2006). Edgin et al. did not, however, find impairments on the phase of the task requiring only inhibitory control. Parent report of everyday EF on the Behavioral Rating Inventory of Executive Function (BRIEF) suggests that ratings of working memory and set-shifting are more impaired than ratings of inhibitory and emotional control, both in younger (ages 4–10 years; Raitano Lee et al., 2011) and older children with DS (ages 7–25 years; Edgin et al., 2011).

It is of interest to note that the Pennington et al. (2003) study, which produced a null result, included measures of planning, inhibition, and working memory, but no set-shifting assessments. Edgin (2003) retested a group of individuals from the Pennington et al. (2003) study on a set-shifting task in which children initially sorted multidimensional cards (i.e., a red rabbit and a blue boat) by one category (i.e., the "Dimensional Change Card Sorting," DCCS task; Zelazo, 2006). After several trials in which they completed sorting by that category, they were asked to sort cards by the other dimension. School-age children and young adults with DS showed substantial deficits on this measure in relation to a matched sample of individuals with William's syndrome. While they performed the initial sorting of the cards with near perfect performance, they were impaired when required to shift their response to the next dimension. These findings replicate earlier findings of Zelazo et al. (1996) in which adults with DS showed more errors when sorting to a second dimension in comparison to a MA-matched control sample. Taken together, these findings consistently point to more impairment on set-shifting and working memory measures as compared to measures of inhibitory control.

Given this pattern, Raitano Lee et al. (2011) hypothesized that individuals with DS may show dissociated deficits, with "cool" EF tasks such as working memory showing greater impairment than EF tasks with an emotional or inhibitory control component ("hot" EF). While "hot" EF has not been specifically measured in past studies of DS beyond parent ratings of emotional control, this could be an interesting avenue of research, given the pattern of findings reported here. A "hot" versus "cool" distinction would suggest dysfunction in the dorsolateral rather than ventromedial prefrontal cortex (Metcalfe and Mischel, 1999; Zelazo, 2006) and is also consistent with findings showing relatively preserved amygdala volumes in DS (Pinter et al., 2001).

One important next step is to determine the profile of prefrontal functions in DS, especially in young children and assess the ways in which these processes may underscore cognitive difficulties across domains. At least one study suggests that the pattern of dissociation described in older children may evolve over time. In contrast to the majority of results in older children, toddlers with DS did show difficulties in inhibitory control when presented with an attractive reward, displaying a shorter latency to touch the rewards than MA-matched controls (Kopp et al., 1983). Therefore, it is quite possible that very young children with DS do "grow out" of prefrontal difficulties relating to inhibitory control while deficits in attention, working memory, and set-shifting tend to persist. Given that the targets of certain pharmacological interventions are the attentional and executive systems (i.e., LDOPS; Salehi et al., 2009), an important step will be to clarify the profile of these functions across development.

Prefrontal function in DS mouse models

In comparison to mouse model work on the MTL, there are far fewer studies examining the profile of prefrontal functions. However, there are neuropathological mechanisms that could influence both hippocampal and prefrontal function, and the bulk of studies that have been conducted show clear differences on prefrontal tasks. It has been noted that one of the most pronounced neuropathological features in the Ts65Dn mouse model of DS is the degeneration of cholinergic basal forebrain (CBF) neurons. CBF neurons have two major projection systems, including (1) projections to the hippocampus from the medial septal nucleus and (2) projections from the nucleus basalis to the frontal cortex (Moon et al., 2010). In mouse models, these neurons are intact at birth but show atrophy by 6 months of age. Another potential neurobiological target gleaned from mouse models is the locus coeruleus, whose early degeneration has been shown to lead to underdevelopment of the noradrenergic inputs to the frontal cortex and hippocampus (Salehi et al., 2009).

Indeed, Ts65Dn mice do display some behaviors that suggest difficulties with frontal function. For instance, they have documented deficits in spatial working memory (Escorihuela et al., 1995), a higher overall activity level (e.g., repetitive jumping), and lower levels of anxiety, with these differences apparent in both their home cage and a novel environment (Coussons-Read and Crnic;, 1996; Turner et al., 2001). Tc1 mice (Morice et al., 2008) have deficits in reversal learning, which could be analogous to the human tests of set-shifting. In a detailed examination of attentional behavior using a novel set of tests, Moon et al. (2010) showed that attentional deficits in Ts65Dn could be improved with prenatal choline supplementation. The mice performed a series of tasks requiring that they poke their nose into one of five holes based on the appearance of a light cue from one of the holes. Successful responses resulted in a liquid reward. In various versions of the task, difficulty was varied to ascertain the mouse's level of attention. On a baseline visual discrimination task, a light from one hole was illuminated after 2s. The light remained on until the mouse responded or until 32s had elapsed. For the first two attention tasks, the task was identical except for cue duration. The mouse was required to make a response to a light from one of the five holes, but the light was present for only 1 or 2s. Another two attention tasks added a variable precue delay, placing greater demands on sustained attention. The final task was the Reward Omission task, in which some trials were unrewarded and the mouse's emotional response to the absence of an expected reward (i.e., activity level through jumping) was recorded. Collectively, these tests likely assess the range of prefrontal cortical functions in humans, including inhibitory control, working memory, and sustained attention.

Consistent with the information reported for humans, the test assessing simple visual discrimination was not significantly different in the Ts65Dn. However, all the tests of sustained attention yielded deficits in the Ts65Dn mouse, with some rescue of function after choline supplementation in pregnancy. The authors also reported significantly increased activity levels in Ts65Dn mice after errors on the task. This overreaction to making an error could reflect impaired "hot" executive function, as discussed above. Therefore, unlike humans with DS, mouse models seem to show uniform impairments across a variety of executive tasks.

Merging mouse and human prefrontal function findings

While data from humans suggest dissociations in the profile of prefrontal functions, the findings in the mouse models suggest overall impairment. Given that the profile of executive function deficits may change across development, it is possible that the dissociation between tasks tapping inhibitory control and other EFs reflects a maturational change that is simply not achieved in the mouse models before the onset of neurodegeneration around 6 months (Granholm et al., 2000). Another explanation may relate to cross-species differences in the structure of the frontal cortex. The prefrontal cortex has been shown to be less structurally segregated in the rodent as compared to the primate and human (i.e., less well-developed dorsolateral prefrontal cortex), a fact that could explain less differentiated functional outcomes associated with dysfunction in this region in the mouse (Preuss, 1995).

Taken together, however, these findings suggest some avenues for future research, including (1) a more detailed analysis of frontal function in humans with DS and mouse models, given that these impairments are present and could have a substantial influence on learning in multiple domains; (2) an analysis of the profile of EFs across development (i.e., does the pattern shift from global to dissociated impairment?); and (3) the development of tests, in both the mouse and the human, that measure these functions in a specific manner. The measures of Moon et al. (2010) are a step in the right direction and utilize paradigms that could be quite similar in mice and humans. Given the evidence for set-shifting difficulties in humans with DS, it will be important to develop a close analog of those measures as well.

Cerebellum

The cerebellum is one of the most affected neural structures in DS (Pinter et al., 2001), with behavioral deficits found in this domain in both mouse models and humans (Frith and Frith, 1974; Olson et al., 2004b). Traditional measures of cerebellar function have been extensively studied in humans with DS, particularly aspects of motor control (Table 5). Eyeblink conditioning is one of the purest measures of cerebellar function available

Table 5. Studies of cerebellar tasks in humans with DS

Study	Findings
Frith and Frith (1974)	DS no improvement on rotor tracking task after 5min test. DS slower than ID and MA on a finger tapping task
Rast and Harris (1985)	DS<CA in Motor Assessment of Infants task
Connolly and Michael (1986)	DS<ID on gross motor and fine motor skill composite scores on the Bruininks Oseretsky Test of Motor Proficiency
Woodruff-Pak et al. (1994)	Older DS (m=47.7) <younger DS (m=27.6), ID (fragile X) <CA on eye blink classical conditioning task
Latash et al. (2002)	DS<CA on both maximal single- and multifinger force production tasks and multifinger tasks
Stedron (2004)	DS<MA on rhythmic finger tapping, DS=MA on eye blink conditioning
Virji-Babul et al. (2006)	DS<MA on complex perceptual discriminations of point-light displays of human action
Lam et al. (2009)	DS<CA (slower) on Fitts' task
Edgin et al. (2010a)	DS<MA on finger sequencing, DS=MA on visuomotor precision, reaction time

MA, mental age match; CA, chronological age match.

(McCormick and Thompson, 1984). However, eyeblink conditioning has led to inconsistent results in the DS literature, with deficits found in older adults but not younger children (Stedron, 2004; Woodruff-Pak et al., 1994). As seen in Table 5, complex motor tasks, such as finger sequencing or rhythmic finger tapping, have been shown to be impaired in humans with DS, suggesting deficits in specific aspects of motor control governed by the cerebellum.

While early work on the cerebellum focused on its role in motor coordination, it is now clear that this structure has a broad range of functions, including links with language, visuospatial cognition, and working memory (Stoodley and Schmahmannm, 2009). The cerebellum is a highly modularized structure, forming separate loops between its various lobules and specific cortical and subcortical regions. Recent studies have shown that the lobes of the cerebellar hemispheres may show differing developing trajectories, with more protracted development in the inferior portions (lobules VI–VII) than anterior sections (lobes I–V) (Tiemeier et al., 2010). While the anterior portion of the cerebellar hemispheres shows associations with motor tasks, the inferior lobes of the cerebellum are more closely related to aspects of higher-level cognition, including prefrontal function and auditory memory (Marvel and Desmond, 2010). Given the robust nature of language and verbal STM deficits in DS (Edgin et al., 2010b), an examination of nontraditional links between the cerebellum and cognitive processes may be particularly important in this syndrome. Equally important will be more detailed neuroimaging studies of this structure in humans with DS, ideally using a similar approach to the study of Tiemeier et al. (2010) in which variation in subregions was closely examined in typically developing children.

Cerebellum in mouse models

A summary of the physiological cerebellar profile for selected mouse models is found in Table 6. In the Ts65Dn, overall cerebellar volume is reduced to 88.1% of the volume observed in euploid, with significant reductions in both granule and Purkinje cell density (Baxter et al., 2000). As can be seen in the table, very few studies have examined the morphological development of individual cerebellar lobules. Given the data in the humans suggesting functional segregation by subregion, this analysis may be an important step forward in understanding cerebellar profiles in the mouse.

Table 6. Differences in cerebellar phenotype among mouse models of DS

Mouse model	Normalized cerebellar volume (compared to euploid)	Granule cell density	Purkinje cell density	Lobular density
Dp(16)1Yey/+, Dp(17)1Yey/+, Dp(10)1Yey/+(Yu et al., 2010)	N/M	N/M	N/M	N/M
Dp(16)1Yu (also see Ts1Yu, Dp (16)1Yey) (Li et al., 2007)	N/M	N/M	N/M	N/M
Ts65Dn (Baxter et al., 2000)	88.1%	76%	89.5%	Midline 81.7%, parasagittal section 80.6% of euploid
Ts1Cje (Olson et al., 2004b)	88.8%	91.2%	98.3%	N/M
Ts1Rhr (Olson et al., 2007)	95%	102%	101%	N/M
Ts1Yah (Pereira et al., 2009)	N/M	N/M	N/M	N/M
Tc1 (Moldrich et al., 2007; O'Doherty et al., 2005)	N/M	84–85%	N/M	Lobe VIII reported to be reduced to approximately 84% of wild type

N/M, not measured.

Cerebellar function in mouse models has historically been assessed through motor tasks such as the rotarod, observations of posture and gait (Crawley, 2007), and adapted eyeblink conditioning paradigms (Chen et al., 1996). Despite global reductions in volume in the Ts65Dn, mixed results have been observed when these animals were tested on motor skills tasks assessing functions typically associated with the cerebellum in mice, with some studies actually finding enhanced function, while others found impairment (Baxter et al., 2000; Costa et al., 1999; Hyde et al., 2001). In comparison, the Tc1 model displays significant deficits on tasks such as the rotarod and static rod as well as a significantly different exploratory behavior profile in an open-field exploration environment (Galante et al., 2009). However, studies of the physiological cerebellar profile of Tc1 have indicated a granule cell density reduction that, while significant, is less than the reduction observed in Ts65Dn (85% vs. 76%; Moldrich et al., 2007). Further, certain aspects of Purkinje cell function are not significantly impaired relative to euploid mice (Galante et al., 2009), suggesting that the motor deficits found in Tc1 may correspond to specific cerebellar impairments.

Merging human and mouse cerebellar findings

Given these results it is clear that this region shows substantial morphological differences in both humans and mouse models. However, the behavioral findings have not been consistent in either species. One reason for inconsistencies in outcome may be because of the complex nature of this structure and its cognitive correlates. Because of the extensive connections between the cerebellum, subcortical, and cortical structures, it is difficult to find tasks that tap only "cerebellum." Another piece of evidence regarding this point comes from our recent work, in which we found a substantial correlation between measures of cerebellar and prefrontal function on the Arizona Cognitive Test Battery (Edgin et al., 2010a). Given mounting indications that there are substantial links between cerebellum and higher-level cognition, it may be too simplistic to continue to assess the global function of this region solely with motor-based outcome assessments.

The linguistic brain

Difficulties with language are a central feature of the cognitive profile in DS. Therefore, it will be important for us to be able to accurately assess language function when interventions are developed for humans. Similar to the domains of neuropsychological function already reviewed, language functions in DS are not uniformly impaired. In this case, however, we cannot lean on mouse model research to help us define the specific domains of language functions compromised in DS.

Children with DS appear to show deficits in language production from an early age. The pattern of language impairment includes a longer transition period from babbling to words, reduced intelligibility of speech, and less word production and word variety resulting in a shorter mean length of utterance (Abbeduto et al., 2007; Berglund et al., 2001; Chapman et al., 1998). When tested in receptive vocabulary, individuals with DS have been shown to perform at the level of MA controls. However, not all studies have suggested this pattern is consistent throughout development. For instance, in a longitudinal study following 17–19-month-old children with DS in comparison to typically-developing 9-month-olds during a 14–21-month period, Cardoso-Martins et al. (1985) found that vocabulary development in DS lagged behind the children's level of cognitive development. Therefore, language is one additional domain in which we could use an understanding of developmental progressions in order to better target treatments.

The majority of studies in older children suggest specific impairment in morphosyntactic capacities (i.e., grammar development; Abbeduto et al., 2007). Morphology concerns aspects of language that convey meaning, such as words and affixes. Morphosyntax involves the way in which these units of speech, or morphemes, are combined together to convey meaning. It is in this area of language that one sees the most dramatic impairments in DS.

In typical development and the development of those with DS, research on the neurological underpinnings of language processing has highlighted the involvement of a number of brain regions. Research using fMRI has expanded the regions implicated in the phonological processing and lexical-semantic aspects of language beyond what were formerly known as "Broca's" and "Wernicke's" areas (inferior frontal, and superior temporal gyri) to include the angular gyrus, certain prefrontal regions (inferior, superior, and portions of the middle frontal gyri along with part of the anterior cingulate), some temporoparietal regions (Binder et al., 1997), as well as parts of the cerebellum and basal ganglia (Booth et al., 2007).

A recent structural neuroimaging study in individuals with DS (Menghini et al., 2011) suggested that receptive morphosyntax was related to the gray matter density of the cerebellum, while productive morphosyntax related to the gray matter density of the inferior and middle temporal gyrus. Another recent functional neuroimaging study reported differences in regional activation between individuals with DS and typically developing age-matched controls during passive story listening. Losin et al. (2009) noted that individuals with DS showed almost the identical pattern of activation when listening to the words of stories read forward or backward, unlike a sample of CA-matched controls that showed more activation in classic receptive language areas (i.e., superior and middle temporal gyri) in the forward condition. Further, individuals with DS did not show activation in these language areas but had greater activation in the cingulate gyrus and parietal lobes, suggesting either a compensatory strategy or the engagement of more diffuse neural mechanisms. While the neurological profile underlying these deficits is quite complex, it is clear that core neuropsychological functions such as those supported by the MTL, prefrontal cortex, and cerebellum may also play a role in this important cognitive domain.

Although mice are not considered a "linguistic" species, recent research has suggested that there are certain behaviors in mice that might help us to better understand language in humans. One behavior that has received some attention, in this regard, is mouse ultrasonic vocalization (USVs), with some studies showing that DS mouse models as well as mouse models of other disorders with communicative deficits are altered in this behavior (Holtzman et al., 1996). Holtzman et al. found that USV behavior was delayed by 4 days in Ts65Dn mouse pups, and there have been a number of studies indicating specific pup and adult USV irregularities in mice with alterations in genes such as FOX P2 and oxytocin gene OXT (i.e., implicated in language and social behavior) as well as neuroligin-4 and neuroligin-3, which have been associated with autism spectrum disorder (ASD; see Fischer and Hammerschmidt, 2011 for review). While it is important to note that USVs are delayed in the Ts65Dn, these vocalizations do not map onto the specific language impairments observed in humans with DS (i.e., morphosyntax). They may instead be more reflective of the animal's drive to communicate. In mouse models of autism, this may be a good proxy for communication impairments, but in DS, USVs are not likely to be as useful.

In addition to USVs, previous studies have indicated that auditory learning tasks (i.e., oddball discrimination) can be completed in mice. In one such task (Peiffer et al., 2003), mice are presented with a pattern of two-tone auditory sequences separated by constant intervals. During the sequence presentation, a startle-eliciting stimulus (a white noise burst producing an acoustic startle reflex in mice) interrupts one of the between-sequence intervals. In the uncued (control) condition, a control two-tone (high/low) sequence is presented repetitively prior to the startle-eliciting stimulus, while in the cued (oddball) condition, an oddball (low/high) sequence is inserted as a "cue" for the startle-eliciting stimulus amongst the high/low sequences. The acoustic startle reflex of the mice is recorded in each condition, and it is expected that if the mice are able to discriminate the oddball sequence, then their startle reflex should be significantly attenuated in the cued condition relative to the uncued condition. Using this reflex modification paradigm, Peiffer et al. (2003) were able to find significant oddball detection differences among male BXSB/MpJ mice with cortical ectopias resembling the pathology seen in dyslexia, which suggests that it may be useful for assessing auditory processing deficits in relation to neurodevelopment.

Other paradigms used in rodents have made use of auditory "go/no-go" tasks, in which rodents are rewarded for responding to short tone sequences containing an oddball tone in contrast to sequences of identical tones. Differences on this task correlate with age-related changes in the primary auditory cortex of rats (de Villers-Sidani et al., 2010). Given these results, it is possible that the addition of auditory learning tasks could allow for the testing of the effects of drugs on multiple neural systems and some additional aspects of the DS phenotype prior to trials in humans.

Recommendations for cognitive assessment moving forward

Using data generated in mouse models of DS, in the past 5 years several drug targets have been identified for cognitive intervention (see Chapter 10 for additional details). Many of these pharmacological agents are currently being tested or will go to trial in humans in the coming years. Returning to our assumptions regarding the bridge between animal models and human work, we can expect success in this transfer only to the extent that the cognitive processes affected in the mouse and the human are as similar as possible. Further, given that some important behaviors (e.g., language or adaptive behavior) cannot be measured in the same way in mice, we want to be certain that the neurological processes modified by any drug will have an effect on important cognitive domains in humans with DS. Based on the data reviewed here, we recommend the following research directions moving forward: (1) better definition of the phenotype in both mice and humans; (2) examination of the functional subregions of affected neural systems; (3) work on cross-species assessment batteries that are targeted at specific neural structures; (4) a closer examination of developmental profiles, rather than just isolated snapshots, in both species; (5) a definition of the impact of compromised neural structures on the broader cognitive profile, including outcomes, such as language and adaptive behavior; (6) a better recognition of the role of interactions across brain systems that may be important in explaining the profile (i.e., the presence of cerebellar–prefrontal interactions); and (7) the joint exploration of the influence of environmental stimulation and pharmacotherapy on cognitive outcomes.

While there have been considerable advances in our knowledge of the cognitive phenotype in both species, we still require more precise definition of the cognitive deficits relating to impaired function in the MTL, prefrontal cortex, and cerebellum, including any dissociations in the function of subregions of these areas. The data reviewed here suggest that dissociations in cognitive functions do exist in humans (i.e., specific language and prefrontal profile), but the findings are less clearly differentiated in mice. In humans, there is evidence for deficits on tasks that tap both the hippocampus proper and the surrounding MTL. The findings in mice clearly show deficits in hippocampal-dependent memory, with mixed evidence for short-term recognition memory impairments. While the profile of prefrontal function is clearly differentiated by domain in older children with DS, the findings in mice seem more uniform, albeit the number of studies examining prefrontal function are fewer in number. The findings regarding cerebellar function show similar inconsistencies in outcomes, with most studies of motor control in humans showing deficits and studies in mice not always showing dsyfunction. We also reviewed the pattern of language deficits in individuals with DS, noting clear dissociations in impairment, with syntax being the strongest area of impairment. While there is no direct analog of syntax in mice, we review some methods to test auditory processing that might prove useful. Moving forward, the development of assessment batteries that can be easily completed in both humans and mice will be essential. In addition, more work is required to understand the neurological status of the specific subregions of affected brain structures. For instance, given that the cerebellum is so highly modularized, it is important to understand the morphology of the specific lobules of the human and mouse cerebellum. As can be seen in Table 6, very little work has been done along these lines to date.

One problem with currently administered cognitive tests is that the tasks are too complex to target specific structures. For instance, the MWM and the human variant (i.e., the computer generated arena) tap a complex set of skills. In the mouse, performance can be affected by swim speed. Similarly in humans, the task is operated by a joystick control, a method that can lead to perseverative responding. Tasks should be developed that are simple as possible in their cognitive demands. However, this goal may not always be fully realized, and thus, statistical control for extraneous demands will often be necessary.

Another need is a clarification of the inconsistencies in measured behavior across both humans and mouse models. Some of the differences between mouse and human studies are difficult to intrepret because measures have differed across studies. As much as we need to develop consistent measures for humans, mice of various genetic alterations also need to be tested on the same paradigms. Further, studies of pharmacological intervention in mice will be better placed to make claims about how these drugs may affect the human if they utilize more than one measure of function (i.e., expanding beyond the MWM), including measures from a broader range of affected cognitive domains. Further, while the Ts65Dn is the most easily utilized and well-understood model, it cannot be ignored that the cognitive profile in this model may differ from the newer models carrying the full trisomy.

As emphasized earlier, there has been little work describing the longitudinal development of the cognitive profile in DS. An understanding of this trajectory is essential in targeting treatments to affect the primary source(s) of cognitive deficits in DS. The evidence reviewed here suggests that the functions targeted by pharmacological treatment (i.e., hippocampal-dependent memory, medial- temporal lobe, and prefrontal function) may have some influence in the broader cognitve profile when measured at a single time point in later childhood, but more work is needed to understand the progression of these deficits across development. Some surprises may be found if we examine this trajectory closely, such as the potential for the cerebellum to play a role in the development of verbal memory deficits or language dysfunction.

Closing remarks

In summary, our understanding of the cognitive profile and neuropathological mechanisms associated with DS has advanced rapidly in the past decade, leading to tests of pharmacological interventions for cognitive rehabilitation that are increasing in number. Crosstalk between studies of cognition in humans and mice is indeed possible, and essential, to maximize the possibility that drugs may positively affect humans at trial. It is necessary that cross-species comparisons are conducted with care to ensure continued development of drug treatments in this population. In our work, we have shown that cognitive assessment can indeed be valid and reliable in humans with DS (Edgin et al., 2010a). Now, we require additional test development resulting in measures that are targeted toward specific neurological structures and can be administered across a range of ages. Only through a more complete understanding of the developmental profile of cognitive function in DS will we be able to target early cognitive deficits in a way that could have the greatest impact in changing the developmental trajectory for each individual.

Acknowledgments

We thank the families who made this work possible. Our work is supported in part by grants from the Down Syndrome Research and Treatment Foundation, DSRTF (to L. N. and J. O. E.); the Thrasher Research Fund (to L. N. and J. O. E.); the Lejeune Foundation (to J. O. E.); and the University of Arizona Foundation (to L. N.). We are also grateful to Bruce Pennington for allowing us to reanalyze some data from our past projects conducted with him.

References

Abbeduto, L., Warren, S., & Conners, F. (2007). Language development in Down syndrome: From the prelinguistic period to the acquisition of literacy. *Mental Retardation and Developmental Disabilities Research Reviews*, *13*, 247–261.

Alloway, T. P. (2009). Working memory, but not IQ, predicts subsequent learning in children with learning difficulties. *European Journal of Psychological Assessment*, *25*, 92–98.

Arnold, S. E., & Trojanowski, J. Q. (1996). Human fetal hippocampal development: I. Cytoarchitecture, myeloarchitecture, and neuronal morphologic features. *Journal of Comparative Neurology*, *367*, 274–292.

Baddeley, A. D. (1986). *Working memory*. London/New York: Oxford University Press.

Bakker, A., Kirwan, C. B., Miller, M., & Star, C. E. L. (2008). Pattern separation in the human hippocampal CA3 and dentate gyrus. *Science*, *319*, 1640–1642.

Banks, W. P., & Prinzmetal, W. (1976). Configural effects in visual information processing. *Perception & Psychophysics*, *19*, 361–367.

Bar-Peled, O., Israeli, M., Ben-Hur, H., Hoskins, I., Groner, Y., & Biegon, A. (1991). Developmental patterns of muscarinic receptors in normal and Down's syndrome fetal brain—an autoradiographic study. *Neuroscience Letters*, *133*, 154–158.

Baxter, L. L., Moran, T. H., Richtsmeier, J. T., Troncoso, J., & Reeves, R. H. (2000). Discovery and genetic localization of Down syndrome cerebellar phenotypes using the Ts65Dn mouse. *Human Molecular Genetics*, *9*(2), 195–202.

Becker, L. E., Armstrong, D. L., & Chan, F. (1986). Dendritic atrophy in children with Down's syndrome. *Annals of Neurology*, *20*, 520–526.

Benda, C. E. (1971). Mongolism. . In: J. Minckler (Ed.), *Pathology of the nervous system* (Vol. 2). New York: McGraw-Hill 1867.

Berglund, E., Eriksson, M., & Johansson, I. (2001). Parental reports of spoken language skills in children with Down syndrome. *Journal of Speech, Language, and Hearing Research*, *44*, 179–191.

Binder, J. R., Frost, J. A., Hammeke, T. A., Cox, R. W., Rao, S. M., & Prieto, T. (1997). Human brain language areas identified by functional magnetic resonance imaging. *Journal of Neuroscience*, *17*(1), 353–362.

Blackwood, W., & Corsellis, J. A. N. (1976). *Greenfield's neuropathology*. Chicago: Yearbook Medical pp. 420–421.

Booth, J. R., Wood, L., Lu, D., Houk, J. C., & Bitan, T. (2007). The role of the basal ganglia and cerebellum in language processing. *Brain Research*, *1133*(1), 136–144.

Braudeau, J., Delatour, B., Duchon, A., Pereira, P. L., Dauphinot, L., de Chaumont, F., et al. (2011). Specific targeting of the GABA-A receptor α5 subtype by a selective inverse agonist restores cognitive deficits in Down syndrome mice. *Journal of Psychopharmacology*, *8*, 1030–1042.

Brooksbank, B. W. L., Walker, D., Balazs, R., & Jorgensen, O. S. (1989). Neuronal maturation in the fetal brain in Down syndrome. *Early Human Development*, *18*, 237–246.

Brown, J. H., Johnson, M. H., Paterson, S. J., Gilmore, R., Longhi, E., & Karmiloff-Smith, A. (2003). Spatial representation and attention in toddlers with Williams syndrome and Down syndrome. *Neuropsychologia*, *41*, 1037–1046.

Bussey, T. J., & Saksida, L. M. (2005). Object memory and perception in the medial temporal lobe: An alternative approach. *Current Opinion in Neurobiology*, *15*, 730–737.

Bussey, T. J., Saksida, L. M., & Murray, E. A. (2003). Impairments in visual discrimination after perirhinal cortex lesions: Testing "declarative" vs "perceptual-mnemonic" views of perirhinal cortex function. *European Journal of Neuroscience*, *17*, 649–660.

Cardoso-Martins, C., Mervis, B., & Mervis, C. (1985). Early vocabulary acquisition by children with Down syndrome. *American Journal of Mental Deficiency*, *90*, 177–184.

Carlesimo, G. A., Marotta, L., & Vicari, S. (1997). Long-term memory in mental retardation: Evidence for a specific impairment in subjects with Down's syndrome. *Neuropsychologia*, *35*, 71–79.

Chakrabarti, L., Scafidi, J., Gallo, V., & Haydar, T. F. (2011). Environmental enrichment rescues postnatal neurogenesis defect in the male and female Ts65Dn mouse model of Down syndrome. *Developmental Neuroscience*, *33*(5), 428–441.

Chapman, R. S., Seung, H.-K., Schwartz, S. E., & Kay-Raining Bird, E. (1998). Language skills of children and adolescents with Down syndrome: II. Production deficits. *Journal of Speech, Language, and Hearing Research*, *41*(4), 861–873.

Chen, C., Bao, S., Lockard, J. M., Kim, J. J., & Thompson, R. F. (1996). Impaired classical eyeblink conditioning in cerebellar-lesioned and purkinje cell degeneration (pcd) mutant mice. *Journal of Neuroscience*, *16*, 2829–2838.

Connolly, B. H., & Michael, B. T. (1986). Performance of retarded children, with and without Down syndrome, on the Bruininks Oseretsky Test of Motor Proficiency. *Physical Therapy*, *66*, 344–348.

Contestabile, A., Fila, T., Ceccarelli, C., Bonasoni, P., Bonapace, L., Santini, D., et al. (2007). Cell cycle alteration and decreased cell proliferation in the hippocampal dentate gyrus and in the neocortical germinal matrix of fetuses with Down syndrome and in Ts65Dn mice. *Hippocampus*, *17*, 665–678.

Cornoldi, C., & Vecchi, T. (2003). *Visuo-spatial working memory and individual differences.* Hove: Psychology Press.

Costa, A. C., Walsh, K., & Davisson, M. T. (1999). Motor dysfunction in a mouse model for Down syndrome. *Physiology and Behavior*, *68*, 211–220.

Coussons-Read, M. E., & Crnic, L. S. (1996). Behavioral assessment of the Ts65Dn mouse, a model for Down syndrome: Altered behavior in the elevated plus maze and open field. *Behavior Genetics*, *26*, 7–13.

Crawley, J. N. (2007). *What's wrong with my mouse?: Behavioral phenotyping of transgenic and knockout mice.* Hoboken: Wiley.

Crnic, L. S., & Pennington, B. F. (2000). Down syndrome: Neuropsychology and animal models. *Progress in Infancy Research*, *1*, 69–111.

Crome, L., Cowie, V., & Slater, E. (1966). A statistical note on cerebellar and brain-stem weight in Mongolism. *Journal of Mental Deficiency Research*, *10*, 69–72.

D'Esposito, M., Detre, J. A., Alsop, D. C., Shin, R. K., Atlas, S., & Grossman, M. (1995). The neural basis of the central executive system of working memory. *Nature*, *378*, 279–281.

Das, I., & Reeves, R. H. (2011). The use of mouse models to understand and improve cognitive deficits in Down syndrome. *Disease Models & Mechanisms*, *4*(5), 596–606.

Davidson, M. C., Amso, D., Anderson, L. C., & Diamond, A. (2006). Development of cognitive control and executive functions from 4 to 13 years: Evidence from manipulations of memory, inhibition, and task switching. *Neuropsychologia*, *44*(11), 2037–2078.

Davisson, M. T., Schmidt, C., & Akeson, E. C. (1990). Segmental trisomy of murine chromosome 16: A new model system for studying Down syndrome. *Progress in Clinical and Biological Research*, *360*, 263–280.

de Villers-Sidani, E., Alzghoul, L., Zhou, X., Simpson, K. L., Lin, R. C. S., & Merzenich, M. M. (2010). Recovery of functional and structural age-related changes in the rat primary auditory cortex with operant training. *Proceedings of the*

National Academy of Sciences of the United States of America, 107(31), 13900–13905.

Dierssen, M., & Ramakers, G. J. (2006). Dendritic pathology in mental retardation: from molecular genetics to neurobiology. *Genes, Brain and Behavior*, *5*(Suppl. 2), 48–60.

Durston, S., Thomas, K., Yang, Y., Ulug, A., Zimmerman, R., & Casey, B. (2002). A neural basis for the development of inhibitory control. *Developmental Science*, *5*, 9–16.

Edgin, J. O. (2003). *A neuropsychological model for the development of the cognitive profiles in mental retardation syndromes: Evidence from Down syndrome and Williams syndrome.* [doctoral dissertation]Denver, CO: University of Denver.

Edgin, J. O., Mason, G. M., Allman, M. J., Capone, G. T., DeLeon, I., Maslen, C., et al. (2010). Development and validation of the Arizona Cognitive Test Battery for Down syndrome. *Journal of Neurodevelopmental Disorders*, *2*, 149–164.

Edgin, J. O., Pennington, B. F., & Mervis, C. B. (2010). Neuropsychological components of intellectual disability: The contributions of immediate, working, and associative memory. *Journal of Intellectual Disability Research*, *54*, 406–417.

Edgin, J. O., Spanò, G., & Nadel, L. (2011). Advances in clinical endpoints for neurocognitive rehabilitation in Down syndrome. In J. A. Rondal, J. Perera & D. Spiker (Eds.), *Neurocognitive rehabilitation of Down syndrome, the early years*. Cambridge: Cambridge University Press.

Ellis, N. R., Woodley-Zanthos, P., & Dulaney, C. L. (1989). Memory for spatial location in children, adults, and mentally retarded persons. *American Journal of Mental Retardation*, *93*, 521–526.

Engidawork, E., & Lubec, G. (2003). Molecular changes in fetal Down syndrome brain. *Journal of Neurochemistry*, *84*, 895–904.

Engle, R. W., Kane, M. J., & Tuholski, S. W. (1999). Individual differences in working memory capacity and what they tell us about controlled attention, general fluid intelligence and functions of the prefrontal cortex. In A. Miyake & P. Shah (Eds.), *Models of working memory: Mechanisms of active maintenance and executive control* (pp. 102–134). New York: Cambridge University Press.

Escorihuela, R. M., Fernández-Teruel, A., Vallina, I. F., Baamonde, C., Lumbreras, M. A., Dierssen, M., et al. (1995). A behavioral assessment of Ts65Dn mice: A putative Down syndrome model. *Neuroscience Letters*, *199*(2), 143–146.

Fernandez, F., & Garner, C. C. (2007). Object recognition memory is conserved in Ts1Cje, a mouse model of Down syndrome. *Neuroscience Letters*, *421*, 137–141.

Fernandez, F., & Garner, C. C. (2008). Episodic-like memory in Ts65Dn, a mouse model of Down syndrome. *Behavioural Brain Research*, *188*, 233–237.

Fernandez, F., Morishita, W., Zuniga, E., Nguyen, J., Blank, M., Malenka, R. C., et al. (2007). Pharmacotherapy for cognitive impairment in a mouse model of Down syndrome. *Nature Neuroscience*, *10*, 411–413.

Ferrer, I., Guionnet, N., Cruz-Sanchez, F., & Tunon, T. (1990). Neuronal alterations in patients with dementia: A Golgi study on biopsy samples. *Neuroscience Letters*, *114*, 11–16.

Fidler, D. J., Hepburn, S., & Rogers, S. (2006). Early learning and adaptive behaviour in toddlers with Down syndrome: Evidence for an emerging behavioural phenotype? *Down Syndrome Research and Practice*, *9*, 37–44.

Fidler, D. J., & Nadel, L. (2007). Education and children with Down syndrome: Neuroscience, development, and intervention. *Mental Retardation and Developmental Disabilities Research Reviews*, *13*, 262–271.

Fischer, J., & Hammerschmidt, K. (2011). Ultrasonic vocalizations in mouse models for speech and socio-cognitive disorders: Insights into the evolution of vocal communication. *Genes, Brain, and Behavior*, *10*, 17–27.

Flórez, J., del Arco, C., Gonzalez, A., Pascual, J., & Pazos, A. (1990). Autoradiographic studies of neurotransmitter receptors in the brain of newborn infants with Down syndrome. *American Journal of Medical Genetics Supplement*, *7*, 301–305.

Friedman, N. P., Miyake, A., Young, S. E., DeFries, J. C., Corley, R. P., & Hewitt, J. K. (2008). Individual differences in executive functions are almost entirely genetic in origin. *Journal of Experimental Psychology: General*, *137*, 201–225.

Frith, U., & Frith, C. D. (1974). Specific motor disabilities in Down's syndrome. *Journal of Child Psychology and Psychiatry*, *15*, 293–301.

Galante, M., Jani, H., Vanes, L., Daniel, H., Fisher, E. M., Tybulewicz, V. L., et al. (2009). Impairments in motor coordination without major changes in cerebellar plasticity in the Tc1 mouse model of Down syndrome. *Human Molecular Genetics*, *18*, 1449–1463.

Gardiner, K. J. (2010). Molecular basis of pharmacotherapies for cognition in Down syndrome. *Trends in Pharmacological Sciences*, *31*, 66–73.

Ginsberg, S. D., Alldred, M. J., & Che, S. (2012). Gene expression levels assessed by CA1pyramidal neuron and regional hippocampal dissections in Alzheimer's disease. *Neurobiology of Disease*, *45*(1), 99–107.

Golden, J. A., & Hyman, B. T. (1994). Development of the superior temporal neocortex is anomalous in trisomy 21. *Journal of Neuropathology and Experimental Neurology*, *53*, 513–520.

Granholm, A. C., Sanders, L. A., & Crnic, L. S. (2000). Loss of cholinergic phenotype in basal forebrain coincides with cognitive decline in a mouse model of Down's syndrome. *Experimental Neurology*, *161*, 647–663.

Guedj, F., Sébrié, C., Rivals, I., Ledru, A., Paly, E., Bizot, J. C., et al. (2009). Green tea polyphenols rescue of brain defects induced by overexpression of DYRK1A. *PLoS One*, *4*(2), e4606. doi:10.1371/journal.pone.0004606.

Hanson, J. E., Blank, M., Valenzuela, R. A., Garner, C. C., & Madison, D. V. (2007). The functional nature of synaptic circuitry is altered in area CA3 of the hippocampus in a mouse model of Down's syndrome. *Journal of Physiology*, *579*, 53–67.

Hayes, S. M., Nadel, L., & Ryan, L. (2007). The effect of scene context on episodic object recognition: Parahippocampal cortex mediates memory encoding and retrieval success. *Hippocampus*, *17*(9), 873–889.

Heuer, E., & Bachevalier, J. (2011). Effects of selective neonatal hippocampal lesions on tests of object and spatial recognition memory in monkeys. *Behavioral Neuroscience*, *125* (2), 137–149.

Holtzman, D. M., Santucci, D., Kilbridge, J., Chua-Couzens, J., Fontana, D. J., Daniels, S. E., et al. (1996). Developmental abnormalities and age-related neurodegeneration in a mouse model of Down syndrome. *Proceedings of the National Academy of Sciences of the United States of America*, *93*, 13333–13338.

Hyde, L. A., & Crnic, L. S. (2002). Reactivity to object and spatial novelty is normal in older Ts65Dn mice that model Down syndrome and Alzheimer's disease. *Brain Research*, *945*, 26–30.

Hyde, L. A., Frisone, D. F., & Crnic, L. S. (2001). Ts65Dn mice, a model for Down syndrome, have deficits in context discrimination learning suggesting impaired hippocampal function. *Behavioural Brain Research*, *118*, 53–60.

Insausti, A. M., Megıas, M., Crespo, D., Cruz-Orive, L. M., Dierssen, M., Vallina, I. F., et al. (1998). Hippocampal volume and neuronal number in Ts65Dn mice: A murine model of Down syndrome. *Neuroscience Letters*, *253*, 175–178.

Jacobs, R., Harvey, A. S., & Anderson, V. (2011). Are executive skills primarily mediated by the prefrontal cortex in childhood? Examination of focal brain lesions in childhood. *Cortex*, *47*(7), 808–824.

Jung, M. W., Wiener, S. I., & McNaughton, B. L. (1994). Comparison of spatial firing characteristics of units in dorsal and ventral hippocampus of the rat. *Journal of Neuroscience*, *14*, 7347–7356.

Kesner, R. P., & Goodrich-Hunsaker, N. J. (2010). Developing an animal model of human amnesia: The role of the hippocampus. *Neuropsychologia*, *48*, 2290–2302.

Kleschevnikov, A. M., Belichenko, P. V., Villar, A. J., Epstein, C. J., Malenka, R. C., & Mobley, W. C. (2004). Hippocampal long-term potentiation suppressed by increased inhibition in the Ts65Dn mouse, a genetic model of Down syndrome. *Journal of Neuroscience*, *24*, 8153–8160.

Kopp, C., Krakow, J., & Johnson, K. (1983). Strategy production by young Down syndrome children. *American Journal of Mental Deficiency*, *88*, 164–169.

Lam, M. Y., Hodges, N. J., Virji-Babul, N., & Latash, M. L. (2009). Evidence for slowing as a function of index of difficulty in young adults with Down syndrome. *American Journal on Intellectual and Developmental Disabilities*, *114*(6), 411–426.

Lanfranchi, S., Cornoldi, C., & Vianello, R. (2004). Verbal and visuospatial working memory deficits in children with Down syndrome. *American Journal of Mental Retardation*, *109*, 456–466.

Lanfranchi, S., Jerman, O., Dal Pont, E., Alberti, A., & Vianello, R. (2010). Executive function in adolescents with Down syndrome. *Journal of Intellectual Disability Research*, *54*, 308–319.

Latash, M. L., Kang, N., & Patterson, D. (2002). Finger coordination in persons with Down syndrome: Atypical patterns of coordination and the effects of practice. *Experimental Brain Research*, *6*(3), 345–355.

Li, Z., Yu, T., Morishima, M., Pao, A., LaDuca, J., Conroy, J., et al. (2007). Duplication of the entire 22.9 Mb human chromosome 21 syntenic region on mouse chromosome 16 causes cardiovascular and gastrointestinal abnormalities. *Human Molecular Genetics*, *16*(11), 1359–1366.

Lorenzi, H. A., & Reeves, R. H. (2006). Hippocampal hypocellularity in the Ts65Dn mouse originates early in development. *Brain Research*, *1104*, 153–159.

Losin, E. A. R., Rivera, S. M., O'Hare, E. D., Sowell, E. R., & Pinter, J. D. (2009). Abnormal fMRI activation pattern during story listening in individuals with Down syndrome. *American Journal on Intellectual and Developmental Disabilities*, *114*(5), 369–380.

Lott, I. T., & Dierssen, M. (2010). Cognitive deficits and associated neurological complications in individuals with Down's syndrome. *Lancet Neurology*, *9*, 623–633.

Mangan, P. A. (1992). *Spatial memoiy abilities and abnormal development of the hippocampal formation in Down syndrome.* Unpublished doctoral dissertation, University of Arizona, Tucson.

Martinez-Cue, C., Baamonde, C., Lumbreras, M. A., Paz, J., Davisson, M. T., Schmitt, C., et al. (2002). Differential effects of environmental enrichment on behavior and learning of male and female Ts65Dn mice, a model for Down syndrome. *Behavioural Brain Research*, *134*, 185–200.

Marvel, C. L., & Desmond, J. E. (2010). Functional topography of the cerebellum in verbal working memory. *Neuropsychology Review*, *20*, 271–279.

McCormick, D. A., & Thompson, R. F. (1984). Cerebellum: Essential involvement in the classically conditioned eyelid response. *Science*, *223*, 296–299.

Menghini, D., Costanzo, F., & Vicari, S. (2011). Relationship between brain and cognitive processes in Down syndrome. *Behavior Genetics*, *41*, 381–393.

Mesulum, M. M. (2002). The human frontal lobes: Transcending the default mode through contingent encoding. In D. T. Stuss & R. T. Knight (Eds.), *Principles of frontal lobe function* (pp. 8–30). Oxford: Oxford University Press.

Metcalfe, J., & Mischel, W. (1999). A hot/cool-system analysis of delay of gratification: Dynamics of willpower. *Psychological Review*, *106*, 3–19.

Miyake, A., Friedman, N. P., Emerson, M. J., Witzki, A. H., & Howerter, A. (2000). The unity and diversity of executive functions and their contributions to complex "frontal lobe" tasks: A latent variable analysis. *Cognitive Psychology*, *41*, 49–100.

Moldrich, R. X., Dauphinot, L., Laffaire, J., Rossier, J., & Potier, M. C. (2007). Down syndrome gene dosage imbalance on cerebellum development. *Progress in Neurobiology*, *82*, 87–94.

Moon, J., Chen, M., Gandhy, S. U., Strawderman, M., Levitsky, D. A., Maclean, K. N., et al. (2010). Perinatal choline supplementation improves cognitive functioning and emotion regulation in the Ts65Dn mouse model of Down syndrome. *Behavioral Neuroscience*, *124*(3), 346–361.

Morice, E., Andreae, L. C., Cooke, S. F., Vanes, L., Fisher, E. M., Tybulewicz, V. L., et al. (2008). Preservation of long-term memory and synaptic plasticity despite short-term impairments in the Tc1 mouse model of Down syndrome. *Learning and Memory*, *15*, 492–500.

Morris, R. (1984). Developments of a water-maze procedure for studying deficits in the water maze spatial learning in the rat. *Journal of Neuroscience Methods*, *11*, 47–60.

Nadel, L. (1986). Down syndrome in neurobiological perspective. In C. J. Epstein (Ed.), *The neurobiology of Down syndrome* (pp. 239–251). New York: Raven Press.

Nadel, L. (1994). Multiple memory systems: What and why. An update. In D. Schacter & E. Tulving (Eds.), *Memory systems* (pp. 39–63). Cambridge, MA: MIT Press.

Nadel, L. (1999). Down syndrome in cognitive neuroscience perspective. In H. Tager-Flusberg (Ed.), *Neurodevelopmental disorders: Contributions to a new framework from the cognitive neurosciences* (pp. 187–221). Cambridge, MA: MIT Press.

Nadel, L. (2003). Down's syndrome: A genetic disorder in biobehavioral perspective. *Genes, Brain, and Behavior*, *2*(3), 156–166.

Nadel, L., & Hardt, O. (2011). Update on memory systems and processes. *Neuropsychopharmacology*, *36*(1), 251–273.

Norman, D. A., & Shallice, T. (1986). Attention to action: Willed and automatic control of behavior. In R. J. Davidson, G. E. Schwartz & D. Shapiro (Eds.), *Consciousness and self-regulation: Advances in research and theory* (Vol. 4), (pp. 1–18). New York: Plenum Press.

O'Doherty, A., Ruf, S., Mulligan, C., Hildreth, V., Errington, M. L., Cooke, S., et al. (2005). An aneuploid mouse strain carrying human chromosome 21 with Down syndrome phenotypes. *Science*, *309*, 2033–2037.

Olson, L. E., Richtsmeier, J. T., Leszl, J., & Reeves, R. H. (2004). A chromosome 21 critical region does not cause specific Down syndrome phenotypes. *Science*, *306*, 687–690.

Olson, L. E., Roper, R. J., Baxter, L. L., Carlson, E. J., Epstein, C. J., & Reeves, R. H. (2004). Down syndrome mouse models Ts65Dn, Ts1Cje, and Ms1Cje/Ts65Dn exhibit variable severity of cerebellar phenotypes. *Developmental Dynamics*, *230*, 581–589.

Olson, L. E., Roper, R. J., Sengstaken, C. L., Peterson, E. A., Aquino, V., Galdzicki, Z., et al. (2007). Trisomy for the Down syndrome "critical region" is necessary but not sufficient for brain phenotypes of trisomic mice. *Human Molecular Genetics*, *16*(7), 774–782.

Pazos, A., del Olmo, E., Diaz, A., del Arco, C., Rodriguez-Puertas, R., Pascual, J., et al. (1994). Serotonergic (5-HT1A and 5-HT1D) and muscarinic cholinergic receptors in DS brains: An autoradiographic analysis. In: *Poster presented at international Down syndrome research conference, Charleston, South Carolina, April 1994.*

Peiffer, A. M., Rosen, G. D., & Fitch, R. H. (2003). Sex differences in rapid auditory processing deficits in ectopic BXSB/MpJ mice. *NeuroReport*, *13*(17), 2277–2280.

Pennington, B. F., Moon, J., Edgin, J., Stedron, J., & Nadel, L. (2003). The neuropsychology of Down syndrome: Evidence for hippocampal dysfunction. *Child Development*, *74*, 75–93.

Pereira, P. L., Magnol, L., Sahún, I., Brault, V., Duchon, A., Prandini, P., et al. (2009). A new mouse model for the trisomy of the Abcg1-U2af1 region reveals the complexity of the combinatorial genetic code of Down syndrome. *Human Molecular Genetics*, *18*(24), 4756–4769.

Pinter, J. D., Eliez, S., Schmitt, J. E., Capone, G. T., & Reiss, A. L. (2001). Neuroanatomy of Down's syndrome: A high-resolution MRI study. *The American Journal of Psychiatry*, *158*, 1659–1665.

Pletcher, M. T., Wiltshire, T., Cabin, D. E., Villanueva, M., & Reeves, R. H. (2001). Use of the comparative physical and sequence mapping to annotate mouse chromosome 16 and human chromosome 21. *Genomics*, *74*, 45–54.

Preuss, T. M. (1995). Do rats have a prefrontal cortex? The Rose-Woolsey-Akert program reconsidered. *Journal of Cognitive Neuroscience*, *7*(1), 1–24.

Raitano Lee, N., Fidler, D. J., Blakeley-Smith, A., Daunhauer, L., Robinson, C., & Hepburn, S. L. (2011). Caregiver report of executive functioning in a population-based sample of young children with down syndrome. *American Journal on Intellectual and Developmental Disabilities*, *116* (4), 290–304.

Rast, M. M., & Harris, S. R. (1985). Motor control in infants with Down syndrome. *Developmental Medicine and Child Neurology*, *27*, 675–685.

Reeves, R. H., Irving, N. G., Moran, T. H., Wohn, A., Kitt, C., Sisodia, S. S., et al. (1995). A mouse model for Down syndrome exhibits learning and behaviour deficits. *Nature Genetics*, *11*, 177–184.

Robinson, B. F., Mervis, C. B., & Robinson, B. W. (2003). The roles of verbal short-term memory and working memory in the acquisition of grammar by children with Williams syndrome. *Developmental Neuropsychology*, *23*, 13–31.

Roper, R. J., Baxter, L. L., Saran, N. G., Klinedinst, D., Beachy, P. A., & Reeves, R. H. (2006). Defective cerebellar response to mitogenic Hedgehog signaling in Down syndrome mice. *Proceedings of the National Academy of Sciences of the United States of America*, *103*(5), 1452–1456.

Roubertoux, P. L., & Carlier, M. (2010). Mouse models of cognitive disabilities in trisomy 21 (Down syndrome). *American Journal of Medical Genetics. Part C, Seminars in Medical Genetics*, *154C*, 400–416.

Rowe, J., Lavender, A., & Turk, V. (2006). Cognitive executive function in Down's syndrome. *British Journal of Clinical Psychology*, *45*, 5–17.

Sago, H., Carlson, E. J., Smith, D. J., Kilbridge, J., Rubin, E. M., Mobley, W. C., et al. (1998). Ts1Cje, a partial trisomy 16 mouse model for Down syndrome, exhibits learning and behavioral abnormalities. *Proceedings of the National Academy of Sciences of the United States of America*, *95*(11), 6256–6261.

Saksida, L. M., Bussey, T. J., Buckmaster, C. A., & Murray, E. A. (2006). No effect of hippocampal lesions on perirhinal cortex-dependent feature-ambiguous visual discriminations. *Hippocampus*, *16*, 421–430.

Salehi, A., Faizi, M., Colas, D., Valletta, J., Laguna, J., Takimoto-Kimura, R., et al. (2009). Restoration of norepinephrine modulated contextual memory in a mouse model of Down syndrome. *Science Translational Medicine*, *1*(7), 7–17.

Schmidt-Sidor, B., Wisniewski, K., Shepard, T. H., & Sersen, E. A. (1990). Brain growth in Down syndrome subjects 15 to 22 weeks of gestational age and birth to 60 months. *Clinical Neuropathology*, *9*, 181–190.

Seregaza, Z., Roubertoux, P. L., Jamon, M., & Soumireu-Mourat, B. (2006). Mouse models of cognitive disorders in trisomy 21: A review. *Behavior Genetics*, *36*, 387–404.

Shamloo, M., Belichenko, P. V., & Mobley, W. C. (2010). Comprehensive behavioral assays to enhance phenotype to genotype linkages and therapeutic screening in mouse models of Down syndrome. *Future Neurology*, *5*, 467–471.

Siarey, R. J., Stoll, J., Rapoport, S. I., & Galdzicki, Z. (1997). Altered long-term potentiation in the young and old Ts65Dn mouse, a model for Down Syndrome. *Neuropharmacology*, *36*(11–12), 1549–1554.

Sommer, T., Rose, M., Weiller, C., & Buchel, C. (2005). Contributions of occipital, parietal and parahippocampal cortex to encoding of object-location associations. *Neuropsychologia*, *43*, 732–743.

Stedron, (2004). *Cerebellar function in Down syndrome.* [doctoral dissertation] Denver, CO: University of Denver.

Stoodley, C. J., & Schmahmannm, J. D. (2009). Functional topography in the human cerebellum: A meta-analysis of neuroimaging studies. *NeuroImage*, *44*, 489–501.

Takashima, S., Ieshima, A., Nakamura, H., & Becker, L. E. (1989). Dendrites, dementia and Down syndrome. *Brain & Development*, *11*, 131–133.

Tiemeier, H., Lenroot, R. K., Greenstein, D. K., Tran, L., Pierson, R., & Giedd, J. N. (2010). Cerebellum development during childhood and adolescence: A longitudinal morphometric MRI study. *NeuroImage*, *49*(1), 63–70.

Turner, C. A., Presti, M. F., Newman, H. A., Bugenhagen, P., Crnic, L., & Lewis, M. H. (2001). Spontaneous stereotypy in an animal model of Down syndrome: Ts65Dn mice. *Behavior Genetics*, *31*, 393–400.

Uecker, A., Mangan, P. A., Obrzut, J. E., & Nadel, L. (1993). Down syndrome in neurobiological perspective: An emphasis on spatial cognition. *Journal of Clinical Child Psychology*, *22*, 266–276.

Ungerleider, L. G., & Mishkin, M. (1982). Two cortical visual systems. In D. J. Ingle, M. A. Goodale & R. J. W. Mansfield (Eds.), *Analysis of visual behavior* (pp. 549–586). Cambridge, MA: MIT Press.

Vicari, S. (2006). Motor development and neuropsychological patterns in persons with Down syndrome. *Behavior Genetics*, *36*, 355–364.

Vicari, S., Bellucci, S., & Carlesimo, G. A. (2005). Visual and spatial longterm memory: Differential pattern of impairments in Williams and Down syndromes. *Developmental Medicine and Child Neurology*, *47*(5), 305–311.

Virji-Babul, N., Kerns, K., Zhou, E., Kapur, A., & Shiffrar, M. (2006). Perceptual-motor deficits in children with Down syndrome: Implications for intervention. *Down Syndrome Research and Practice*, *10*(2), 74–82.

Visu-Petra, L., Benga, O., Tincas, I., & Miclea, M. (2007). Visual-spatial processing in children and adolescents with Down's syndrome: A computerized assessment of memory skills. *Journal of Intellectual Disability Research*, *51*, 942–952.

Wang, P. P., Doherty, S., Rourke, S. B., & Bellugi, U. (1995). Unique profile of visuoperceptual skills in a genetic syndrome. *Brain and Cognition*, *29*, 54–65.

White, N. S., Alkire, M. T., & Haier, R. J. (2003). A voxel-based morphometric study of nondemented adults with Down syndrome. *NeuroImage*, *20*(1), 393–403.

Wiebe, S., Sheffield, T., Nelson, J., Clark, C., Chevalier, N., & Espy, K. (2010). The structure of executive function in 3-year-olds. *Journal of Experimental Child Psychology*, *108* (3), 436–452.

Wisniewski, K. E., & Kida, E. (1994). Abnormal neurogenesis and synaptogenesis in Down syndrome. *Developmental Brain Dysfunction*, *7*, 289–301.

Wisniewski, K., & Schmidt-Sidor, B. (1989). Postnatal delay of myelin formation in brains from Down syndrome infants and children. *Clinical Neuropathology*, *8*, 55–62.

Woodruff-Pak, D. S., Papka, M., & Simon, E. (1994). Eyeblink classical conditioning in Down's Syndrome, Fragile X syndrome and normal adults over and under age 35. *Neuropsychology*, *8*, 14–24.

Yu, T., Li, Z., Jia, Z., Clapcote, S. J., Liu, C., Li, S., et al. (2010). A mouse model of Down syndrome trisomic for all human chromosome 21 syntenic regions. *Human Molecular Genetics*, *19*, 2780–2791.

Zelazo, P. D. (2006). The Dimensional Change Card Sort (DCCS): A method of assessing executive function in children. *Nature Protocols*, *1*, 297–301.

Zelazo, P. D., Burack, J., Benedetto, E., & Frye, D. (1996). Theory of mind and rule use in individuals with Down Syndrome: A test of the uniqueness and specificity claims. *Journal of Child Psychology and Psychiatry, and Allied Disciplines*, *37*, 479–484.

Zola-Morgan, S., Squire, L. R., Amaral, D. G., & Suzuki, W. A. (1989). Lesions of perirhinal and parahippocampal cortex that spare the amygdala and hippocampal formation produce severe memory impairment. *Journal of Neuroscience*, *9*(12), 4355–4370.

M. Dierssen and R. de la Torre (Eds.)
Progress in Brain Research, Vol. 197
ISSN: 0079-6123

CHAPTER 8

Perturbation of dendritic protrusions in intellectual disability

Josien Levenga[†] and Rob Willemsen[‡,*]

[†] *Department of Physiology and Neuroscience, New York University, School of Medicine, New York, NY, USA*
[‡] *CBG-Department of Clinical Genetics, Erasmus MC, Rotterdam, The Netherlands*

Abstract: Intellectual disability (ID) affects 1–3% of the general population and is defined by an intelligence quotient score under 70 and the presence of two or more adaptive behaviors. Learning and memory involves the change in the transmission efficacy at the synapse (synaptic plasticity). Synaptic plasticity is the ability of the connection, or synapse, between two functional neurons to change in strength. Many molecular mechanisms are involved in the change in synaptic strength, which can result in changes in spine morphology. Spines are specialized dendritic protrusions and their change in morphology is implicated in learning and memory. In several cases of ID, the link between spine abnormalities (abnormal in number, size, and shape) and ID is well described, including nonsyndromic ID and Down, Fragile X, and Rett syndromes. This chapter discusses the underlying molecular mechanisms of this altered spine phenotype.

Keywords: spines; mental retardation; Down syndrome; fragile X syndrome; Rett syndrome.

Introduction

Our brain consists of different cell types which are contributing to our ability to learn, think, and remember. The molecular basis of learning and memory has been related to the brain's ability to change its anatomical shape, a process named plasticity (Kandel, 2001). Neuronal cells are believed to be the cells that retrieve information and can store information. Although the neuron is the principal functional unit, it is not the only cell type in the brain. Other cell types include oligodendrocytes, astrocytes, and microglia, which have an important role to support neurons in their functioning. Learning and memory is not restricted to a single brain region. Different brain regions have been identified which are responsible for specific types of memories, including implicit and explicit memory. To store memories and to learn, neurons have to communicate with each other in complex

*Corresponding author.
Tel.: +31 107043152; Fax: +31 107044736
E-mail: r.willemsen@erasmusmc.nl

DOI: 10.1016/B978-0-444-54299-1.00008-X

networks using synapses (chemical and electrical). It appears that a "typical" neuron may contain 1000–10,000 synapses and that each synapse contains over 1000 protein components (Grant et al., 2005). Most neuronal communication occurs with chemical synapses. At a chemical synapse, the plasma membrane of the signal-passing neuron (the *presynaptic* neuron) comes into close apposition with the membrane of the target (*postsynaptic*) neuron. The presynaptic neuron releases synaptic vesicles that contain a chemical substance called a neurotransmitter. Neurotransmitters bind to specific receptors located at an electron-dense structure at the postsynaptic cell membrane, named postsynaptic density. Binding of the neurotransmitter to a receptor can affect the postsynaptic cell in two different ways; they can excite or inhibit the receiving neuron. Glutamate is one of the most abundant excitatory neurotransmitters in the brain. Most glutamatergic excitatory synapses occur on small protrusions that are located along the dendrite. These protrusions are called spines and they play an important role in learning and memory processes since they are the main site to receive stimulation of neighboring neurons (Kasai et al., 2010). The first observations of dendritic spines in brain tissue originate from 1891 by Santiago Ramon y Cajal using Golgi's staining method. Spines are highly dynamic structures and their morphology can change very rapidly upon different types of stimulation, a process termed synaptic plasticity (Vanderklish and Edelman, 2002; von Bohlen Und Halbach, 2009; Yang et al., 2009). Typically, spines consist of a head that is connected to the dendrite by a neck. The dendritic spine density can reach up to 50 spines per 10-μm stretch of a neuron's dendrite. Spines can change their morphology within seconds by remodeling the architecture of their actin cytoskeleton. The size of the dendritic spine corresponds to the size of the synapse, which corresponds to the strength of the synapse. Typically, spines with strong synaptic contacts have a large spine head. Spines can be divided into two main classes: (1) immature spines or filopodia and (2) mature spines (Fig. 1). Filopodia are long and thin protrusions that typically

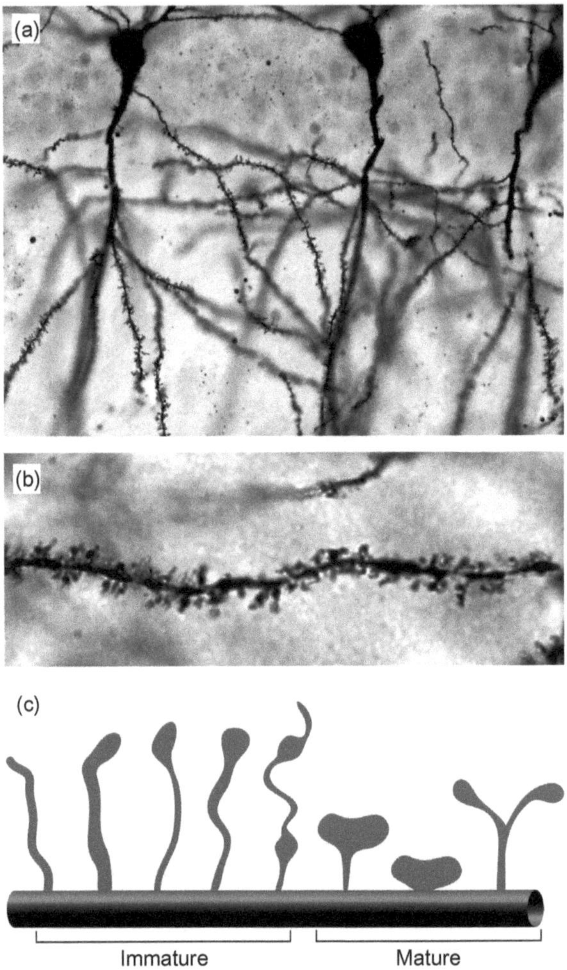

Fig. 1. Morphology of dendritic spines. Photomicrograph of murine hippocampal neurons stained with Golgi impregnation staining method (a). The Golgi technique selectively impregnates single neurons with silver chromate and allows visualization of dendrites and axons, including dendritic spines (b). Graphical representation of dendritic spine morphologies defined as mature or immature (c). (For color version of this figure, the reader is referred to the Web version of this chapter.)

lack "heads" (or have very small heads) while mature spines have a distinct head and neck. Based on these different morphologies, spines are also named mushroom-like, stubby, thin, and branched.

Spines are highly dynamic after specific stimulation and during development (see Box 1). The

BOX 1

LEARNING AND MEMORY

It is believed that learning and memory involves long-term potentiation (LTP) and long-term depression (LTD), which are considered to be the major cellular mechanisms underlying learning and memory (Kandel, 2001). LTP is mainly described as strengthening of the connection between a presynaptic neuron and postsynaptic neuron for a longer period. Glutamate is one of the neurotransmitters that can induce LTP. Due to the binding of glutamate to the *N*-methyl-D-aspartic (NMDA) receptor, the receptor pore is opened and Ca^{2+} flows into the cell. The rise of Ca^{2+} in the cell triggers short lasting activation of proteins, like CamKII. Active CamKII can phosphorylate α-amino-3-hydroxy-5-methyl-4-isoxazolepropionic acid (AMPA) receptor to modulate insertion of more AMPA receptors at the postsynaptic membrane. Thus, LTP results in more AMPA receptors in the postsynaptic membrane which improves the sensitivity to signals from the presynaptic neurons and thus strengthens the connection (Matsuzaki et al., 2004). LTD is the antithesis of LTP and is seen as the weakening of the synapse, reflected by a reduced number of ion receptors at the postsynaptic membrane (see for review Malenka and Bear, 2004). Different types of LTD can be distinguished, like NMDA receptor, metabotropic glutamate receptor (mGluR) or endocannabinoid-dependent LTD. Most studies examined the NMDA receptor-dependent LTD in the hippocampus. NMDA receptor LTD is induced after a small, slow rise in postsynaptic Ca^{2+}. Another important type of LTD is induced by group I mGluRs. It was demonstrated that a paired-pulse low-frequency stimulation (PP-LFS) protocol or stimulation of group I mGluRs by DHPG ((RS)-3,5-dihydroxyphenylglycine) results in LTD that is independent of NMDARs (Huber et al., 2001). This form of LTD is dependent on local protein synthesis at the synapse and also results in a net loss of AMPA receptors (Schuman et al., 2006; Steward, 2002). The gain or loss of AMPA receptors at the postsynaptic membrane is related to the strength to the synapse and also to the morphology of the spines.

number of spines on dendrites and the morphology of the spines can change dramatically. The number of neurons and the number of synapses are at a maximum in early postnatal life. The synapses are weak during development, but as a response to environmental inputs, activity and experience, the immature synaptic connections will be strengthened. Contrary, inactive synapses become weaker and will eventually be eliminated ("use-it-or-lose-it"). Thus, elimination of spines and change of morphology seem to be essential to establish neuronal circuits in early life (Fig. 2). In many intellectual disability (ID) syndromes, it has been found that the morphology and/or number of excitatory dendritic spines are aberrant compared to age-matched control individuals using Golgi-stained brain sections (Figs. 1 and 2). The first link between ID and aberrant spines dates already from the early 70s (Purpura, 1974). Purpura showed that ID was associated with a significant increase of abnormally long, thin spines on dendrites of cortical neurons in children with ID. Later, many of the ID syndromes have been linked to mutations in (synaptic) proteins that are probably directly or indirectly involved in the stabilization of the spine structure, including the Rho GTPase signaling pathway, epigenetic pathways, and local mRNA translation at the synapse (Kiebler and Bassell, 2006; Ramakers, 2002; van Bokhoven and Kramer, 2010). This inability to preserve the spine structure probably results in altered information processing and consequently learning and memory storage defects leading to cognitive impairment. In this chapter, we discuss the progress that is made about our understanding of synaptic plasticity in specific ID syndromes and the proteins that play a role in the spine structure and synaptic plasticity.

Syndromes of ID and spine abnormalities

ID is defined as a developmental disability with onset during childhood, characterized by significant

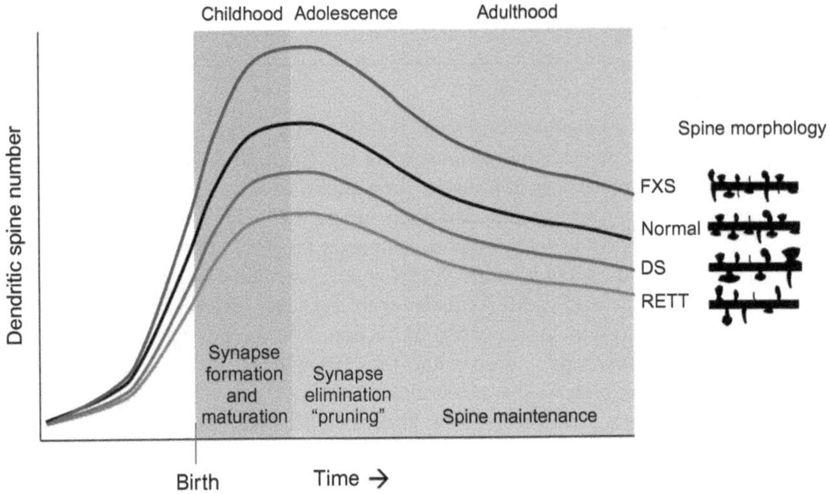

Fig. 2. Spine dynamics and maintenance during life. Excitatory spines undergo extensive changes during life. During neonatal development and after birth, spines are formed and shaped. During adolescence, a process called pruning will cause elimination of (weak) connections between neurons (the spines) that are considered unnecessary. After adolescence, the spines have to be maintained to be functional until death. During development, the formation and number of spines can be altered due to a genetic cause resulting in ID and/or behavioral abnormalities. Patients with FXS show an increased number of spines and more immature spines. On the other hand, patients with DS have less spines but with increased spine heads (mature). The other phenotype is shown in patients with Rett syndrome who have also decreased number of spines with a decreased spine head (immature). Figure according to (Penzes et al., 2011). (For color version of this figure, the reader is referred to the Web version of this chapter.)

impairment of intellectual functioning and adaptive skills causing major limitations to living a normal independent life. Previous definitions of ID include mental retardation; however, for a number of reasons, this term has been changed to ID (Schalock et al., 2007). ID affects approximately 2–3% of the general population and the severity of the ID are based on intelligence quotient (IQ) scores, a measure of general intellectual functioning (Ropers, 2006). ID can be caused by genetic and nongenetic factors. To date, over 80 genes located on the X chromosome and over 300 genes on autosomes have been identified that give rise to ID. Moreover, it has been predicted that more than 1400 human genes on the autosomes can cause ID in a monogenic model of disease (Chelly et al., 2006). Interestingly, a recent study suggests that the majority of ID cases occurred by spontaneous mutations in paternal sperm or maternal egg cells (Vissers et al., 2010). Inherited ID distinguishes syndromic ID as intellectual deficits associated with other medical and behavioral signs and symptoms, whereas nonsyndromic ID refers to intellectual deficits that appear without other abnormalities. The most common causes of nongenetic ID include problems at birth (ischemia), exposure to toxins, infectious disease during pregnancy, and malnutrition. In the next section, we focus on spine abnormalities in three common ID syndromes.

Down syndrome

The most common genetic cause of ID in humans is Down syndrome (DS). Next to ID, individuals with DS have a typical facial appearance and often other congenital problems, for example, heart problems. Studies of brains from patients with DS, either postmortem observations or volumetric MRI studies, have shown reduced brain weight. Particularly, the cerebellum, frontal, and temporal lobes have shown reduced brain volumes (Pinter et al., 2001; Wisniewski, 1990). All symptoms found in DS have been assumed to arise from an excessive amount of genetic material but they may also depend on epigenetic modifications or be the indirect consequence of structural/functional changes. The gene dosage hypothesis proposed that the 1.5-fold increase in gene expression of the genes located on HSA21 is responsible for the brain abnormalities and ID in patients with DS. However, it appears that trisomy 21 not only lead to increased gene expression but also to decreased gene expression of genes on the disomic chromosomes. This might result from regulatory mechanisms that are imbalanced in DS. For example, microRNAs that are located on HSA21 could be overexpressed in DS resulting in decreased expression of target mRNAs (Kuhn et al., 2010). Neuropathological studies of the cellular structure of the brain using postmortem brain tissue of patients with DS (without Alzheimer's pathology) have shown dendritic spine alterations. The number of spines was often reduced and many spines have an increased spine head area in brains of patients with DS (Chapleau et al., 2009). The perturbed spine morphology and number might logically be a result of deregulated gene expression, leading to altered synaptic plasticity. This makes trisomy 21 a complex syndrome to study dendritic spine phenotypes because multiple proteins may contribute to this particular phenotype. Therefore, several animal models were generated to study this syndrome in more detail (see Table 1 and Chapter 9 of this volume).

Using an animal model, it is possible to study the impact of imbalanced gene dosage during developmental stages, from the embryonic phase till adulthood. Several transgenic mouse models for single or multiple genes were developed using BAC or YAC technology to study the contribution of each individual gene located on HSA21. Of course, these models do not exactly replicate the DS phenotype. Therefore, other models were generated that better match to genetically trisomy. In mice, the murine genes that are homologous to the genes located on HSA21 are spread across three chromosomes, namely, chromosome 16, 10, and 17 (Dierssen et al., 2009). Because the genes responsible for DS in humans are spread over different chromosomes in mice, it makes it nearly impossible to generate a mouse model that completely replicates DS. However, the first description of the Ts65Dn mouse model increased the knowledge underlying the molecular mechanisms involved in DS. This mouse model carries a distal part of murine chromosome 16 and shares multiple phenotypes with human DS patients, including memory impairments and cortical dysfunction (Dierssen et al., 2003; Lott and Dierssen, 2010 and see Chapter 10 this volume). In addition, a reduced neuronal density, reduced number of excitatory synapses, and reduction in synaptic length have been found in specific subregions of the hippocampus (Kurt et al., 2004). Interestingly, enriched environment had no effect on spine morphology or number of layer III pyramidal neurons in Ts65Dn mice, while the enriched environment had a clear effect on spine number and morphology in control animals (Dierssen et al., 2003). Moreover, several experiments have shown deficits in field excitatory postsynaptic responses after inducing long-term potentiation (LTP) in hippocampal slices from Ts65Dn mice with different stimulation paradigms (Costa and Grybko, 2005; Siarey et al., 1999, 1997 and see Chapter 10). Besides an altered LTP, induction of long-term depression (LTD) in hippocampal slices from Ts65Dn mice was significantly enhanced compared to control diploid hippocampal slices. Overall, this result suggests a defect in both synaptic plasticity and in the development of neuronal circuitry.

The second mouse model for DS that has been developed was named Ts1Cje mice and is a trisomic mouse model for a reduced size of the distal part of mouse chromosome 16 (Belichenko et al., 2007). Ts1Cje mice were less severely affected compared to the Ts65Dn mice. Also, the altered spine morphology was less severe then in Ts65Dn mice, that is, the numbers of spines

Table 1. Animal models and spine abnormalities

MR syndrome	Genetic cause	Animal model	Spine phenotype	References
Down syndrome	Trisomy HSA21	Ts65Dn	Decreased number of spines, enlarged spine heads	Belichenko et al. (2009c); Kurt (2000); Kurt et al. (2004)
		Ts1Cje	Decreased number of spines, enlarged spine heads but less severe compared to Ts65Dn	Belichenko et al. (2007, 2009c)
		Ts1Rhr	Decreased number of spines, enlarged spine heads but less severe compared to Ts65Dn and Ts1Cje	Belichenko et al. (2009c)
		Ts1Yu	To be studied	
		Tc1	To be studied	O'Doherty et al. (2005)
Fragile X syndrome	Expanded CGG repeat in 5′ UTR of *FMR1* gene	*Fmr1* KO mouse	Layer V pyramidal neurons: —increased number of spines	Galvez and Greenough (2005); McKinney et al. (2005)
			—more immature spines	Galvez and Greenough (2005); Irwin et al. (2002); Nimchinsky et al. (2001)
			Hippocampus: —increased number of spines —more immature spines	Grossman et al. (2006a,b); Levenga et al. (2011a,b)
Rett syndrome	Mutations in *MECP2* gene	*Mecp2* knockout mice ("Bird strain")	Lower spine density, decreased spine head size, and increased spine neck length	Belichenko et al. (2009c); Fukuda (2005)
		Mecp2 truncated protein ("Jaenisch strain")	Impaired dendritic spine density, decrease spine head size, increased spine neck length, and altered spine distribution	Belichenko et al. (2009a, b); Smrt et al. (2007)

were decreased, but the average spine volumes were increased. Also in this mouse model, it was shown that LTP was decreased, while LTD was enhanced in hippocampal slices from Ts1Cje mice (Siarey et al., 2005). This suggests that the genes located on the telomeric end of the murine chromosome are important for learning and memory and are involved in synaptic plasticity.

The frequently cited Down syndrome critical region (DSCR) hypothesizes that a specific region on HSA21 is responsible for the symptoms in DS, although a single DSCR probably does not exist. However, it is still interesting to study which dosage-sensitive genes within this region contribute to the DS phenotype. Therefore, a new mouse model was generated named Ts1Rhr that carries a trisomic segment that corresponds closely to the DSCR (Belichenko et al., 2009a). Ts1Rhr mice also recapitulate the decreased dendritic spine density and increased size of spine heads in fascia dentata, although less marked compared to the Ts65Dn and Ts1Cje mice. In addition, this mouse model also shows decreased LTP in hippocampal slices (Belichenko et al., 2009a).

In 2005, another mouse model, named Tc1 mice, has been generated that carries almost whole HSA21 (O'Doherty et al., 2005). Tc1 mice display a pleiotropic and variable "human"-like phenotype. Like the other mouse models, it displays deficits in synaptic plasticity, congenital heart defects, and hippocampal-dependent learning defects. Although a rigorously description of the spine morphology has not yet been performed, it can be expected that these mice will also show similar spine phenotypes as the Ts65Dn mouse model since the Tc1 mice do show altered synaptic plasticity shown by a decreased LTP in hippocampal slices. Finally, Li and coauthors generated the most recent trisomic DS model by using a larger part of murine chromosome 16 that corresponds to HSA21 (Li et al., 2007). Several DS phenotypes are displayed in this mouse model; however, the spine phenotype has still to be studied.

Most DS mouse models show similar phenotypes, however, in different degrees. Overall, all DS mouse models show defects in spine morphology and/or number of spines and have synaptic plasticity defects, illustrated by altered LTP and LTD. These results contribute to unravel the exact mechanisms involved in DS, and these mouse models can be used for drug intervention studies. However, it is important to realize that these studies have their limitations. The question remains whether specific genes have a higher contribution to the phenotype.

Since there are more than 243 known genes located on HSA21, the spine phenotype in DS is a result of a complex interaction of many proteins. Their impact on synaptic plasticity, resulting in altered spine density and morphology, is of great interest and still under study. Two candidate genes, located in the "DSCR", can be envisioned as affecting synapse formation and function: *Dyrk1A* and *Kcnj6* (Girk2). Expression levels of *Dyrk1A* are increased in Ts65Dn, Ts1Cje, and Ts1Rhr mice (Amano et al., 2004; Ramakrishna et al., 2005; Saran et al., 2003). Dyrk1A is expressed not only in neurons during fetal and postnatal life but also in adults and aging subjects. Therefore, Dyrk1A seems to play an important role during both neurodevelopment and aging. Dyrk1A has been found in the nucleus and in the soluble and insoluble cytoplasmic fractions and can be phosphorylated which probably determined the intercellular transportation of Dyrk1a. Dyrk1a on its turn can phosphorylate Tau protein on multiple sites, suggesting that Dyrk1a is involved in the tau pathology seen in early onset of Alzheimer's disease (AD) in patients with DS. Interestingly, non-DS AD brain can also show increased Dyrk1a expression. In addition, dendritic spine number is also decreased in AD. To investigate the Dyrk1a effects alone, a Dyrk1a transgenic mouse has been generated. This mouse model also shows a number of neurological phenotypes, including altered neuronal morphology and impairments in hippocampal-dependent learning (Ahn et al., 2006; Altafaj et al., 2001; Branchi et al., 2004). Furthermore, Dyrk1a heterozygous mice have a clear phenotype on dendritic arborization, showing that pyramidal neurons are less branched and also the spine number is significantly reduced (Benavides-Piccione et al., 2005). Overall, the data suggest that Dyrk1a might be a key player in the altered spine morphology found in patients with DS.

The other candidate gene is *Kcnj6*, which encodes for Girk2. Girk2 is a subunit of a G protein-coupled inwardly rectifying potassium channel that serves as an effector for postsynaptic $GABA_B$, $5HT_{1a}$, muscarine m2, and certain other receptors. *Kcnj6* is present in three copies in patients with DS and in all the mouse models. In the DS mouse models, LTP is decreased, while LTD is enhanced (Siarey et al., 2005). This suggests an overactive GABA inhibitory input on pyramidal cells. Enhanced levels of Girk2 might cause hyperpolarization of neurons, since currents through the Girk2-containing channels drive resting membrane potential to the equilibrium potential of K^+. Interestingly, when one of the three Kcnj6 copies was removed in the Ts65Dn mice, it could rescue some of the DS phenotypes suggesting that the function of Girk2 channels is indeed altered (Cramer et al., 2010).

Of course, the spine phenotype is not only linked to these two genes. Therefore, studies were performed to rank candidate genes for the DS phenotype, based on the variability in brain regional expression of genes from various Ts65Dn mice (Sultan et al., 2007). A few pathways have been identified that were altered owing to trisomy 21, including NFAT-dependent transcription, the

SHH signaling pathway, and the Notch signaling pathway. In conclusion, the altered spine phenotype found in DS results from a complex interaction of overexpression of several genes and their interactors.

Fragile X syndrome

Fragile X syndrome (FXS) is the most common ID syndrome caused by one single gene mutation. It affects approximately 1:4000 males and 1:7000 females in the population (for recent review see Willemsen et al., 2011). FXS is caused by lack of one single protein, the fragile X mental retardation protein (FMRP), owing to a mutation in the 5′ untranslated region (UTR) of the *FMR1* gene. The 5′ UTR of *FMR1* contains a CGG trinucleotide repeat that normally has a length between 5 and 55. In patients with FXS, this repeat has expanded above 200 CGG repeats (full mutation). As a result of this expansion, the promoter region is extensively methylated directing silencing of the *FMR1* gene. Therefore, patients with FXS lack FMRP in their brain and consequently patients with FXS will develop ID and behavioral abnormalities (Willemsen et al., 2004). Based on IQ scores, the severity of their MR ranges from mild (50<IQ>70) to severe (IQ<50). Besides ID, patients with FXS are often confronted with epilepsy, sleep disturbances, ADHD, and autistic features (Hagerman, 2002).

In 1994, an FXS animal model has been created by replacing the wild-type murine *Fmr1* gene with a nonfunctional *Fmr1* gene in which a neomycin resistance cassette was placed in exon 5, using homologous recombination in embryonic stem (ES) cells employing conventional transgenic ES technology (Bakker et al., 1994). This *Fmr1* knockout (KO) mouse model lacks FMRP expression and develops similar phenotypes as reported in patients, including learning and behavioral deficits and functional and morphological changes of spines (Oostra and Nelson, 2006). Regarding the spine phenotype, several studies have demonstrated an altered spine morphology and often an increased spine density in the different brain areas of *Fmr1* KO mice compared to control individuals, like as in brains of patients with FXS (Comery et al., 1997; Cruz-Martin et al., 2010; Galvez and Greenough, 2005; Grossman et al., 2006b, 2010; Hinton et al., 1991; Irwin et al., 2000, 2001, 2002; Levenga et al., 2011a; McKinney et al., 2005; Nimchinsky et al., 2001). In addition, *Fmr1* KO-cultured hippocampal neurons have more spines with an immature appearance (named filopodia) compared to wild-type neurons (Antar et al., 2006; De Vrij et al., 2008; Levenga et al., 2010). The abnormal protrusion morphology in cultured hippocampal neurons enabled *in vitro* rescue studies using specific drug treatments (De Vrij et al., 2008; Levenga et al., 2011b; Su et al., 2010).

Recently, it has become clear that the alterations in the brain due to lack of FMRP is not observed in all brain regions. For example, overexpression of specific proteins in the *Fmr1* KO brain can be brain region-specific. Schuett et al. analyzed the abundance of synaptic proteins in wild-type and *Fmr1* KO mice (Schuett et al., 2009). They observed that SAPAP1 and Shank3 expression were only elevated in the neocortical synaptosomes of the *Fmr1* KO mice, while SAPAP2 and SAPAP3 expression were only elevated in the hippocampal synaptosomes. For the spines, the debate is concentrated at which developmental time-point the abnormal protrusion morphology starts in *Fmr1* KO mice and in which brain areas. Although several research groups have studied the protrusion morphology in *Fmr1* KO mice, generally the pyramidal neurons of the cortex were studied and not the hippocampal neurons in adult *Fmr1* KO mice. Since the hippocampus is crucial for learning and memory, it is important to study the spine morphology in this brain region of *Fmr1* KO mice as well.

The protrusion morphology in the hippocampus seems to be variable. Grossman et al. demonstrated more longer and fewer shorter spines in the CA1 region of young adult (9–13 weeks)

Fmr1 KO mouse compared to wild-type mice (Grossman et al., 2006a). However, the density of protrusions in the CA1 region of *Fmr1* KO mouse did not differ from wild types. More recently, they studied the development of dendritic spines in the dentate gyrus of the *Fmr1* KO mice (Grossman et al., 2010). This study showed a higher protrusion density in the *Fmr1* KO mice during development (P15–P60) and an overall more immature morphological phenotype (higher proportion of thin headed spines). Another study using hippocampal slices of 7 days old *Fmr1* KO mice showed significant differences in protrusion morphology, including increased protrusion length and reduced spine head area (Bilousova et al., 2008). Recently, an FXS-related protrusion morphology phenotype could be demonstrated in hippocampi of relatively old *Fmr1* KO mice *ex vivo*. This study focused their analysis on the CA1 and CA3 region of the hippocampus. Unexpectedly, only the CA1 pyramidal neurons showed a clear immature protrusion phenotype in *Fmr1* KO mice, while no significant differences could be observed in the CA3 region (Levenga et al., 2011a). Thus, the immature protrusion morphology found in *Fmr1* KO mice seems not only to be age-specific, but also specific for specific brain subregions. Interestingly, the immature protrusion morphology phenotype in the CA1 hippocampal region could be linked to levels of mGluR5 expression. Of course, it remains possible that during postnatal development, pyramidal neurons in both areas show altered protrusion morphology, which only in CA3 region disappears during adulthood.

The molecular mechanisms underlying the altered protrusion phenotype in FXS are another area of focus for research. A model (mGluR theory) was proposed in which FMRP normally functions as a repressor of translation of specific mRNAs at the synapse. Upon mGluR5 stimulation by release of glutamate, FMRP is the "brake" on the protein-synthesis-dependent functions of mGluR5 activation, including strengthening of the synapse. In the absence of FMRP, mGluR5 activation is exaggerated and the synaptic strength is weakened (Bear et al., 2004). According to the mGluR theory, owing to the excessive protein synthesis after group I mGluR stimulation, excessive numbers of AMPA receptors are internalized which results in the immature protrusion phenotype in FXS. Several studies have shown that prolonged treatment of wild-type hippocampal neurons with DHPG, a group I mGluR agonist, results in increased number of long, thin dendritic spines (filopodia), resembling the protrusion phenotype found in FXS. However, evidence is missing identifying which proteins are responsible for the excessive AMPA receptor internalization in FXS (Abu-Elneel et al., 2008; Vanderklish and Edelman, 2002). Activity-regulated cytoskeleton-associated protein (Arc) (also termed as Arg3.1) is an immediate-early gene induced in response to sensory experience, learning, LTP, spatial exploration, and novelty and might be a good candidate for the excessive AMPA receptor internalization (Gusev et al., 2005; Guzowski et al., 2006). It has been suggested that *Arc* is an mRNA target of FMRP and that the protein levels of *Arc* are increased in the absence of FMRP (Zalfa et al., 2003). Park et al. demonstrated that *Arc* synthesis is required for mGluR-LTD induction, and Waung et al. showed that *Arc* is important for mGluR-dependent LTD by increasing the AMPA receptor internalization rate (Park and Tang, 2009; Waung et al., 2008). Therefore, Arc may play a major role in the excessive AMPA receptor internalization and altered protrusion morphology. Striatal enriched phosphatase (STEP) is another interesting candidate in AMPA receptor internalization (Zhang et al., 2008). STEP can phosphorylate AMPA and NMDA receptors to mediate internalization of these receptors. STEP mRNA is present in the dendrites and is synthesized after group I mGluR stimulation. STEP expression might be elevated in *Fmr1* KO mice, and in this way induces increased AMPA receptor internalization. Recently, a STEP/*Fmr1* double KO mouse model has been generated and preliminary data showed that this mouse model is rescued in open field phenotype (P. Lombroso, personal communication). Whether STEP is indeed an mRNA target of FMRP is thus far only hypothesized and needs to be confirmed.

Excessive AMPA receptor internalization alone cannot induce the immature appearance of the protrusions in FXS, since the cytoskeletal structure is also important for the protrusion morphology. The actin cytoskeleton of neurons plays an important role in cell polarity, neurite outgrowth, dendritic morphology, arborization and maintenance, and synapse formation. It has been reported that the cytoskeletal structure does not only play a role in FXS, but also in other forms of ID, including LIMK1, oligophrenin-1, PAK3, and ARHGEF6 (Newey et al., 2005; Ramakers, 2002). Small GTPases, such as Rho, Rac, and Cdc42, form a large family of proteins characterized by their ability to bind and hydrolyze GTP. They act as molecular switches affecting various biological activities and regulating growth and migration, and their action is not restricted to neurons and synapses alone, although their role in synapses seems to be important for the pathophysiology of ID (Boda et al., 2010). RhoA, Rac1, and Cdc42 are best known for their characteristic effects on the actin cytoskeleton, but more recently, a role has been identified in microtubule organization (Gundersen et al., 2004). Because microtubules and actin filaments make up the structural framework of dendrites and protrusions, the function of these networks in dendritic development and protrusion morphogenesis has been extensively studied in a variety of model systems.

Several lines of evidence have linked FMRP to GTPase signaling. First, *Fmr1* KO mice show a general impairment in LTP that can be rescued by activation of Ras/PI3K cascade (Hu et al., 2008). Second, in *Drosophila*, it has been shown that dFmrp affects dendritic development by regulating the actin cytoskeleton through a translational suppression of Rac1 and profilin (Reeve et al., 2005). In addition, CYFIP acts as a RAC1 effector that antagonizes *dFMR1* function (Schenck et al., 2003). Third, FMRP appears to be a negative regulator of PP2Ac mRNA translation, and absence of FMRP results in increased expression of PP2Ac leading to alterations in actin remodeling in fibroblast cell lines (Castets et al., 2005). Finally, another downstream effector of Rac, p21-activated kinases (PAK), a family of serine-threonine kinases that consists of at least three members, PAK1, PAK2, and PAK3, has been associated with FXS. Inhibition of PAK in the *Fmr1* KO can rescue cellular and behavioral FXS phenotypes, including the abnormal protrusion morphology, locomotor activity, and anxiety (Hayashi et al., 2007). Furthermore, FMRP seems to play an important role in microtubule stability. FMRP represses the translation of microtubule-associated protein 1b (*Map1b*) mRNA during active synaptogenesis in neonatal brain development. In the absence of FMRP, elevated Map1b protein expression leads to abnormally increased microtubule stability, thereby hindering normal development of dendritic protrusion (Lu et al., 2004).

In conclusion, lack of FMRP expression leads to abnormal protrusion morphology in different brain areas, and this abnormal protrusion morphology seems to be a hallmark of many other ID disorders, syndromic or nonsyndromic, as well. However, whether this phenotype is a cause or a consequence of ID and if we can use this feature as a translational end point to investigate the therapeutic effects of pharmacological interventions are still open questions that need to be addressed in the near future.

Rett syndrome

Rett syndrome is a progressive postnatal neurodevelopmental disorder and is one of the most common causes of ID in females. It affects approximately 1:10,000 females worldwide and is caused by loss-of-function mutations in the gene encoding methyl-CpG binding protein 2 (MECP2) (Chahrour and Zoghbi, 2007). MECP2 protein is involved in epigenetic changes of DNA via transcriptional repression through DNA methylation. Thus, lack of MECP2 expression may cause overexpression of target genes. However, evidence for overexpression of target genes is not unambiguous, although abnormal expression levels of a number of genes important for synaptogenesis have been reported (Guy et al., 2011). The MECP2 gene is located on the X chromosome and affects primarily girls who usually developed normally in the first 6–18 months of life, but show regression in psychomotor development, autonomic dysfunction, and regression in brain growth. Additional characteristics are autistic features, sleep impairments, panic-like attacks. After initial regression, the condition stabilizes and female patients may recover some skills and usually survive till adulthood, although incidences of sudden death are higher compared to control individuals. Boys, who carry MECP2 mutations, die soon after birth because of their single X chromosome.

The brains of patients with Rett syndrome show decreased brain volume with an increased cell density in several brain areas, especially in the cerebral cortex (Armstrong, 2002). Interestingly, in Rett syndrome, the weight of the brain is significantly reduced but does not decrease significantly with age. This suggests that atrophy does not account for the small size. The absence of gross macroscopical neuroanatomic abnormalities in Rett syndrome suggests that dysfunction occurs at the microcircuit level. Indeed, at the cellular level, the neuronal soma is smaller and Golgi staining of postmortem brain sections has shown a reduced dendritic arborization, reduced number of mature spines, and defects in spine morphology (Armstrong, 2002; Belichenko et al., 1994; Chapleau et al., 2009; Schule et al., 2008). Quantitative analyses revealed significantly lower dendritic spine density in secondary and tertiary apical dendrites of CA1 pyramidal neurons from individuals with Rett compared to unaffected (non-ID) individuals using DiOlistic labeling of neurons (Chapleau et al., 2009).

To study Rett syndrome in more detail, several animal models have been generated (Chahrour and Zoghbi, 2007; Chen et al., 2001; Shepherd and Katz, 2011). Mecp2 knockout mice show neurodevelopmental and behavioral phenotypes similar to those described for patients with Rett syndrome (Smrt et al., 2007). Importantly, the conditional deletion of Mecp2 in postmitotic neurons recapitulates most phenotypes observed in the Mecp2 knockout mice, suggesting that Rett syndrome is caused by a specific defect in MeCP2 function in mature neurons (Guy et al., 2011). Studies of postnatal brain development in Mecp2 knockout mice also showed that Mecp2 was not important for the generation of immature neurons but rather for neuronal maturation. Electrophysiological analysis of neocortical neurons and circuits revealed reduced excitatory postsynaptic currents in Mecp2 mutant mice (Dani et al., 2005). In addition, morphological analyses demonstrated a reduced number of synapses in the hippocampus, abnormalities in dendritic arborization, and spines (number and delayed maturation) (Belichenko et al., 2009b,c; Chao et al., 2007; Kishi and Macklis, 2004). Altogether, these specific phenotypes point to deficits in experience-dependent synaptic plasticity. To stabilize synaptic connections, many proteins are involved; however, in Rett syndrome, two proteins might play a role in the impaired synaptic plasticity, brain-derived-neurotrophic-factor (BDNF), and distal-less homeobox 5 (DLX5). BDNF is essential for learning and memory. Two research groups have discovered that Mecp2 regulates the expression of the gene encoding for BDNF. They reported that Mecp2 can bind selectively to Bdnf promoter III and represses expression of BDNF (Chen et al., 2003). Furthermore, depolarization of the cell membrane, causing the influx of calcium ions, consequently results in phosphorylation of Mecp2, which is then released from the Bdnf promoter, facilitating Bdnf transcription. Misregulation of this process seems to be involved in Rett syndrome.

The other gene, DLX5, was found to be transcribed two times higher in Mecp2 null mice compared to wild-type mice. The protein encoded by this gene is thought to regulate the production of enzymes that synthesize GABA and normally this gene is imprinted. Failure of this imprinting process might result in neurological disorders. A loss of imprinting of DLX5 in Rett syndrome patients might result in altered GABAergic synaptic plasticity. Although the gene that is responsible for Rett syndrome has been identified, the exact molecular mechanisms underlying impaired synaptic plasticity remain poorly understood (Bapat and Galande, 2005).

Concluding remarks

All the ID syndromes described in this chapter have many things in common and one of them is the altered spine morphology and/or spine number, resulting in altered communication between neurons. However, interestingly, the spine alterations are different for each individual ID syndrome. Therefore, it seems that in each syndrome different molecular mechanisms are involved in the aberrant spine morphology, but all are leading to ID. The altered spine morphology might be the cause or the result of deficits in synaptic plasticity, leading to problems in intellectual performance. To understand the molecular mechanisms underlying these ID syndromes, animal models have been generated. These *in vivo* models have remained indispensable to our knowledge and they might help us to find a therapeutic target that can improve the life of individuals with ID.

Acknowledgments

The authors wish to thank their colleagues from Erasmus MC for their contributions. Artwork from Tom de Vries-Lentsch is greatly appreciated. Financial support was provided by ZonMw (912-07-022) and CureFXS E-Rare (EU/FIS PS09102673).

References

Abu-Elneel, K., Ochiishi, T., Medina, M., Remedi, M., Gastaldi, L., Caceres, A., et al. (2008). A delta-catenin signaling pathway leading to dendritic protrusions. *Journal of Biological Chemistry*, *283*, 32781–32791.

Ahn, K. J., Jeong, H. K., Choi, H. S., Ryoo, S. R., Kim, Y. J., Goo, J. S., et al. (2006). DYRK1A BAC transgenic mice show altered synaptic plasticity with learning and memory defects. *Neurobiology of Disease*, *22*, 463–472.

Altafaj, X., Dierssen, M., Baamonde, C., Marti, E., Visa, J., Guimera, J., et al. (2001). Neurodevelopmental delay, motor abnormalities and cognitive deficits in transgenic mice overexpressing Dyrk1A (minibrain), a murine model of Down's syndrome. *Human Molecular Genetics*, *10*, 1915–1923.

Amano, K., Sago, H., Uchikawa, C., Suzuki, T., Kotliarova, S. E., Nukina, N., et al. (2004). Dosage-dependent over-expression of genes in the trisomic region of Ts1Cje mouse model for Down syndrome. *Human Molecular Genetics*, *13*, 1333–1340.

Antar, L. N., Li, C., Zhang, H., Carroll, R. C., & Bassell, G. J. (2006). Local functions for FMRP in axon growth cone motility and activity-dependent regulation of filopodia and spine synapses. *Molecular and Cellular Neuroscience*, *32*, 37–48.

Armstrong, D. D. (2002). Neuropathology of Rett syndrome. *Mental Retardation and Developmental Disabilities Research Reviews*, *8*, 72–76.

Bakker, C. E., Verheij, C., Willemsen, R., Vanderhelm, R., Oerlemans, F., Vermey, M., et al. (1994). Fmr1 knockout mice: A model to study fragile X mental retardation. *Cell*, *78*, 23–33.

Bapat, S., & Galande, S. (2005). Association by guilt: Identification of DLX5 as a target for MeCP2 provides a molecular link between genomic imprinting and Rett syndrome. *Bioessays*, *27*, 676–680.

Bear, M. F., Huber, K. M., & Warren, S. T. (2004). The mGluR theory of fragile X mental retardation. *Trends in Neurosciences*, *27*, 370–377.

Belichenko, P. V., Oldfors, A., Hagberg, B., & Dahlstrom, A. (1994). Rett syndrome: 3-D confocal microscopy of cortical pyramidal dendrites and afferents. *Neuroreport*, *5*, 1509–1513.

Belichenko, P. V., Kleschevnikov, A. M., Salehi, A., Epstein, C. J., & Mobley, W. C. (2007). Synaptic and cognitive abnormalities in mouse models of Down syndrome: Exploring genotype-phenotype relationships. *The Journal of Comparative Neurology*, *504*, 329–345.

Belichenko, N. P., Belichenko, P. V., Kleschevnikov, A. M., Salehi, A., Reeves, R. H., & Mobley, W. C. (2009a). The "Down syndrome critical region" is sufficient in the mouse model to confer behavioral, neurophysiological, and synaptic phenotypes characteristic of Down syndrome. *Journal of Neuroscience*, *29*, 5938–5948.

Belichenko, N. P., Belichenko, P. V., & Mobley, W. C. (2009b). Evidence for both neuronal cell autonomous and nonautonomous effects of methyl-CpG-binding protein 2 in the cerebral cortex of female mice with Mecp2 mutation. *Neurobiology of Disease*, *34*, 71–77.

Belichenko, P. V., Wright, E. E., Belichenko, N. P., Masliah, E., Li, H. H., Mobley, W. C., et al. (2009c). Widespread changes in dendritic and axonal morphology in Mecp2-mutant mouse models of Rett syndrome: Evidence for disruption of neuronal networks. *The Journal of Comparative Neurology*, *514*, 240–258.

Benavides-Piccione, R., Dierssen, M., Ballesteros-Yanez, I., Martinez de Lagran, M., Arbones, M. L., Fotaki, V., et al. (2005). Alterations in the phenotype of neocortical pyramidal cells in the Dyrk1A+/- mouse. *Neurobiology of Disease*, *20*, 115–122.

Bilousova, T., Dansie, L., Ngo, M., Aye, J., Charles, J. R., Ethell, D. W., et al. (2008). Minocycline promotes dendritic spine maturation and improves behavioral performance in the fragile X mouse model. *Journal of Medical Genetics*, *46*, 94–102.

Boda, B., Dubos, A., & Muller, D. (2010). Signaling mechanisms regulating synapse formation and function in mental retardation. *Current Opinion in Neurobiology*, *20*, 519–527.

Branchi, I., Bichler, Z., Minghetti, L., Delabar, J. M., Malchiodi-Albedi, F., Gonzalez, M. C., et al. (2004). Transgenic mouse *in vivo* library of human Down syndrome critical region 1: Association between DYRK1A overexpression, brain development abnormalities, and cell cycle protein alteration. *Journal of Neuropathology and Experimental Neurology*, *63*, 429–440.

Castets, M., Schaeffer, C., Bechara, E., Schenck, A., Khandjian, E. W., Luche, S., et al. (2005). FMRP interferes with the Rac1 pathway and controls actin cytoskeleton dynamics in murine fibroblasts. *Human Molecular Genetics*, *14*, 835–844.

Chahrour, M., & Zoghbi, H. Y. (2007). The story of Rett syndrome: From clinic to neurobiology. *Neuron*, *56*, 422–437.

Chao, H. T., Zoghbi, H. Y., & Rosenmund, C. (2007). MeCP2 controls excitatory synaptic strength by regulating glutamatergic synapse number. *Neuron*, *56*, 58–65.

Chapleau, C. A., Calfa, G. D., Lane, M. C., Albertson, A. J., Larimore, J. L., Kudo, S., et al. (2009). Dendritic spine pathologies in hippocampal pyramidal neurons from Rett syndrome brain and after expression of Rett-associated MECP2 mutations. *Neurobiology of Disease*, *35*, 219–233.

Chelly, J., Khelfaoui, M., Francis, F., Cherif, B., & Bienvenu, T. (2006). Genetics and pathophysiology of mental retardation. *European Journal of Human Genetics*, *14*, 701–713.

Chen, R. Z., Akbarian, S., Tudor, M., & Jaenisch, R. (2001). Deficiency of methyl-CpG binding protein-2 in CNS neurons results in a Rett-like phenotype in mice. *Nature Genetics*, *27*, 327–331.

Chen, W. G., Chang, Q., Lin, Y., Meissner, A., West, A. E., Griffith, E. C., et al. (2003). Derepression of BDNF transcription involves calcium-dependent phosphorylation of MeCP2. *Science*, *302*, 885–889.

Comery, T. A., Harris, J. B., Willems, P. J., Oostra, B. A., Irwin, S. A., Weiler, I. J., et al. (1997). Abnormal dendritic spines in fragile X knockout mice: Maturation and pruning deficits. *Proceedings of the National Academy of Sciences of the United States of America*, *94*, 5401–5404.

Costa, A. C., & Grybko, M. J. (2005). Deficits in hippocampal CA1 LTP induced by TBS but not HFS in the Ts65Dn mouse: A model of Down syndrome. *Neuroscience Letters*, *382*, 317–322.

Cramer, N. P., Best, T. K., Stoffel, M., Siarey, R. J., & Galdzicki, Z. (2010). GABAB-GIRK2-mediated signaling in Down syndrome. *Advances in Pharmacology*, *58*, 397–426.

Cruz-Martin, A., Crespo, M., & Portera-Cailliau, C. (2010). Delayed stabilization of dendritic spines in fragile X mice. *Journal of Neuroscience*, *30*, 7793–7803.

Dani, V. S., Chang, Q., Maffei, A., Turrigiano, G. G., Jaenisch, R., & Nelson, S. B. (2005). Reduced cortical activity due to a shift in the balance between excitation and inhibition in a mouse model of Rett syndrome. *Proceedings of the National Academy of Sciences of the United States of America*, *102*, 12560–12565.

De Vrij, F. M. S., Levenga, J., Van der Linde, H. C., Koekkoek, S. K., De Zeeuw, C. I., Nelson, D. L., et al. (2008). Rescue of behavioral phenotype and neuronal protrusion morphology in FMR1 KO mice. *Neurobiology of Disease*, *31*, 127–132.

Dierssen, M., Benavides-Piccione, R., Martinez-Cue, C., Estivill, X., Florez, J., Elston, G. N., et al. (2003). Alterations of neocortical pyramidal cell phenotype in the Ts65Dn mouse model of Down syndrome: Effects of environmental enrichment. *Cerebral Cortex*, *13*, 758–764.

Dierssen, M., Herault, Y., & Estivill, X. (2009). Aneuploidy: From a physiological mechanism of variance to Down syndrome. *Physiological Reviews*, *89*, 887–920.

Fukuda, T., Yamashita, Y., Nagamitsu, S., Miyamoto, K., Jin, J. J., Ohmori, I., et al. (2005). Methyl-CpG binding protein 2 gene (MECP2) variations in Japanese patients with Rett syndrome: pathological mutations and polymorphisms. *Brain Development*, *27*, 211–217.

Galvez, R., & Greenough, W. T. (2005). Sequence of abnormal dendritic spine development in primary somatosensory cortex of a mouse model of the fragile X mental retardation syndrome. *American Journal of Medical Genetics. Part A*, *135*, 155–160.

Grant, S. G., Marshall, M. C., Page, K. L., Cumiskey, M. A., & Armstrong, J. D. (2005). Synapse proteomics of multiprotein complexes: En route from genes to nervous system diseases. *Human Molecular Genetics*, *14*, R225–R234.

Grossman, A. W., Aldridge, G. M., Weiler, I. J., & Greenough, W. T. (2006a). Local protein synthesis and spine morphogenesis: Fragile X syndrome and beyond. *Journal of Neuroscience*, *26*, 7151–7155.

Grossman, A. W., Elisseou, N. M., McKinney, B. C., & Greenough, W. T. (2006b). Hippocampal pyramidal cells in adult Fmr1 knockout mice exhibit an immature-appearing profile of dendritic spines. *Brain Research*, *1084*, 158–164.

Grossman, A. W., Aldridge, G. M., Lee, K. J., Zeman, M. K., Jun, C. S., Azam, H. S., et al. (2010). Developmental characteristics of dendritic spines in the dentate gyrus of Fmr1 knockout mice. *Brain Research*, *1355*, 221–227.

Gundersen, G. G., Gomes, E. R., & Wen, Y. (2004). Cortical control of microtubule stability and polarization. *Current Opinion in Cell Biology*, *16*, 106–112.

Gusev, P. A., Cui, C., Alkon, D. L., & Gubin, A. N. (2005). Topography of Arc/Arg3.1 mRNA expression in the dorsal and ventral hippocampus induced by recent and remote spatial memory recall: Dissociation of CA3 and CA1 activation. *Journal of Neuroscience*, *25*, 9384–9397.

Guy, J., Cheval, H., Selfridge, J., & Bird, A. (2011). The role of MeCP2 in the brain. *Annual Review of Cell and Developmental Biology*, *10*, 631–652.

Guzowski, J. F., Miyashita, T., Chawla, M. K., Sanderson, J., Maes, L. I., Houston, F. P., et al. (2006). Recent behavioral history modifies coupling between cell activity and Arc gene transcription in hippocampal CA1 neurons. *Proceedings of the National Academy of Sciences of the United States of America*, *103*, 1077–1082.

Hagerman, R. J. (2002). The physical and behavioural phenotype. In R. J. Hagerman & P. Hagerman (Eds.), *Fragile-X syndrome: Diagnosis, treatment and research* (pp. 3–109). Baltimore: The Johns Hopkins University Press.

Hayashi, M. L., Rao, B. S., Seo, J. S., Choi, H. S., Dolan, B. M., Choi, S. Y., et al. (2007). Inhibition of p21-activated kinase rescues symptoms of fragile X syndrome in mice. *Proceedings of the National Academy of Sciences of the United States of America*, *104*, 11489–11494.

Hinton, V. J., Brown, W. T., Wisniewski, K., & Rudelli, R. D. (1991). Analysis of neocortex in three males with the fragile X syndrome. *American Journal of Medical Genetics*, *41*, 289–294.

Hu, H., Qin, Y., Bochorishvili, G., Zhu, Y., van Aelst, L., & Zhu, J. J. (2008). Ras signaling mechanisms underlying impaired GluR1-dependent plasticity associated with fragile X syndrome. *Journal of Neuroscience*, *28*, 7847–7862.

Huber, K. M., Roder, J. C., & Bear, M. F. (2001). Chemical induction of mGluR5- and protein synthesis–dependent long- term depression in hippocampal area CA1. *Journal of Neurophysiology*, *86*, 321–325.

Irwin, S. A., Galvez, R., & Greenough, W. T. (2000). Dendritic spine structural anomalies in fragile-X mental retardation syndrome. *Cerebral Cortex*, *10*, 1038–1044.

Irwin, S. A., Patel, B., Idupulapati, M., Harris, J. B., Crisostomo, R. A., Larsen, B. P., et al. (2001). Abnormal dendritic spine characteristics in the temporal and visual cortices of patients with fragile-X syndrome: A quantitative examination. *American Journal of Medical Genetics*, *98*, 161–167.

Irwin, S. A., Idupulapati, M., Gilbert, M. E., Harris, J. B., Chakravarti, A. B., Rogers, E. J., et al. (2002). Dendritic spine and dendritic field characteristics of layer V pyramidal neurons in the visual cortex of fragile-X knockout mice. *American Journal of Medical Genetics*, *111*, 140–146.

Kandel, E. R. (2001). The molecular biology of memory storage: A dialogue between genes and synapses. *Science*, *294*, 1030–1038.

Kasai, H., Fukuda, M., Watanabe, S., Hayashi-Takagi, A., & Noguchi, J. (2010). Structural dynamics of dendritic spines in memory and cognition. *Trends in Neurosciences*, *33*, 121–129.

Kiebler, M. A., & Bassell, G. J. (2006). Neuronal RNA granules: Movers and makers. *Neuron*, *51*, 685–690.

Kishi, N., & Macklis, J. D. (2004). MECP2 is progressively expressed in post-migratory neurons and is involved in neuronal maturation rather than cell fate decisions. *Molecular and Cellular Neuroscience*, *27*, 306–321.

Kuhn, D. E., Nuovo, G. J., Terry, A. V., Jr., Martin, M. M., Malana, G. E., Sansom, S. E., et al. (2010). Chromosome 21-derived microRNAs provide an etiological basis for aberrant protein expression in human Down syndrome brains. *Journal of Biological Chemistry*, *285*, 1529–1543.

Kurt, M. A., Davies, D. C., Kidd, M., Dierssen, M., & Flórez, J. (2000). Synaptic deficit in the temporal cortex of partial trisomy 16 (Ts65Dn) mice. *Brain Research*, *858*, 191–197.

Kurt, M. A., Kafa, M. I., Dierssen, M., & Davies, D. C. (2004). Deficits of neuronal density in CA1 and synaptic density in the dentate gyrus, CA3 and CA1, in a mouse model of Down syndrome. *Brain Research*, *1022*, 101–109.

Levenga, J., de Vrij, F. M., Oostra, B. A., & Willemsen, R. (2010). Potential therapeutic interventions for fragile X syndrome. *Trends in Molecular Medicine*, *16*, 516–527.

Levenga, J., de Vrij, F. M., Buijsen, R. A., Li, T., Nieuwenhuizen, I. M., Pop, A., et al. (2011a). Subregion-specific dendritic spine abnormalities in the hippocampus of Fmr1 KO mice. *Neurobiology of Learning and Memory*, *95*, 467–472.

Levenga, J., Hayashi, S., de Vrij, F. M., Koekkoek, S. K., van der Linde, H. C., Nieuwenhuizen, I., et al. (2011b). AFQ056, a new mGluR5 antagonist for treatment of fragile X syndrome. *Neurobiology of Disease*, *42*, 311–317.

Li, Z., Yu, T., Morishima, M., Pao, A., LaDuca, J., Conroy, J., et al. (2007). Duplication of the entire 22.9 Mb human chromosome 21 syntenic region on mouse chromosome 16 causes cardiovascular and gastrointestinal abnormalities. *Human Molecular Genetics*, *16*, 1359–1366.

Lu, R., Wang, H., Liang, Z., Ku, L., O'Donnell, W. T., Li, W., et al. (2004). The fragile X protein controls microtubule-associated protein 1B translation and microtubule stability in brain neuron development. *Proceedings of the National Academy of Sciences of the United States of America*, *101*, 15201–15206.

Malenka, R. C., & Bear, M. F. (2004). LTP and LTD: An embarrassment of riches. *Neuron*, *44*, 5–21.

Matsuzaki, M., Honkura, N., Ellis-Davies, G. C., & Kasai, H. (2004). Structural basis of long-term potentiation in single dendritic spines. *Nature*, *429*, 761–766.

McKinney, B. C., Grossman, A. W., Elisseou, N. M., & Greenough, W. T. (2005). Dendritic spine abnormalities in the occipital cortex of C57BL/6 Fmr1 knockout mice. *American Journal of Medical Genetics. Part B, Neuropsychiatric Genetics*, *136B*, 98–102.

Newey, S. E., Velamoor, V., Govek, E. E., & Van Aelst, L. (2005). Rho GTPases, dendritic structure, and mental retardation. *Journal of Neurobiology*, *64*, 58–74.

Nimchinsky, E. A., Oberlander, A. M., & Svoboda, K. (2001). Abnormal development of dendritic spines in fmr1 knock-out mice. *Journal of Neuroscience*, *21*, 5139–5146.

O'Doherty, A., Ruf, S., Mulligan, C., Hildreth, V., Errington, M. L., Cooke, S., et al. (2005). An aneuploid mouse strain carrying human chromosome 21 with Down syndrome phenotypes. *Science*, *309*, 2033–2037.

Oostra, B., & Nelson, D. L. (2006). Animal models of fragile X syndrome: Mice and flies. In R. D. Wells & T. Ashizawa (Eds.), *Genetic instabilities and neurological diseases* (pp. 175–194). Amsterdam: Elsevier.

Park, C. S., & Tang, S. J. (2009). Regulation of microRNA expression by induction of bidirectional synaptic plasticity. *Journal of molecular neuroscience*, *38*, 50–56.

Penzes, P., Cahill, M. E., Jones, K. A., VanLeeuwen, J. E., & Woolfrey, K. M. (2011). Dendritic spine pathology in neuropsychiatric disorders. *Nature Neuroscience*, *14*, 285–293.

Pinter, J. D., Eliez, S., Schmitt, J. E., Capone, G. T., & Reiss, A. L. (2001). Neuroanatomy of Down's syndrome: A high-resolution MRI study. *The American Journal of Psychiatry*, *158*, 1659–1665.

Purpura, R. P. (1974). Dendritic spine dysgenesis and mental retardation. *Science*, *186*, 1126–1128.

Ramakers, G. J. (2002). Rho proteins, mental retardation and the cellular basis of cognition. *Trends in Neurosciences*, *25*, 191–199.

Ramakrishna, N., Meeker, C., Li, S., Jenkins, E. C., Currie, J. R., Flory, M., et al. (2005). Polymerase chain reaction method to identify Down syndrome model segmentally trisomic mice. *Analytical Biochemistry*, *340*, 213–219.

Reeve, S. P., Bassetto, L., Genova, G. K., Kleyner, Y., Leyssen, M., Jackson, F. R., et al. (2005). The Drosophila fragile x mental retardation protein controls actin dynamics by directly regulating profilin in the brain. *Current Biology*, *15*, 1156–1163.

Ropers, H. H. (2006). X-linked mental retardation: Many genes for a complex disorder. *Current Opinion in Genetics and Development*, *16*, 260–269.

Saran, N. G., Pletcher, M. T., Natale, J. E., Cheng, Y., & Reeves, R. H. (2003). Global disruption of the cerebellar transcriptome in a Down syndrome mouse model. *Human Molecular Genetics*, *12*, 2013–2019.

Schalock, R. L., Luckasson, R. A., Shogren, K. A., Borthwick-Duffy, S., Bradley, V., Buntinx, W. H., et al. (2007). The renaming of mental retardation: Understanding the change to the term intellectual disability. *Intellectual and Developmental Disabilities*, *45*, 116–124.

Schenck, A., Bardoni, B., Langmann, C., Harden, N., Mandel, J. L., & Giangrande, A. (2003). CYFIP/Sra-1 controls neuronal connectivity in Drosophila and links the Rac1 GTPase pathway to the fragile X protein. *Neuron*, *38*, 887–898.

Schuett, J., Falley, K., Richter, D., Kreienkamp, H. J., & Kindler, S. (2009). Fragile X mental retardation protein regulates the levels of scaffold proteins and glutamate receptors in postsynaptic densities. *Journal of Biological Chemistry*, *284*, 25479–25487.

Schule, B., Armstrong, D. D., Vogel, H., Oviedo, A., & Francke, U. (2008). Severe congenital encephalopathy caused by MECP2 null mutations in males: Central hypoxia and reduced neuronal dendritic structure. *Clinical Genetics*, *74*, 116–126.

Schuman, E. M., Dynes, J. L., & Steward, O. (2006). Synaptic regulation of translation of dendritic mRNAs. *Journal of Neuroscience*, *26*, 7143–7146.

Shepherd, G. M., & Katz, D. M. (2011). Synaptic microcircuit dysfunction in genetic models of neurodevelopmental disorders: Focus on Mecp2 and Met. *Current Opinion in Neurobiology*, *21*, 827–833.

Siarey, R. J., Stoll, J., Rapoport, S. I., & Galdzicki, Z. (1997). Altered long-term potentiation in the young and old Ts65Dn mouse, a model for Down Syndrome. *Neuropharmacology*, *36*, 1549–1554.

Siarey, R. J., Carlson, E. J., Epstein, C. J., Balbo, A., Rapoport, S. I., & Galdzicki, Z. (1999). Increased synaptic depression in the Ts65Dn mouse, a model for mental retardation in Down syndrome. *Neuropharmacology*, *38*, 1917–1920.

Siarey, R. J., Villar, A. J., Epstein, C. J., & Galdzicki, Z. (2005). Abnormal synaptic plasticity in the Ts1Cje segmental trisomy 16 mouse model of Down syndrome. *Neuropharmacology*, *49*, 122–128.

Smrt, R. D., Eaves-Egenes, J., Barkho, B. Z., Santistevan, N. J., Zhao, C., Aimone, J. B., et al. (2007). Mecp2 deficiency leads to delayed maturation and altered gene expression in hippocampal neurons. *Neurobiology of Disease*, *27*, 77–89.

Steward, O. (2002). mRNA at synapses, synaptic plasticity, and memory consolidation. *Neuron*, *36*, 338–340.

Su, T., Fan, H. X., Jiang, T., Sun, W. W., Den, W. Y., Gao, M. M., et al. (2010). Early continuous inhibition of group 1 mGlu signaling partially rescues dendritic spine abnormalities in the Fmr1 knockout mouse model for fragile X syndrome. *Psychopharmacology*, *215*, 291–300.

Sultan, M., Piccini, I., Balzereit, D., Herwig, R., Saran, N. G., Lehrach, H., et al. (2007). Gene expression variation in Down's syndrome mice allows prioritization of candidate genes. *Genome Biology*, *8*, R91.

van Bokhoven, H., & Kramer, J. M. (2010). Disruption of the epigenetic code: An emerging mechanism in mental retardation. *Neurobiology of Disease*, *39*, 3–12.

Vanderklish, P. W., & Edelman, G. M. (2002). Dendritic spines elongate after stimulation of group 1 metabotropic glutamate receptors in cultured hippocampal neurons. *Proceedings of the National Academy of Sciences of the United States of America*, *99*, 1639–1644.

Vissers, L. E., de Ligt, J., Gilissen, C., Janssen, I., Steehouwer, M., de Vries, P., et al. (2010). A *de novo* paradigm for mental retardation. *Nature Genetics*, *42*, 1109–1112.

von Bohlen Und Halbach, O. (2009). Structure and function of dendritic spines within the hippocampus. *Annals of Anatomy*, *191*, 518–531.

Waung, M. W., Pfeiffer, B. E., Nosyreva, E. D., Ronesi, J. A., & Huber, K. M. (2008). Rapid translation of Arc/Arg3.1 selectively mediates mGluR-dependent LTD through persistent increases in AMPAR endocytosis rate. *Neuron*, *59*, 84–97.

Willemsen, R., Oostra, B. A., Bassell, G. J., & Dictenberg, J. (2004). The fragile X syndrome: From molecular genetics to neurobiology. *Mental Retardation and Developmental Disabilities Research Reviews*, *10*, 60–67.

Willemsen, R., Levenga, J., & Oostra, B. (2011). CGG repeat in the FMR1 gene: Size matters. *Clinical Genetics*, *80*, 214–225.

Wisniewski, K. E. (1990). Down syndrome children often have brain with maturation delay, retardation of growth, and cortical dysgenesis. *American Journal of Medical Genetics. Supplement*, *7*, 274–281.

Yang, G., Pan, F., & Gan, W. B. (2009). Stably maintained dendritic spines are associated with lifelong memories. *Nature*, *462*, 920–924.

Zalfa, F., Giorgi, M., Primerano, B., Moro, A., Di Penta, A., Reis, S., et al. (2003). The fragile X syndrome protein FMRP associates with BC1 RNA and regulates the translation of specific mRNAs at synapses. *Cell*, *112*, 317–327.

Zhang, Y., Venkitaramani, D. V., Gladding, C. M., Kurup, P., Molnar, E., Collingridge, G. L., et al. (2008). The tyrosine phosphatase STEP mediates AMPA receptor endocytosis after metabotropic glutamate receptor stimulation. *Journal of Neuroscience*, *28*, 10561–10566.

M. Dierssen and R. de la Torre (Eds.)
Progress in Brain Research, Vol. 197
ISSN: 0079-6123

CHAPTER 9

The *in vivo* Down syndrome genomic library in mouse

Yann Herault[†,‡,§,*], Arnaud Duchon[†], Emilie Velot[†], Damien Maréchal[†] and Véronique Brault[†]

[†] *Institut de Génétique et de Biologie Moléculaire et Cellulaire, Translational medicine and Neurogenetics program, IGBMC, CNRS, INSERM, Université de Strasbourg, UMR7104, UMR964, Illkirch, Strasbourg, France*
[‡] *Transgenèse et Archivage Animaux Modèles, TAAM, CNRS, UPS44, Orléans, France*
[§] *Institut Clinique de la Souris, ICS, GIE CERBM, Illkirch, Strasbourg, France*

Abstract: Mouse models are key elements to better understand the genotype–phenotype relationship and the physiopathology of Down syndrome (DS). Even though the mouse will never recapitulate the whole spectrum of intellectual disabilities observed in the DS, mouse models have been developed over the recent decades and have been used extensively to identify homologous genes or entire regions homologous to the human chromosome 21 (Hsa21) that are necessary or sufficient to induce DS cognitive features. In this chapter, we review the principal mouse DS models which have been selected and engineered over the years either for large genomic regions or for a few or a single gene of interest. Their analyses highlight the complexity of the genetic interactions that are involved in DS cognitive phenotypes and also strengthen the hypothesis on the multigenic nature of DS. This review also addresses future research challenges relative to the making of new models and their combination to go further in the characterization of candidates and modifier of the DS features.

Keywords: locomotion; learning; memory; trisomy 21; genetic dosage.

Introduction

In recent years, a myriad of Down syndrome (DS) mouse models have been generated and analyzed for several reasons. All the models identified or generated so far were isolated from a genetic basis. There are many advantages using the mouse as a model organism to study DS. First, inbred strains of mice are available with individuals sharing more than 98% of identical genetic loci. Thus, the reproducibility of results from one individual to another is improved, and

*Corresponding author.
Tel.: +33-388-655657; Fax: +33-388-655690
E-mail: herault@igbmc.fr

DOI: 10.1016/B978-0-444-54299-1.00009-1

phenotypic assessment can be on a large number of genetically identical individuals. The genetic composition is well defined with stable biological data, provided that environmental conditions are strictly controlled (breeding condition, food, day–night cycle, etc.). The efforts put into the sequencing of the mouse genome have greatly accelerated the human–mouse comparative research. Using a C57BL/6J female, a comparative analysis was made between the 2.5-Gb mouse genome and 2.9-Gb of the human genome, covering it almost entirely. Over 90% of genes are in regions of synteny between the two species (Venter et al., 2001), and almost 99% of human genes have mouse equivalents (Tecott, 2003). In addition, using different inbred strains of mice, one can really assess the contribution of the genetic environment and of the genetic background to the variability and penetrance of the DS phenotypes. Thus, the impact of trisomy can vary in different genetic contexts and has been proposed to explain the individual variability in this syndrome (Deitz and Roper, 2011). The long arm of human chromosome 21 contains 33.7Mb and approximately 291 annotated genes in hg19.refGene (Pruitt et al., 2003) of which 181 have a counterpart in the mouse genome according to the Mammalian Orthology section of the Mouse Genome Informatics (The Jackson Laboratory, Bar Harbor, Maine). These genes are found in three regions of synteny located on three different chromosomes in the mouse, with, from the centromere to the telomere of Hsa21, the largest region found on Mmu16 (about 37Mb or 119 orthologous genes between *Lipi* and *Zfp295*), then a small region on Mmu17 (1.1Mb or 20 genes between *Umodl1* and *Hsf2bp*), and finally a region on chromosome Mmu10 (2.3Mb with 42 genes between *Cstb* and *Prmt2*) (Dierssen et al., 2009, see also Chapters 2 and 5). Based on this well-preserved homology, many DS mouse models have been developed, taking into consideration that, even if the expressivity of phenotypes may be species dependent, the genetic processes disturbed by dose effects should be kept between species. The following sections of this chapter will focus on the different mouse models for DS studies. As shown in Table 1, we chose first to categorize the models with one class carrying a freely segregating extra-chromosome, and a second class carrying a supernumerary copy as tandem duplication or in combination. Then additional models with overexpression of candidate genes are introduced and detailed (Table 2). Principal phenotypes observed in mouse models bearing segmental trisomies are reported in Table 3.

Models with an extra-chromosome

Only a few mouse models carry an additional freely segregating chromosome as observed in most of the case of DS patients in which three copies of Human chromosome 21 is found. Besides the increase in gene dosage, the presence of an additional chromosome can have additional consequences, disrupting the nuclear organization or the segregation of chromosome during meiosis. Shapiro has also stressed that the presence of additional chromosome could partly perturb the overall control of gene expression leading to common phenotypes observed in different trisomies. Thus, these models carrying a whole extra-chromosome are not only interesting for defining useful candidates for therapies, but they might also contribute to stress the importance of chromosomal region in inducing specific phenotypes and also to understand further pathogenic mechanisms.

The best known and most widely used mouse model in the study of DS is the Ts(17^{16})65Dn (Ts65Dn; Fig. 1). The model was selected after X-ray induced irradiation of mouse stocks which led to a reciprocal translocation between the telomeric end of Mmu16 and the centromeric part of Mmu17 (Davisson et al., 1993). Thus, this mouse model has a partial trisomy of Mmu16 that extends from upstream of *Mrpl39* to *Zfp295*, that is, 122 genes, including the DS critical region (DSCR), but also 5% of nonhomologous genes to Hsa21 belonging to Mmu17. The breakpoint has been mapped more precisely recently and

Table 1. Description of the most studied mouse models for Down syndrome, including list of all the common denominations, the origins and details of the region targeted in each specific model

Symbol	Synonym	MGI id	Official nomenclature	Origin	Genetic interval	Size (Mb) [a]	Number of genes [b]
Tc1	Tc(Hsa21)91-1Emcf, Tc1	3814712	Tc(Hsa21)1TybEmcf	Transchromosome	All Hsa 21[c]	48.1	
Ts16				Robertsonian translocation	All Mmu16	98.3	679
Ts65Dn	Ts(16C3–4;17A2)65Dn	3512067	Ts(17^{16})65Dn	Segmental translocation	*Mrpl39–Zfp295*	13.5	122
Ts1Cje	Ts(12^{16})1Cje, Ts108Cje	3623022	Ts(16C-tel)1Cje	Segmental translocation	*Sod1–Mx1* [d]	7.5	67
Ts2Cje	Ts(Rb(12.17^{16}))2Cje	3531161	Rb(12.$Ts17^{16}$65Dn)2Cje	Segmental translocation	*Mrpl39–Zfp295*	13.5	122
Dp(16) Yey	Dp(16)1Yey, Dp(16) 1Yu, Dp1Yu, Ts1Yey	3714529	Dp(16Lipi-Zfp295)1Yey	Segmental duplication	*Lipi–Zfp295*	22.6	140
Dp(10) Yey	Ts2Yey	4461397	Dp(10Prmt2-Pdxk)1Yey	Segmental duplication	*Prmt2–Pdxk*	2.3	53
Dp(17) Yey	Ts3Yey	4461398	Dp(17Abcg1-Rrp1b) 1Yey	Segmental duplication	*Abcg1–Rrp1b*	1.0	18
Ts1Yah	Dp(17)1Yah	4367905	Dp(17Abcg1-Cbs)1Yah	Segmental duplication	*Abcg1 – U2af1*	0.6	13
Ts43H	Ts(16;17)43H			Translocation	*Umodl1-Cbs*	30	<310
Ts1Rhr		3487283	Dp(16Cbr1-ORF9)1Rhr	Segmental duplication	*Cbr1–Orf9*	4.1	31
Ms1Rhr		3487281	Del(16Cbr1-ORF9)1Rhr	Segmental deletion	*Cbr1–Orf9*	4.1	31
Df(17) 1Yey	Del(17)1Yey, Df (Abcg1-Rrp1b)1Yey	4818361	Del(17Abcg1-Rrp1b) 1Yey	Segmental deletion	*Abcg1–Rrp1b*	1.0	18
Df(10) 1Yey	Del(10)1Yey, Df (Prmt2-Pdxk)1Yey	4818358	Del(10Prmt2-Pdxk)1Yey	Segmental deletion	*Prmt2–Pdxk*	2.3	53
Ms1Yah		4868462	Del(10Prmt2-Col6a1) 3Yah	Segmental deletion	*Prmt2–Col6a1*	0.5	12
Ms4Yah		4868385	Del(10Prmt2-Cstb)4Yah	Segmental deletion	*Prmt2–Cstb*	2.2	52

[a]The size was calculated by difference between gene interval coordinate (NCBI37/mm9).
[b]Number of gene was known, mouse protein-coding and nonprotein-coding genes taken from the NCBI RNA reference sequences collection (RefSeq).
[c]This mouse line contained several microdeletion and duplication on Hsa21 transchromosome.
[d]In this line, Sod1 gene was knockout.

Table 2. Mouse models for candidate genes

Symbol	Name	Synonym	MGI id	Reference
Tg(SOD1)10Cje	Transgene insertion 10, Charles J Epstein	218/10, TgHS-218/10	3053950	Epstein et al. (1987)
Tg(SOD1)2Gur	Transgene insertion 2, Mark E Gurney	N1029, N29, SOD1WT, tg-SOD1, WT SOD1,	2448770	Gurney et al. (1994)
Tg(ACTB-SOD3)1Tido	Transgene insertion 1, Tim D Oury	EC-SOD	3773992	Oury et al. (1992)
Tg(huS100B)5Daoh	Transgene insertion 5, Alexander Marks	S-100beta, S100B TG, TghuS100B	2386685	Friend et al. (1992a)
Tg(Sh100b)				Reeves et al. (1994)
TgSim2				Ema et al. (1999)
TgN(*Sim2*)Sagu				Chrast et al. (2000)
Tg(APP751)	Tg(Eno2-APP*751) 10Cord	NSE: beta-APP75	3784388	Quon et al. (1991)
Tg(AppYAC)	Tg(APP)8.9Btla	APPYAC, hABetaPP, hAPP-YAC, Py8.9, Tg(APP) C9.24Jgh	2447591	Lamb et al. (1993)
Tg(Pfkl)224Yg	Transgene insertion 224, Yorum Groner	PFKL minigene	3789607	Elson et al. (1994)
Tg Ets2				Sumarsono et al. (1996)
Tg(DYRK1A)36Wjs	Transgene insertion 36, Woo-Joo Song	DYRK1A BAC TG	4867514	Ahn et al. (2006)
Tg(MT1A-Dyrk1a)Xest	Transgene insertion, Xavier Estivill	TgDyrk1A	3814576	Altafaj et al. (2001)
Tg(RCAN1)M9Mapr	Transgene insertion M9, Melanie A Pritchard	RCAN1ox, Tg(RCAN1) 1Mapr	3809545	Keating et al. (2008)
TgYAC152F7		*DSCRC, TTC3, DSCR, DYRK1A*		Adams et al. (2004), Branchi et al. (2004b), Chabert et al. (2004), Sebrie et al. (2008), Smith et al. (1995, 1997b)
TgYAC230E8		*SETD4, CBR1, CBR3, DOPEY2, MORC3, CHAF1B, CLDN14*		Smith and Rubin (1997), Smith et al. (1995)
TgPCP4				Mouton-Liger et al. (2011)

clearly has shown the presence of an additional trisomy for 60 genes located on the Mmu17 centromeric region (Duchon et al., 2011b; Reinholdt et al., 2011). The first phenotype observed in this model is the increase of postnatal lethality that affects 17% of the trisomic newborns due to cardiovascular defects similar to DS patients (Moore, 2006). The Ts65Dn animals who survive display phenotypes that are comparable to those observed in patients with early developmental delay (Holtzman et al., 1996), a reduced weight and a small size, (Blazek et al., 2010) and facial anomalies (Hill et al., 2007; Olson et al., 2007).

A large number of studies have described deficits in different kinds of memories in Ts65Dn mice. For example, the spatial working memory is affected in several alternation tests (Belichenko et al., 2007; Fernandez et al., 2007), like the Y-maze (Chang and Gold, 2008; Faizi et al., 2011) or the radial maze (Bimonte-Nelson et al., 2003; Demas et al., 1998). Spatial reference memory assessed in the Morris water maze was also

Table 3. Main behavioral and cognitive disabilities found in DS mouse models

Mice	Main phenotypes observed	References
Tc1	*Behavioral tasks affected*	
	Locomotor deficits	O'Doherty et al. (2005)
	Hyperactivity	Galante et al. (2009a)
	Spatial working memory (MWM)	Morice et al. (2008)
	Explicit declarative memory (NOR; short-term)	Morice et al. (2008)
	Reduced LTP (hippocampal DG)	Morice et al. (2008)
	Structural alterations	
	Decreased size of the cerebellum and of the granule cell density	O'Doherty et al. (2005)
	Cerebellar circuits alteration	Galante et al. (2009a)
Ts16	*Behavioral tasks affected*	
	No behavioral test published	
	Structural alterations	
	Reduced brain weight	Haydar et al. (1996, 2000)
	Transient delay in the radial expansion of the cortical wall and persistent reduction in cortical volume of the neocortex	Haydar et al. (1996)
	Decreased number of neuron precursors in the neocortex	Haydar et al. (2000)
Ts65Dn	*Behavioral tasks affected*	
	Explicit declarative memory (NOR)	Fernandez and Garner (2008), Fernandez et al. (2007)
	Short-term working memory (spontaneous alternation)	Belichenko et al. (2007a), Bimonte-Nelson et al. (2003), Chang and Gold (2008), Demas et al. (1996), Faizi et al. (2011), Fernandez et al. (2007)
	Reference memory (MWM hidden platform test); spatial learning (MWM probe trial); cognitive flexibility (MWM reverse platform test)	Escorihuela et al. (1998), Martinez-Cue et al. (2005), Olson et al. (2007b), Reeves et al. (1995), Rueda et al. (2010), Sago et al. (2000), Stasko and Costa (2004)
	Contextual memory (fear conditioning)	Costa et al. (2008), Salehi et al. (2009)
	Hyperactivity	Coussons-Read and Crnic (1996), Faizi et al. (2011), Stewart et al. (2007)
	Decreased LTP (hippocampal CA1 and DG)	Kleschevnikov et al. (2004a), Siarey et al. (1997)
	Increased LTD (hippocampal CA1)	Siarey et al. (1999)
	Structural alterations	
	Reduced brain volume with delayed expansion of the neocortex (embryonic period)	Aldridge et al. (2007), Chakrabarti et al. (2007), Holtzman et al. (1996)
	Decreased cell density in the cortex, specifically in layer IV-VI present postnatally	Chakrabarti et al. (2007)
	Reduction of the volume of the ventricles due to reduced neural precursors proliferation in the hippocampal and cortical ventricular zones	Bianchi et al. (2010)
	Reduced thickness of the hippocampal wall at embryonic stage	Chakrabarti et al. (2007)
	Decreased neurogenesis in hippocampus	Clark et al. (2006)
	Decreased number of granule cells in the dentate gyrus at postnatal life stage	Bianchi et al. (2010), Contestabile et al. (2007), Insausti et al. (1998), Lorenzi and Reeves (2006)
	Deduction of the number, increase in size, and structural alteration of excitatory synapses in the hippocampus and temporal cortex	Belichenko et al. (2004), Kurt et al. (2000, 2004), Popov et al. (2011)

(*Continued*)

Table 3. Main behavioral and cognitive disabilities found in DS mouse models (*Continued*)

Mice	Main phenotypes observed	References
	Hypertrophy and increased number of the astrocytes in the hippocampus	Holtzman et al. (1996)
	Cerebellum hypoplasia with decreased density of granule and Purkinje cells	Aldridge et al. (2007), Baxter et al. (2000b), Moldrich et al. (2007), Roper et al. (2006b)
	Degeneration of basal forebrain cholinergic neurons	Holtzman et al. (1996)
	Increased number of inhibitory neurons	Chakrabarti et al. (2007, 2010)
Ts1Cje	*Behavioral tasks affected*	
	Short-term working memory (spontaneous alternation)	Belichenko et al. (2007a), Sago et al. (1998)
	Reference memory (MWM hidden platform test); spatial learning (MWM probe trial); cognitive flexibility (MWM reverse platform test)	Sago et al. (1998)
	Hypoactivity	Sago et al. (1998)
	Decreased LTP (hippocampal CA1 and DG)	Belichenko et al. (2007a), Siarey et al. (2005)
	Increased LTD (hippocampal CA1)	Siarey et al. (2005)
	Structural alterations	
	Reduced brain volume (embryonic period)	Ishihara et al. (2010)
	Enlargement of ventricles' size	Ishihara et al. (2010)
	Decreased neurogenesis	Hewitt et al. (2010), Ishihara et al. (2010)
	Decreased hippocampal spine density	Belichenko et al. (2007, 2009a)
	Reduced cerebellar volume with decreased granule cell density	Olson et al. (2004b)
	Increased number of astrocytes	Hewitt et al. (2010)
Ts2Cje	*Behavioral tasks affected*	
	ND	
	Structural alterations	
	Reduced brain volume (embryonic period); enlarged ventricles; decreased neurogenesis	Ishihara et al. (2010)
Ts1Rhr	*Behavioral tasks affected*	
	Explicit declarative memory (NOR; long-term)	Belichenko et al. (2009a)
	Short-term working memory (spontaneous alternation)	Belichenko et al. (2009a)
	Decreased LTP (DG)	Belichenko et al. (2009a)
	Structural alterations	
	Larger brain weight with increased thickness of the motor cortex and a greater volume of the posterior hippocampus (7.5months)	Belichenko et al. (2009a)
	Decreased hippocampal spine density	Belichenko et al. (2007, 2009a)
	Reduced volume of the cerebrum and cerebellum (4 months)	Belichenko et al. (2009a)
Dp(16)Yey (Ts1Yey)	*Behavioral tasks affected*	
	Reference memory (MWM hidden platform test)	Yu et al. (2010c)
	Contextual memory (fear conditioning)	Yu et al. (2010c)
	Decreased LTP (hippocampal CA1)	Yu et al. (2010c)
	Structural alterations	
	No report	
Dp(10)Yey (Ts2Yey)	No phenotype sofar reported	
Dp(17)Yey (Ts3Yey)	*Behavioral tasks affected*	
	Increased LTP (hippocampal CA1)	Yu et al. (2010c)
	Structural alterations	
	No report	

Table 3. Main behavioral and cognitive disabilities found in DS mouse models (*Continued*)

Mice	Main phenotypes observed	References
Ts1Yah	*Behavioral tasks affected*	
	Explicit declarative memory (NOR)	Lopes Pereira et al. (2009)
	Short-term working memory (spontaneous alternation)	Lopes Pereira et al. (2009)
	Reference memory (MWM hidden platform test) (better performance)	Lopes Pereira et al. (2009)
	Increased LTP (hippocampal CA1)	Lopes Pereira et al. (2009)
	Structural alterations	
	No report	
Ts43H	*Behavioral tasks affected*	
	Spatial learning deficits (MWM)	Vacik et al. (2005a)
	Structural alterations	
	No report	
Ts1Yey;	*Behavioral tasks affected*	
Ts2Yey;	Reference memory (MWM hidden platform test)	Yu et al. (2010d)
Ts3Yey	Contextual memory (fear conditioning)	Yu et al. (2010d)
	Structural alterations	
	No report	

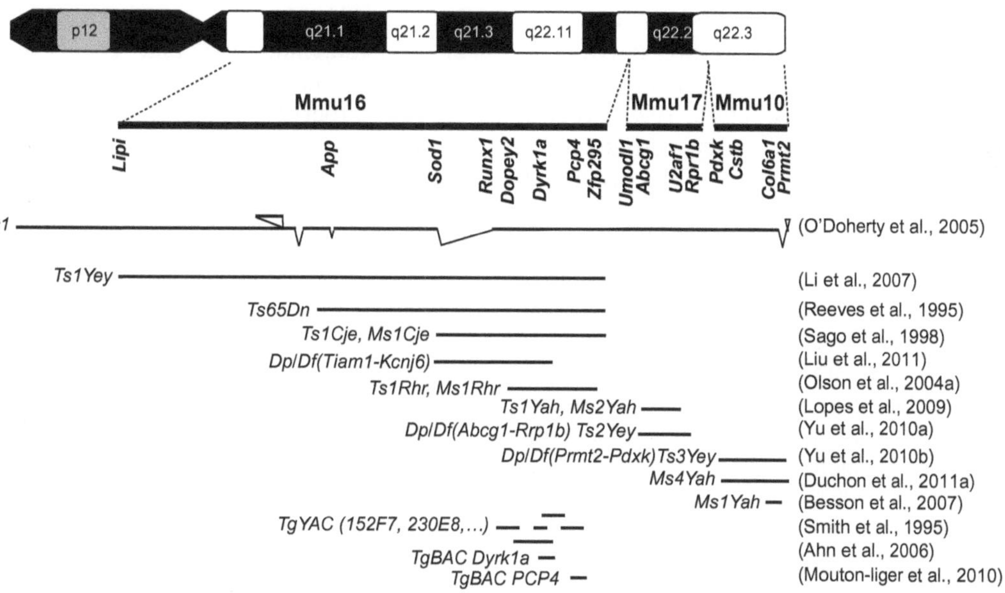

Fig. 1. Schematic representation of the localization of trisomic (Ts) or monosomic (Ms) regions in the different mouse models of Down syndrome. The chromosome Hsa21 freely segregating in the Tc1 is indicated with small internal deletions or duplications. The size of the interval is indicated with respect to the position on the Mmu16, 17, or 10.

affected to different degrees depending on the protocols used (Escorihuela et al., 1998; Olson et al., 2007; Reeves et al., 1995; Rueda et al., 2010; Sago et al., 2000; Stasko and Costa, 2004). But these animals also show deficits in explicit memory in the novel object-recognition test

(Fernandez and Garner, 2008; Fernandez et al., 2007) and in context-discrimination learning (Hyde et al., 2001). They have a limited ability to acquire an operant learning conditioning (Wenger et al., 2004) and have reduced sensitivity to pain reflecting a reduced nociception or an altered nociceptive information processing (Martinez-Cue et al., 1999). Trisomic mice are hyperactive and less anxious than control (Coussons-Read and Crnic, 1996) even in a familiar environment (Faizi et al., 2011). Ts65Dn are able to learn with similar performance as controls in simple tasks when they are not stressed, which is no longer the case when stress is induced (Faizi et al., 2011) or if their environment is enriched (Dierssen et al., 2003; Martinez-Cue et al., 2002). A study of the sociability of these animals showed no difference in social interactions, but a deficit in social memory, which depends on the length of the trial, the inter trial time, the context, and the age of the mice (Faizi et al., 2011). At morphological level, Ts65Dn mice exhibit a deficit in the number of neurons in different brain structures (Insausti et al., 1998; Lorenzi and Reeves, 2006). A study using MRI experiments showed a reduction in the brain and cerebellar volumes in Ts65Dn individuals compared to euploid controls (Baxter et al., 2000), despite an unaltered form (Aldridge et al., 2007). The density of cerebellar granule and Purkinje cells was significantly lower in Ts65Dn mice, a phenotype that was subsequently observed in the cerebellum of DS patients. This phenotype is due to a weaker response of granule cell precursors to the Sonic Hedgehog (Shh) mitotic signal (Moldrich et al., 2007; Roper et al., 2006). It appears that neural precursors exhibit an attenuated response to the Shh pathway due to overexpression of the inhibitor Ptch1 (Protein patched homolog 1). This overexpression is linked to an increase of the intracellular peptide domain of APP (Amyloid Precursor Protein intracellular domain peptide) derived from the cleavage of APP by γ-secretase. This finding provides a link between the *APP* trisomic gene and DS neurogenesis deficits (Trazzi et al., 2011). A degeneration of cholinergic neurons in the brain of Ts65Dn mice is also observed from the age of 4 months. This feature is found in patients with Alzheimer's disease (AD) (Granholm et al., 2000). The degeneration of the cholinergic system and the elevation of the amount of App protein in the hippocampus are involved in neurological deficits of these mice (Seo and Isacson, 2005). Overexpression of the *APP* gene is also linked to increased susceptibility of DS patients to epilepsy and this through mGluR5 (metabotropic glutamate receptor 5) (Westmark et al., 2010). But triplication in the Ts65Dn model seems also to have opposite effects, including an increased number of interneurons in the somatosensory cortex (Perez-Cremades et al., 2010) and the forebrain, indicating that the reduction of cell proliferation is not spread out in all brain structures. This increase of interneurons number alters hippocampal circuitry by altering the balance of excitation/inhibition and could be due to overexpression of the transcription factors Olig1 (oligodendrocyte lineage transcription factor 1) and Olig2 (Chakrabarti et al., 2010). It has also been shown by *in vitro* stimulation of CA1 synapses in the hippocampus that adult Ts65Dn mice exhibit decreased long-term potentiation (LTP see Chapters 8 and 10). This lack of LTP is associated with an increase in GABAergic inhibition or other inhibitory circuit changes in the hippocampus (Costa and Grybko, 2005; Kleschevnikov et al., 2004b), such as the reduction in number of excitatory synapses by neurons not only in the hippocampus but also in the temporal cortex (Kurt et al., 2000, 2004), an increase of the size of the latter (Belichenko et al., 2004), and alterations in their ultrastructure (Popov et al., 2011). Changes in both inhibitory and synaptic proteins have been observed in the composition of excitatory and inhibitory receptors such as $GABA_A$ receptor (Belichenko et al., 2009b). Based on these observations, a series of therapeutic approaches have been tested on mice. In particular, chronic treatments with picrotoxin or pentylenetetrazol (PTZ) reduce γ-aminobutyric acid (GABA) inhibition and restore the LTP and hippocampus-dependent learning, even after cessation of treatment (Fernandez et al., 2007; Rueda et al., 2008). Targeting the same pathway, Braudeau et al. (2011) used a GABA inverse agonist, specific for the α5 subunit of the GABA receptor, to rescue the intellectual disabilities in the Ts65Dn and limit the risk of epilepsy induced by a PTZ treatment. Alternatively, the use of memantine, a noncompetitive antagonist of NMDA (*N*-methyl-D-aspartate) receptor also showed facilitation of learning in these mice (Costa et al., 2008; Rueda et al., 2010). A trial of early treatment with fluoxetine, an antidepressant that inhibits reuptake of serotonin, enhanced cell proliferation. Ultimately, a test of fear conditioning, hippocampus dependent, showed a restoration of memory in treated subjects (Bianchi et al., 2010). Similarly, supplementation of the maternal diet with excess of choline during gestation and lactation improved cognitive function (Moon et al., 2010). A trial to reduce β-amyloid by injection of an α-secretase inhibitor in Ts65Dn mice also resulted in a positive effect on the mice performance in the Morris water maze test, showing the importance of developing treatment seeking to lower levels of β-amyloid (Netzer et al., 2010). Finally, a trial of dual therapy was undertaken with two peptides, NAPVSIPQ and SALLRSIPA, that mimic the action of the proteins: that mimic the action of the activity-dependent neuroprotective protein and the activity-dependent neurotrophic factor.

These two peptides are derived from glial proteins and have shown *in vitro* to give a natural neuroprotection against oxidative damage in cortical neurons of DS patients (Busciglio et al., 2007). This treatment resulted in an increase of NR2B (NMDA receptor 2B) in the hippocampus of Ts65Dn that normally have reduced levels of NR2B and was associated with recovery of learning and retention which returned to its baseline level if treatment was stopped (Incerti et al., 2011).

Although the Ts65Dn model has the advantage to have, like human patients, an extra-chromosome, several drawbacks remain: in particular and above all, the Ts65Dn mice are trisomic for genes on the Mmu17 from the centromere to *Pde10a* that could alter the specific consequences of the trisomic genes from the Mmu16 *Mrpl39-Zfp295* genetic interval (A. Duchon and Y. Herault, personal communication). This additional change might be overcome by a more elaborated strategy in which functional copy number of gene (s) of interest could be altered using knock-out or deletion of specific Mmu16 genes. Furthermore, only a few Ts65Dn males are fertile (Moore, 2006), and the translocated chromosome should be kept on a hybrid genetic background resulting in more variability of the induced phenotypes. Nevertheless, the Ts65Dn mice still have a strong potential using a subtractive strategy, that allow removing of the excessive gene copy number by combining the Ts65Dn with a mice carrying a deletion of the genetic interval homologous to human chromosome 21 (Duchon et al., 2011a).

To gain insight into DS mouse models, the first transchromosomic mouse line carrying an almost complete copy of Hsa21 was obtained in 2005 (O'Doherty et al., 2005). An additional copy of human chromosome even in a mouse context could better reproduce the DS phenotypes by preserving the complex set of regulatory sequences for large genes and the whole integrity of the chromosomal sequence which is found in three independent genomic regions in the mouse. The Tc1 mouse was engineered using the microcell-mediated chromosome transfer (MMCT) technique (Fournier and Ruddle, 1977), a technique used before that with a limited success. Indeed previous report using MMCT showed only chimera with varying contribution of trisomic cells to tissues but still with alterations in behavioral tests (Shinohara et al., 2001). In addition, these mice showed a hypoplasic thymus and cardiac defects. Unfortunately, in this first set of experiments, Hsa21 was not transmitted to the derived progeny from these chimeric mice. The Tc1 model was obtained using a similar technique (O'Doherty et al., 2005) and showed surprisingly no defect in the Morris water maze test (unlike Ts65Dn) while locomotor activity and learning were impaired, with the spatial working memory and the long-term memory also affected. The authors highlighted a lack of synaptic plasticity in the hippocampus. Indeed, LTP was reduced in transchromosomic mice, due to reduced expression of the glutamate receptor type 1 subunit of α-amino-3-hydroxy-5-methyl-4-isoxazolepropionic acid receptors (Morice et al., 2008). Mice also showed a difference in their exploratory pattern in open field, indicating a lack of habituation to novelty. A deficit was also shown in social memory (Duchon et al., 2011a) and in the test of social recognition, unlike the Ts65Dn mouse line (Faizi et al., 2011). For locomotion, the mice also show a lack of balance whereas their muscle strength is preserved (Galante et al., 2009).

Additional advantage is that transchromosomic mice carry an extra-chromosome, but the major drawbacks of this model are the rearrangements present on the Hsa21, certainly caused by the radiations used during its relocation in murine cells (Reynolds et al., 2010).

Models with supernumerary copy

There are two additional DS mouse models that were selected and found to carry translocation of genetic material homologous to human chromosome 21. The oldest mouse model for DS, the Ts16, was isolated in the 1970s. It was originated from crosses between double heterozygous males carrying Robertsonian translocations involving mouse chromosome 16 (Mmu 16) (Gropp et al., 1975). It corresponds, therefore, to a complete trisomy of mouse chromosome 16. Ts16 mice die *in utero* with many phenotypes, affecting notably the heart and the study of the craniofacial development. Even if limited to study embryonic mice development or to *in vitro* cell culture, Ts16 were used to assess the number of acetylcholinesterase -positive neurons in the basal forebrain and showed a significant loss of 40% of these neurons compared to control mice (Sweeney et al., 1989). Moreover, by extending the lifespan of these cholinergic cells which have been transplanted into the hippocampus of young adult disomic mice, it was possible to show that these cells suffered significant atrophy (Holtzman et al., 1992). The study of the signaling pathway of brain-derived neurotrophic factor (BDNF) on Ts16 hippocampal neurons in culture showed that the level of phosphorylation of the receptor tyrosine kinase-dependent BDNF (TrkB) was significantly reduced by 33% in response to BDNF in hippocampal neurons of Ts16 mice. In addition, Ts16 cells have high levels of a truncated kinase-deficient isoform (T1) of the TrkB receptor (TrkB T1). The loss of function of the gene encoding this protein can restore the survival of cortical and hippocampal neurons of Ts16 mice (Dorsey et al., 2002, 2006).

Experiments performed using immortalized cell lines derived in particular from hippocampal and cortical neurons show an alteration in calcium homeostasis. In the same way, the response to several neurotransmitters such as glutamate or nicotine showed an alteration in the kinetics of calcium entrance and exit (Cardenas et al., 1999, 2002b; Caviedes et al., 2006). Cells from cerebral cortex neurons showed approximately a 50% decrease in choline assimilation and a decrease in acetylcholine release after a nicotine treatment. In hippocampal neurons, only depolarization-induced acetylcholine release or nicotine treatment is reduced (Cardenas et al., 2002a) which was associated with abnormal endosomal function due to overexpression of the APP protein in immortalized cell lines of derived from the cerebral cortex after cholesterol treatment (Arriagada et al., 2007).

The main limitation of the Ts16 model is the presence of a complete extracopy of Mmu16. Therefore, in addition to the trisomy of the Mmu16 homologous region to Hsa21, it is also trisomic for the Mmu16 regions homologous to Hsa3, 8, 16, and 22. In addition, several other genes on the distal edge of Hsa21 long arm are essential for DS phenotypes and are not carried by the Mmu16. Thus, Ts16 mice are not trisomic for some of the genes of Hsa21 and are for some other ones that are not involved in the pathogenesis of DS, which makes results interpretation very difficult (Dierssen et al., 2009).

The Ts43H was described as a model trisomic for the Hsa21 homologous region found on the Mmu17 chromosome(Vacik et al., 2005). The model was not completely relevant as its trisomy encompasses more than 310 genes located on the proximal part of Mmu17.

Another model with segmental genome trisomy is the mouse model Ts1Cje. It has a partial trisomy of chromosome 16 extending from murine *Sod1* gene to *Zfp295* gene, that is, 67 genes. This model results from a reciprocal translocation between Mmu12 and Mmu16 (Ts(12^{16})Cje), which occurred during an experiment to inactivate of the *Sod1* gene (Sago et al., 1998). During the translocation process, a small deletion of five genes located at the end of the Mmu12 occurred leading to a monosomy for *Dnach11, Sp4, Sp8, Abcb5, Itgb8, Maccl,* and *Tmem196* (Duchon et al., 2011b). These mice survive to adulthood, and males are not sterile.

Difficulties in learning and memory have been found in particular reference memory in the Morris water maze (Sago et al., 1998) and working memory in a T maze (Belichenko et al., 2007). However, these deficits are less severe than those seen in the Ts65Dn model and are not accompanied by a cholinergic loss (Sago et al., 1998). In contrast to the Ts65Dn model, Ts1Cje mice are hypoactive (Sago et al., 1998) and they do not show any deficit in the test of object recognition (Fernandez and Garner, 2007). Ts1Cje mice have a decrease in LTP and an increase in LTD (long-term depression) in the CA1 and dentate gyrus (DG), but less severe than those observed in the Ts65Dn model (Siarey et al., 2005). These anomalies can be explained by structural abnormalities observed in synapses of the Ts1Cje model, such as a reduction in the density of dendritic spines (Belichenko et al., 2007) and in the production of neurons or a decrease in sites of connections (Hewitt et al., 2010; Ishihara et al., 2010). *In vivo* and *in vitro* measurements in Ts1Cje mouse brain structures are characterized by a sharp decrease in membrane potential and ATP production in cortical astrocytes, as well as a related mitochondrial dysfunction, a hyperphosphorylation of the tau protein without formation of neurofibrillary tangles and an increase of oxidative stress in hippocampal neurons (Ishihara et al., 2009; Shukkur et al., 2006). These data suggest that some genes of the Ts1Cje trisomic segment other than *App* and *Sod1* may be responsible for these alterations. Potential candidates are *Dyrk1A, Olig1,* and *Olig2.*

A study of neurospheres originated from neural progenitor cells of the Ts1Cje mice neocortex, which have been isolated at embryonic day 14, showed a decreased cell proliferation and a high rate of cell death. The expression profile of these neural progenitor cells versus controls showed that 54% of triplicated genes have a ratio significantly higher. In addition, some genes which are not carried by the extra-chromosome, and related to proliferation, cell differentiation, or glial functions, are deregulated. But *Sod1* and *App* genes are not implicated in those phenotypes as they are not triplicated in this model (Hewitt et al., 2010; Moldrich et al., 2009).

The cerebellar volume of Ts1Cje mice is reduced while the number of granule cells and Purkinje cells is not significantly different (Olson et al., 2004b). Transcriptomic studies of the Ts1Cje mice cerebellum showed that Mmu16 genes present in three copies in this model are overexpressed by 50% and that the transcriptome is altered during postnatal development of the cerebellum as in the Ts65Dn model (Dauphinot et al., 2005).

To find an alternative to the limitations of the Ts65Dn model including the problem of fertility in males, a new mouse model, designated Ts[Rb (12.1716)]2Cje, was isolated from spontaneous Robertsonian fusions (Villar et al., 2005). In this mouse model, the triplicated fragment of the Ts65Dn model is found fused to Mmu12. This new model has enlarged ventricles, neurogenesis problems (Ishihara et al., 2010), and an increased oxidative stress as in the Ts1Cje model (Ishihara et al., 2009).

The use of the Cre LoxP technology enabled a dramatic expansion of mouse models. This technique allows the rapid generation of duplications or deletions of regions of interest (Brault et al., 2006, 2007). Thus, new models were created, including models of regions of synteny on the previously unexplored mouse chromosomes 10 and 11. The advantage of this technique is that the modified region can be targeted precisely. The disadvantage, of course, is that these models do not carry physically an extra-chromosome but a segmental tandem duplication for the region of interest.

The first models produced by this technology are the Ts1Rhr and Ms1Rhr models. Indeed, to assess the contribution of the DSCR to the DS phenotypes, two mouse models, trisomic (Ts1Rhr) and monosomic (Ms1Rhr), for this region were created (Olson et al., 2004a; Fig. 1). The DSCR contains 31 genes between *Cbr1* and *Orf9*. Contrary to Ts65Dn mice, the Ts1Rhr adult mice are significantly larger than controls and have a larger skull than the euploid with a rostrocaudal elongation, but they do not display brachycephaly. Surprisingly, changes that were scored for the volume of brain area in the Ts1Rhr were different from that observed in the Ts65Dn model. In addition, no difference was scored in the Morris water maze test, induction of LTP in the CA1, hippocampal volume, or in cerebellum for Ts1Thr mice (Aldridge et al., 2007). Nevertheless, the triplication of the DSCR alone caused memory deficits in the object recognition, associated with changes in LTP in fascia dentata, boosting once again the DSCR theory (Belichenko et al., 2009a). Overall, the data suggest that abnormalities found in DS people and in Ts65Dn are not due to the triplication of one or more genes in a specific region of the Hsa21 but to the interaction between noncontiguous genes.

The Dp(16)1Yey mice present a duplication of 22.9Mb with 140 genes of Mmu16, from *Lipi* to *Zfp295*, that is homologous to the human region 21q11q22.2. Compared to the Ts65Dn model, these mice are trisomic for a larger region extending from *Zfp295* to *Lipi*. The advantage of this model is that it contains only a triplication of Hsa21 homologous genes (Li et al., 2007). This new mouse model has both cardiac and gastrointestinal abnormalities. At cognitive level, these mice show a learning deficit in the Morris water maze test and a decrease in LTP, like Ts65Dn mice (Yu et al., 2010d). This suggests that genes present in the duplicated region contribute to those phenotypes and that genes present in the Ts65Dn region nonhomologous to the Hsa21 do not influence the phenotype.

The Ts1Yah and Dp(17)1Yey models include a duplication of the Hsa21 homologous region on Mmu17, extending respectively from *Abcg1* to *U2af1* for Ts1Yah and *Abcg1* to *Rrp1b* for Dp(17)1Yey. Both models have an increased LTP in the hippocampus measured either *in vivo* or *in vitro* (Lopes Pereira et al., 2009; Yu et al., 2010d). Dp(17)1Yey mice show no deficit in the fear conditioning test. Ts1Yah mice show a deficit of habituation in the open field and a deficit of memory in the tests of object recognition and Y maze, with, in contrast, a trend to be more efficient in learning to find the platform in the Morris water maze test. However, this trend was not found in the Dp(17)1Yey, which may be due to a difference in the achievement's sequence of the test or in the age of mice at the time of the studies (6–8months Ts1Yah vs. 2–4 months for Dp(17)1Yey).

The Dp(10)1Yey has a duplication of the Hsa21 homologous region of Mmu10, from *Prmt2* to *Pdxk*. The study of this model shows no change in LTP in the hippocampus. More experiences of memorization in Morris water maze and fear conditioning tests show that duplication of this region does not cause major deleterious effects (Yu et al., 2010d). These results were very surprising regarding the number of genes (almost 50) found in the region.

Parallel to the designed trisomic mice, corresponding monosomic models were also generated, matching with regions of synteny on

Mmu17 (from *Abcg1* to *Rrp1b*) for the Df(17)1Yey lineage (Yu et al., 2010a), on Mmu10 from *Prmt2* to *Pdxk* for the Df(10)1Yey (Yu et al., 2010a), from *Cstb* to *Prmt2* for the Ms4Yah (Duchon et al., 2011a), and from *Col6a1* to *Prmt2* for Ms1Yah (Besson et al., 2007). Both Df(17)1Yey and Df(10)1Yey lineages showed a deficit in acquisition and retention in the Morris water maze test. In the Df(17)1Yey model, the deficit was due to a lower swimming speed which is not changed in Df(10)1Yey mice. The fear conditioning test revealed a deficit of the associative memory in both mutant lines (Yu et al., 2010a). Ms4Yah mice showed no performance deficit in cognitive tests of working memory, social interaction, or motor coordination (Duchon et al., 2011a). Ms1Yah mice do not present any cognitive defect (Besson et al., 2007).

Very recently, Yu and colleagues have successfully produced a complete trisomic mouse model. This mouse line contains three copies of all the three regions of synteny of Hsa21. They carry the three Dp(16)1Yey, Dp(17)1Yey, and Dp(10)1Yey trisomies and correspond more to a human DS condition. The mice show a reduction of LTP in the hippocampus associated with a deficit in the Morris water maze test and in fear conditioning (Yu et al., 2010b,c). In addition, whereas the Dp(17)1Yey mouse line showed an increase in LTP, the triple trisomic model displayed a complete loss of LTP, which highlight the complexity of genetic interactions between different regions. An additional model trisomic or monosomic for the *Tiam1-Kcnj6* region has been recently used to demonstrate the presence of candidate genes for cardiac phenotypes within the region (Liu et al., 2011).

Recent experience of phenotype rescue was tried to determine whether genes of the distal part of Hsa21, present on Mmu10, could play a role in DS. To do this, Ms4Yah mice were crossed with Tc1 mice. The resulting progeny from this cross was then phenotyped to determine whether double transgenic Tc1Ms4Yah mice presented motor or cognitive Tc1 like phenotypes. The result of this study showed that genes in this region, located between *Cstb* and *Prmt2*, were not contributing to the Tc1-induced phenotypes, suggesting that they don't interfere with the DS physiopathology (Duchon et al., 2011a). With the results observed in the Dp(10)Yey, the most telomeric part of the Hsa21 does not seem to be necessary or sufficient for inducing DS-related behavioral phenotypes.

The panel of trisomic models now available has enabled to map Hsa21 genes or regions with strong effects on their own, but it also has pointed at the complexity of the interactions among different regions leading to DS phenotypes. Hence, it was shown that the trisomy of different non-overlapping regions of Hsa21 can lead to the same phenotype, revealing that a main phenotype can be lead by several regions among this chromosome. This is the case of the *Mrpl39–Zfp295* region triplicated in the Ts65Dn model and of the *Abcg1–U2af1* region triplicated in the Ts1Yah model that both participate to the impairment of the explicit memory as assessed by the NOR (Lopes Pereira et al., 2009). On the other hand, regions sharing common triplicated genes do not necessarily share the same phenotypes. Hence, one of the most complete DS model, the Tc1 model, has normal long-term memory and synaptic plasticity, whereas these phenotypes are affected in the less extended Ts65Dn model (Morice et al., 2008; O'Doherty et al., 2005). Construction of additional models with smaller trisomic regions as well as combining the different models together will help to decipher the complex genetic interactions that are at work in DS.

Models overexpressing single genes or small groups of genes

The complete sequencing and characterization of Hsa21 resulted with an almost complete catalog of Hsa21 genes, some of them with functions already known (Gardiner et al., 2003). Even so, their role in the pathology of DS is not yet clear. In an attempt to identify individual genes or loci on Hsa21 that are responsible for producing the different DS phenotypes, mouse models were created overexpressing single genes or a restricted group of genes. Genes were chosen depending on their localization on Hsa21 (genes present in the DSCR, see Chapter 2) or on the potential role of the corresponding protein (involved in neuropathologies, skeletal, or cardiac developmental anomalies). These single gene gain-of-function models were created by inserting a human or murine transgene under the control of either their own promoter or a chosen promoter, but the second case has the drawback to lead to a spatiotemporal expression of the gene that is different from the endogenous one. Models of single-gene overexpression have revealed genes associated to neuropathological defects such as *Sod1* (Epstein et al., 1987), *S100b* (Gerlai and Roder, 1993, 1996), *App* (Games et al., 1995; Hsiao et al., 1996; Moechars et al., 1998), or *Sim2* (Ema et al., 1999). But phenotypes observed in those models are associated to an uncontrolled overexpression of the transgene that does not allow the establishment of a true correlation with DS characteristics. Transfer in the mouse genome of larger fragments, thanks to the use of YACs and BACs/PACs (respectively yeast or bacterial/phage P1 artificial chromosomes), has the advantage to implicate relatively large inserts with probably all the regulatory elements of the gene needed for a proper regulation of its expression. It has permitted to uncover the role of genes such as *Dyrk1a* whose overexpression leads to learning and behavioral defects (Smith and Rubin, 1997; Smith et al., 1997a) or *App* whose protein is implicated in aggregates formation responsible of AD (Lamb et al., 1993). But these models present several major inconvenient like unspecific insertions of the DNA fragment that can lead to variable expression of the genes present on the YAC and the lack of control of fragment copies number inserted into the genome. We will review in this section the different models of overexpression of genes implicated in neural formation and function. We will focus on the endocellular phenotypes and affected molecular pathways that they revealed.

The first Hsa21 gene that has been overexpressed in a mouse model is the *Sod1* gene coding for the Cu/Zn-superoxide dismutase, a key enzyme in the metabolism of oxygen free radicals. Sod1 catalyzed the dismutation of the superoxide anion to hydrogen peroxide. Mutations in this gene were found in the familial lateral sclerosis (ALS), a neuropathy with progressive degeneration of the motor neurons of the cerebral cortex and the spinal cord. This enzyme may lead to the increased production of active oxygen species in DS and was suggested to be associated with poorer cognitive performance in DS and particularly with premature neuronal aging and Alzheimer-type neurological lesions. Transgenic mice carrying the human *SOD1* gene (Tg(SOD1)10Cje) were constructed by microinjection in fertilized eggs of a 14.5-kb linear fragment of human genomic DNA containing the entire *SOD1* (Epstein et al., 1987) and different lines were established that presented an enzymatic activity increased between 1.6- and 6-fold in the brain. These mice have a diminished number of cells in several brain areas and a decreased LTP in the CA1 pyramidal neurons (Harris-Cerruti et al., 2004; Zang et al., 2004). Overexpression of SOD1 in transgenic mice also triggered signs of premature aging in the neuromuscular junction of the tongue and hindlimb muscles similar to those observed with DS (Avraham et al., 1988, 1991; Groner et al., 1990; Yarom et al., 1988). Degeneration of motor neurons was also found in another transgenic line overexpressing a human SOD1 carrying a G93A substitution, a model for ALS suggesting a contribution of allele to some DS-related phenotypes (Gurney et al., 1994). Proteomic analysis of the cerebellum and cerebral hemisphere of the transgenic mice overexpressing human SOD1 has shown reduced proteasome activity, similar to the decrease observed during normal aging. It could be responsible for early aging in DS patients (Le Pecheur et al., 2005). However, a protective effect has also been described in neurons and astrocytes from brains of transgenic mice using oxidative stress conditions *in vitro* and *in vivo* with overexpression of extracellular SOD1 (EC-SOD) which was found to protect against aging-induced cognitive impairment (Levin et al., 2002).

The *S100B* gene codes for the β subunit of the homodimers or heterodimers of S100 proteins, a large family of calcium binding protein involved in various physiological process (Donato et al., 2009). The S100B homodimer is the major form in the mammalian brain where it is a neurotrophic factor produced by astrocytes from the central nervous system (CNS) and released during development by the action of serotonin and acting on serotonin neurons (Azmitia et al., 1990). It was demonstrated that it can stimulate neurite extension (Marshak et al., 1992) and to play a role in synaptogenesis (Muller et al., 1993), dendritic branching (Yan et al., 1997), and apoptosis (Mariggio et al., 1994). In adult brain, levels of S100B decrease but the protein continues to play a role in cytoskeletal stabilization, neuronal plasticity, and LTP (Nishiyama et al., 2002; Rothermundt et al., 2003). Elevated levels of S100β have been observed in the brain of DS patients, but also of people with AD and conditions of increased inflammation. Transgenic mice were created to overexpress either the murine *S100b* (Reeves et al., 1994) or the human *S100B* gene (Friend et al., 1992) both under the control of their own promoter. The two TgS100b lines, showing two- and sevenfold S100b overexpression had hippocampal astrocytosis and levels of axonal proliferation proportional to the transgene copy number, pointing at the role of S100b as a growth factor in the brain (Reeves et al., 1994). The four Tg(S100B) lines generated for the human gene and containing 10–100 copies of the gene also showed astrocytosis and an increased proliferation of the dendrites in the hippocampus and cerebellum at a young age. The number of dendrites then prematurely decreased with age and was accompanied by a hippocampal-dependent spatial learning and memory deficit in the adult animal (Shapiro and Whitaker-Azmitia, 2004; Whitaker-Azmitia et al., 1997; Winocur et al., 2001). The hypothesis that *S100B* overexpression leads to pathological brain aging in DS was further supported by the finding that the clusterin protein, a multifunctional glycoprotein used as an indicator of aging or cellular stress, is greatly increased in the neurons of young Tg(*S100B*) mice (Gerlai and Roder, 1996; Shapiro et al., 2004). However, Gerlai and Roder (1996) reported a learning deficit in transgenic mice that was already detectable at 3months of age, but with no additional progression while aging. This finding is in contradiction with the hypothesis of a role in progressive neurodegenerative and questions the role of S100B in the neurodegeneration found in DS.

Sim2 is a homolog of Drosophila *single-minded*, whose product is a transcription factor that plays a pivotal role in the CNS midline development (Nambu et al., 1990, 1991). During mouse embryonic development, *Sim2* expression profile is restricted to the ventral diencephalon. Two transgenic lines overexpressing *mSim2* artificially driven by the chicken *β-actin* promoter were constructed and showed increased expression in the hippocampus and amygdale (Ema et al., 1999). Both lines were viable and fertile, with overall normal phenotypes, but had mild impairments in the fear conditioning test that requires nervous influx from the hippocampus and cerebellar amygdale. This impairment is however milder than those observed in the Ts65Dn and Ts1Cje models, suggesting that additional genes in the Hsa21 must contribute to the phenotype. Another transgenic model of *mSim2* overexpression was created by injection of a BAC clone containing the complete mouse *Sim2* gene into fertilized eggs (Chrast et al., 2000). Transgene was shown to have the same spatial pattern of expression as the endogenous *Sim2* gene. Whereas no histopathological phenotype was observed in the brain of these mice, a detailed behavioral analysis revealed an altered anxiety-related exploratory behavior as well as a reduced sensitivity to pain similar to that of the partial trisomy 16 models.

APP is an integral membrane protein found in many tissues and organs and concentrated in the synapse of neurons. It is best known and most commonly studied as the precursor molecule whose proteolysis generates β-amyloid (Aβ) the primary component of the amyloid plaques found in the brain of AD patients, but little is known about its primary function. It has been speculated that it is implicated as a regulator of synapse formation (Priller et al., 2006), neural plasticity (Turner et al., 2003), and iron export (Duce et al., 2010). People with DS develop Alzheimer's like dementia much earlier than the general population and most if not all of them develop brain changes associated with this disease. Several transgenic mouse models for different human or murine, wild type or mutant APP isoforms put under the control of different promoters have been constructed (Games et al., 1995; Higgins et al., 1993; Hsiao et al., 1995; Moechars et al., 1996; Quon et al., 1991). These mice present some of the prominent behavioral and pathological features of AD, like age-related impairment in learning and memory, neuronal loss, gliosis, neuritic changes, amyloid deposition, and an abnormal tau phosphorylation, but none of them have all the defects observed in the human pathology (Richardson and Burns, 2002). Mice expressing the human wild-type APP751 isoform, driven by the neuron-specific enolase promoter, present β-amyloid deposits and a deficit in spatial memory (Moran et al., 1995; Quon et al., 1991). Transgenic FVB/N mice expressing the mutated or wild-type human or mouse APP695 isoform driven by a hamster prion protein gene promoter displayed an age-dependent CNS disorder with early death, neophobia, and impaired spatial alternation (Hsiao et al., 1995). These symptoms were accompanied by diminished glucose utilization in the cerebral cortex and hypertrophic astrocytic gliosis in the limbic system and widespread areas of the neocortex, but spared the sensorimotor cortex, hence resembling the phenotype observed in AD. But no amyloid plaque formation was observed, pointing out at dissociation between the plaque formation and the other deleterious processes related to APP overexpression. A model containing the entire 400kb human APP gene in a YAC did not show any amyloid deposit (Lamb et al., 1993), but, at the homozygous state, showed learning and memory deficits starting at age 9–10 months and cataract at 16months (Frederikse and Ren, 2002; Harris-Cerruti et al., 2004). The physical and biochemical properties of the amyloid plates were different in AD and in transgenic mice (Kuo et al., 2001). Additional results in the model of Hsiao (Hsiao et al., 1995) suggest lipidic peroxidation preceding amyloid plaque formation (Pratico et al., 2001). In order to get all the AD neuropathologic lesions into one mouse model, double and triple genes transgenic models were created. For example, transgenic models for PS1, APP, and Tau developed amyloid plaques together with synaptic deficit (Oddo et al., 2003). Nevertheless, the involvement of APP in the endosomal phenotypes and the degeneration of basal forebrain cholinergic neurons observed in the Ts65Dn mouse model (Salehi et al., 2006) was further supported by the rescue experiment described previously (Cataldo et al., 2003).

PFKL (liver phosphofructokinase) codes for one of the three types (M-muscle, P-platelet, and L-liver) of subunits that form the phosphofructokinase (PFK), a tetrameric enzyme that catalyzes a key step in glycolysis, namely the conversion of D-fructose 6-phosphate to D-fructose 1,6-bisphosphate (Vankeuren et al., 1986; Vora and Francke, 1981). These three subunits randomly associate to form catalytically active homo- or heterotetramers. PFK activity was found to be increased in fibroblasts and erythrocytes of DS patients (Anneren and Gebremedhin, 1987), and the relative amount of PFKL-containing tetramers were increased at the expense of which did not contain PFKL (Vora and Francke, 1981). In addition, changes in glycolytic rates and in the concentrations of glycolytic intermediates and products were found in DS (Elson et al.,

1992). Transgenic mice containing a murine PFKL gene cDNA hybrid construct expressed PFKL in an endogenous tissue-specific manner. In the brain of these mice, PFKL activity revealed a temporal dimension and was doubled during embryogenesis while it was unchanged in the adult (Elson et al., 1994). This model presents an abnormal glucose metabolism, with an enhanced rate in the brain (Peled-Kamar et al., 1998). This phenotype has been described in some young adults with DS. In addition, insulin induced by glucose is also misregulated (Knobler et al., 1997).

Purkinje cell protein 4 (PCP4), also called brain-specific polypeptide 19 (PEP19), is a neurospecific small calmodulin-binding protein, regulating the calmodulin-mediated signal which is critical for synaptic function. The murine *Pcp4* gene is highly expressed in neuroectoderm cells during mouse embryogenesis and in primarily cerebellar Purkinje cells. PCP4 is located in the DSCR and it has been hypothesized that its overexpression contributes to the pathogenesis of the disorder. Levels of PCP4 expression were also found to be altered in other neurodegenerative or neurocognitive pathologies such as alcoholism (Iwamoto et al., 2004), Huntington's disease (Utal et al., 1998), Parkinson's disease (Skold et al., 2006), and AD (Slemmon et al., 1996). Phenotypic analysis of Pep-19/pcp4-null mice revealed that loss of Pcp4 altered dramatically the synaptic plasticity at the excitatory synapses of cerebellar Purkinje cells (Wei et al., 2011). LTP at parallel fiber–Purkinje cell synapses was replaced by LTD. As the result, mutant mice had marked deficit in locomotor tests. A transgenic mouse model containing one copy of human PCP4 revealed that PCP4 overexpression induces precocious neuronal differentiation associated with an increase in the amount of the activated form of CamKIIδ, suggesting that PCP4 can induce the Ca^{2+}–CaM signaling pathway (Mouton-Liger et al., 2011).

The *ETS2* gene codes for the transcription factor protein C-ets-2 of the ETS family that is involved in stem cell development, cell death, and tumorigenesis. During murine development, ETS2 is highly expressed in newly formed cartilage, including skull precursor cells and vertebrate primordia. Sumarsono et al. generated transgenic mice to investigate the consequences of overexpression of Ets2 (Sumarsono et al., 1996). They found that the mice developed neurocranial, viscerocranial, and cervical skeletal abnormalities similar to those observed in trisomy-16 mice and human with DS. But ETS2 overexpression also resulted in neuronal apoptosis (Wolvetang et al., 2003). Increased apoptosis correlated with increased expression of p53 and alterations in downstream factors in the p53 pathway. They additionally found that the brains from the mice had elevated expression of APP and presenilin-1 and increased β-amyloid production and showed that ETS2 transactivated the APP gene via specific Ats-binding sites in the APP promoter. These findings could hence explain why APP expression in DS is three- to fourfold higher than what is expected from the trisomy-dependent 1.5-fold increase in gene load and suggests a role for Ets2 in the pathogenesis of the brain abnormalities seen in DS and AD.

Among the around 30 genes present on the DSCR, the dual-specific tyrosine-(Y)-regulated kinase 1A gene (*Dyrk1A*) has retained most of the attention as there are accumulating data that suggest that DYRK1A has a role in the pathogenic mechanisms underlying the neuropathology of DS. DYRK1A is a serine/threonin kinase and is the mammalian homolog of the *Drosophila melanogaster* minibrain protein kinase, which was shown to be essential for normal postembryonic neurogenesis in the fly (Tejedor et al., 1995). *DYRK1A* is expressed ubiquitously but not evenly in the fetal and adult brain, with a particularly strong expression in the cerebellum, olfactory bulb, and hippocampus (Guimera et al., 1996; Song et al., 1996). It is involved in neurogenesis, dendritic development, and degenerative processes mediated by tau protein. DYRK1A was shown to phosphorylate or interact with many proteins from different pathways. Among targets of DYRK1A, one can list the cAMP responsive element-binding protein (CREB) (Hammerle et al., 2003), the transcription regulator nuclear factor of activated T-cells (NFAT) (Arron et al., 2006) or the endocytotic protein dynamin 1 implicated in vesicle trafficking (Hammerle et al., 2003). Interestingly, DYRK1A haplo-insufficient mice exhibit decreased brain size, motor abnormalities, and cognitive deficits. Experiments demonstrated that they have altered neocortical pyramidal cells which are smaller, less branched, and less spinous (Benavides-Piccione et al., 2005; Fotaki et al., 2002). The expression and kinase activity of DYRK1A were shown to be upregulated in the brain of DS subjects (Liu et al., 2008). Three transgenic mouse models overexpressing *DYRK1A* have been produced. These mice carry the human *DYRK1A* genomic DNA in a BAC or a YAC, or they carry extracopies of murine *Dyrk1a* cDNA (Ahn et al., 2006; Altafaj et al., 2001; Smith et al., 1997a). The *DYRK1A* BAC TG mice exhibit impairment in hippocampal-dependent learning and memory and changes in both LTP and LTD (Ahn et al., 2006). Molecular investigation of this model provided evidence that overexpression of DYRK1A may contribute to the early onset of AD in DS patients through increased phosphorylation of two key target proteins in AD, namely APP (at Thr668) and the microtubule-associated protein tau (at Ser202, Thr212, and Ser404; Ryoo et al., 2008, 2007). A second model for DYRK1A carries a transgene with a 6.7-kb fragment encoding the sMT-la promoter and the *Dyrk1A*-coding sequence. Transgenics have neuro-developmental delay, motor abnormalities, and impairment in spatial learning and cognitive flexibility (Altafaj et al., 2001) some of which are reversed by normalizing specifically dyrk1a dosage in adult mice (Ortiz-abalia et al., 2008), suggesting also a dosage effect in adult function. Microarray analyses of the cerebellum of those mice revealed an upregulation of the NR2A subunit of the NMDA type of glutamate receptors and calcium-imaging experiments indicated a prolonged decay of NMDA-induced calcium transients in the cerebellum (Altafaj et al., 2008). Together, these results reveal that overexpression of *Dyrk1a* alters the NMDA-induced calcium rise in the cerebellar neurons which is probably inducing the locomotor phenotypes. The third model is transgenic for a 180-kb YAC (157F2) containing several DSCR genes (DSCR5, TTC3, DSCR3, and DYRK1A; Smith et al., 1997a). Mice exhibit lower performance in learning and memory, with specific deficits in learning flexibility (Chabert et al., 2004) and hyperactivity. IRM brain morphological analysis revealed an increased brain weight with a specific enlargement of the hypothalamus/thalamus region and increased neuronal size (Sebrie et al., 2008). Biochemical investigation of the model revealed an altered phosphorylation of the transcription factor FKHR, Cyclin b1. and the transcription factors of the CREB family, suggesting that *DYRK1A* overexpression lead to some of the DS phenotypes by perturbing the regulation of the cell cycle (Branchi et al., 2004). The specific role of DYRK1A in the phenotypes observed in this mouse model was confirmed by the use of DYRK1A-specific inhibitors such the Epigallocatechin gallate (a natural polyphenol found in green tea leaves) that restored most of the phenotypic characteristics in these mice (Guedj et al., 2009).

DSCR1, also designed as RCAN1, CSP1, or MCIP1, was discovered by Fuentes and collaborators who identified several putative exons in the 3-Mb DSCR region of the Hsa21 and was hence named Down syndrome candidate region-1 (DSCR1) (Fuentes et al., 1995). Northern blot analysis revealed that a 2.2-kb DSCR1 transcript is expressed at the highest level in adult heart and fetal brain, suggesting that DSCR1 might play a role during CNS development. In the adult brain, DSCR1 is predominantly expressed in neurons of the cerebral cortex, hippocampus, substantia nigra, thalamus, and medulla oblongata. DSCR1 encodes a functional inhibitor of calcineurin, a ubiquitous and multifunctional calcium-activated protein phosphatase that is the most abundant protein in the brain where it regulates other neuronal functions, neurotransmitter release, neurite outgrowth, and neuronal cell death. In particular, DSCR1 was shown to inhibit calcineurin-dependent gene transcription through the inhibition of NFAT translocation to the nucleus and is seen as an endogenous regulator of calcineurin-mediated signaling pathways (Fuentes et al., 2000). In addition, there are hints for implication of DSCR in the regulation of mitochondrial function and oxidative stress, acting as a stress response element: the Drosophila homolog of DSRC1 influences the activity of the mitochondrial ADP/ATP translocator (Chang and Min, 2005), is able to induce the expression of SOD1 (Ermak et al., 2004), also triplicated on Hsa21, and its acute overexpression protects cells from oxidative stress (Ermak et al., 2002). DSCR1 expression is consistently increased 1.8-fold in human fetal tissues bearing trisomy 21 (Baek et al., 2009). Mice deficient for DSCR1 revealed pronounced spatial learning and memory deficits as well as impaired LTP (Hoeffer et al., 2007), making DSCR1 a good candidate to contribute to mental disability in DS. While transgenic mice overexpressing DSCR1 are viable, develop normally and show no overt abnormal phenotype, investigating exocytosis in chromaffin cells, a classic cellular model of neuronal exocytosis revealed that DSCR1 regulates the number and the speed of vesicles undergoing exocytosis (Keating et al., 2008). Dierssen and collaborators have investigated the effect of DSCR1 overexpression on the cognitive functions using transgenic mice TgRCAN1 containing the human isoform splice variant 1 placed under the control of the PDGFb promoter to drive transgene expression to brain areas. Among the transgenic lines that were derived from the injection of the construct, three of them expressed the transgene in brain regions, particularly in the hippocampus and cerebral cortex (Dierssen et al., 2011). While they did not see any motor or sensory impairment, they confirmed the effect of DSCR1 on visuospatial learning and memory seen in knockout mice and similar to the phenotype in DS persons. They conclude that DSCR1 may contribute to specific learning phenotypes seen in DS patients and propose that excess inhibition of endogenous calcineurin, which in turn acts as an inhibitory constraint in the hippocampal phosphorylation/dephosphorylation balance, has a significant impact on learning but not memory. Finally, Arron and collaborators used mathematical modeling to analyze the synergistic impact of both DSCR1 and DYRK1A on the NFAT pathway regulation in DSCR1- and Dyrk1A-overexpressing mice, calcineurin- and Nfatc-deficient mice, and models of DS (Arron et al., 2006). They proposed that the 1.5-fold increase in dosage of DSCR1 and DYRK1A cooperatively destabilizes a regulatory circuit, leading to reduced NFATc activity.

DOPEY2 (C21ORF5) cDNA was isolated during a screening of human brain cDNAs for potential to encode proteins larger than 50kD (Nagase et al., 1999). C21ORF5 expression was detected in all human tissues examined and in all regions of the brain tested (amygdala, corpus callosum, cerebellum, caudate nucleus, hippocampus, substantia nigra, subthalamic nucleus, and thalamus) and was found to be overexpressed in Ts65Dn mice in which it is in three copies and in lymphoblasts of DS patients. Among eight transgenic mouse lines carrying YACs containing human DNA fragments covering the DSCR created by Rachidi and collaborators (Rachidi et al., 2007), two independent lines carrying the YAC 230E8 had cerebellar alteration with significant elongation of the antero-posterior axis, a major neurological defect in culmen and declivus lobules, increased cortical cell density and mild learning defects. YAC 230E8 contains seven genes among which DOPEY2 is the best candidate for the cerebellar alterations considering its high expression in the brain and that its homologous genes in yeast, *Caenorhabditis elegans* and *Drosophila* are involved in morphogenesis. In addition, a study of the expression profile of *Dopey2* using *in situ* hybridization on brain sections of YAC 230E8, mice showed a clear expression at the level of the cortex, the cerebellum, and the hippocampus and revealed an increase of the number of cortical cells overexpressing Dopey2 in the transgenic mice compared to the disomic controls (Rachidi et al., 2005).

Conclusion

The last decade has been enriched tremendously with a variety of new models of DS. The discovery of the first trisomic model in 1990 by Muriel Davisson and coworkers (Davisson et al., 1990) and the development of new chromosomal engineering techniques (Fournier and Ruddle, 1977; Herault et al., 1998; Ramirez-Solis et al., 1995) opened the way forward. Nowadays, the community has access to a comprehensive zoo with more than 12 new models with copy number variation for regions homologous to Hsa21. The new models correspond to increase copy number encompassing sequences for protein coding gene, miRNA (micro ribonucleic acid) or lncRNA (long noncoding ribonucleic acid) and conserved noncoding sequences. The analysis of the new model has started unraveling the complex contribution of the different regions of human chromosome 21 to the DS brain alteration. The myth of a single master gene is unfortunately dead, but the identification of a few targets will allow the definition of a multimodal network of genes interacting together and further explaining the variability of the DS. We could expect that there will be some new models brought on the scene in the new future and that together they will lead to a better knowledge of the DS physiopathology, but even more interestingly to the development of new therapeutic avenues for people.

Acknowledgments

We thank members of the research group, the IGBMC laboratory, the ICS, and the AnEUploidy consortium (www.aneuploidy.org). We are grateful to the National Centre for Scientific Research (CNRS), the INSERM, the University of Strasbourg, the European commission with the AnEUploidy project (LSHG-CT-2006-037627), and the Foundation for Medical Research and foundation Jerome Lejeune for supports.

Abbreviations

BACs	bacterial artificial chromosomes
CNS	central nervous system
DG	dentate gyrus
DS	Down syndrome
Hsa21	Homo sapiens chromosome 21
LncRNA	long noncoding ribonucleic acid
LTP	long-term potentiation
miRNA	micro ribonucleic acid
Mmu10, 16, or 17	Mus musculus chromosome 10, 16, or 17
PACs	phage P1 artificial chromosomes
YACs	yeast artificial chromosomes

References

Adams, D. J., Biggs, P. J., Cox, T., Davies, R., van der Weyden, L., Jonkers, J., et al. (2004). Mutagenic insertion and chromosome engineering resource (MICER). *Nature Genetics*, *36*, 867–871.

Ahn, K. J., Jeong, H. K., Choi, H. S., Ryoo, S. R., Kim, Y. J., Goo, J. S., et al. (2006). DYRKIA BAC transgenic mice show altered synaptic plasticity with learning and memory defects. *Neurobiology of Disease*, *22*, 463–472.

Aldridge, K., Reeves, R. H., Olson, L. E., & Richtsmeier, J. T. (2007). Differential effects of trisomy on brain shape and volume in related aneuploid mouse models. *American Journal of Medical Genetics. Part A*, *143A*, 1060–1070.

Altafaj, X., Dierssen, M., Baamonde, C., Marti, E., Visa, J., Guimera, J., et al. (2001). Neurodevelopmental delay, motor abnormalities and cognitive deficits in transgenic mice overexpressing Dyrk1A (minibrain), a murine model of Down's syndrome. *Human Molecular Genetics*, *10*, 1915–1923.

Altafaj, X., Ortiz-Abalia, J., Fernandez, M., Potier, M. C., Laffaire, J., Andreu, N., et al. (2008). Increased NR2A expression and prolonged decay of NMDA-induced calcium transient in cerebellum of TgDyrk1A mice, a mouse model of Down syndrome. *Neurobiology of Disease*, *32*, 377–384.

Anneren, G., & Gebremedhin, M. (1987). Trace-elements and transport proteins in serum of children with Down-syndrome and of healthy siblings living in the same environment. *Human Nutrition—Clinical Nutrition, 41C*, 291–299.

Arriagada, C., Astorga, C., Atwater, I., Rojas, E., Mears, D., Caviedes, R., et al. (2007). Endosomal abnormalities related to amyloid precursor protein in cholesterol treated cerebral cortex neuronal cells derived from trisomy 16 mice, an animal model of Down syndrome. *Neuroscience Letters, 423*, 172–177.

Arron, J. R., Winslow, M. M., Polleri, A., Chang, C. P., Wu, H., Gao, X., et al. (2006). NFAT dysregulation by increased dosage of DSCR1 and DYRK1A on chromosome 21. *Nature, 441*, 595–600.

Avraham, K. B., Schickler, M., Sapoznikov, D., Yarom, R., & Groner, Y. (1988). Downs-syndrome—Abnormal neuromuscular-junction in tongue of transgenic mice with elevated levels of human CU/ZN-superoxide dismutase. *Cell, 54*, 823–829.

Avraham, K. B., Sugarman, H., Rotshenker, S., & Groner, Y. (1991). Downs-syndrome—Morphological remodeling and increased complexity in the neuromuscular-junction of transgenic CuZn-superoxide dismutase mice. *Journal of Neurocytology, 20*, 208–215.

Azmitia, E. C., Dolan, K., & Whitakerazmitia, P. M. (1990). S-100B but not NGF, EGF, insulin or calmodulin is a CNS serotonergic growth-factor. *Brain Research, 516*, 354–356.

Baek, K. H., Zaslavsky, A., Lynch, R. C., Britt, C., Okada, Y., Siarey, R. J., et al. (2009). Down's syndrome suppression of tumour growth and the role of the calcineurin inhibitor DSCR1. *Nature, 459*, 1126–1130 U1113.

Baxter, L. L., Moran, T. H., Richtsmeier, J. T., Troncoso, J., & Reeves, R. H. (2000). Discovery and genetic localization of Down syndrome cerebellar phenotypes using the Ts65Dn mouse. *Human Molecular Genetics, 9*, 195–202.

Belichenko, N. P., Belichenko, P. V., Kleschevnikov, A. M., Salehi, A., Reeves, R. H., & Mobley, W. C. (2009a). The "Down syndrome critical region" is sufficient in the mouse model to confer behavioral, neurophysiological, and synaptic phenotypes characteristic of Down syndrome. *The Journal of Neuroscience, 29*, 5938–5948.

Belichenko, P. V., Kleschevnikov, A. M., Salehi, A., Epstein, C. J., & Mobley, W. C. (2007). Synaptic and cognitive abnormalities in mouse models of Down syndrome: Exploring genotype-phenotype relationships. *The Journal of Comparative Neurology, 504*, 329–345.

Belichenko, P. V., Klescrevnikov, A. M., Masliah, E., Wu, C. B., Takimoto-Kimura, R., Salehi, A., et al. (2009b). Excitatory-inhibitory relationship in the fascia dentata in the Ts65Dn mouse model of Down syndrome. *The Journal of Comparative Neurology, 512*, 453–466.

Belichenko, P. V., Masliah, E., Kleschevnikov, A. M., Villar, A. J., Epstein, C. J., Salehi, A., et al. (2004). Synaptic structural abnormalities in the Ts65Dn mouse model of Down syndrome. *The Journal of Comparative Neurology, 480*, 281–298.

Besson, V., Brault, V., Duchon, A., Togbe, D., Bizot, J.-C., Quesniaux, V. F. J., et al. (2007). Modeling the monosomy for the telomeric part of human chromosome 21 reveals haploinsufficient genes modulating the inflammatory and airway responses. *Human Molecular Genetics, 16*, 2040–2052.

Bianchi, P., Ciani, E., Guidi, S., Trazzi, S., Felice, D., Grossi, G., et al. (2010). Early pharmacotherapy restores neurogenesis and cognitive performance in the Ts65Dn mouse model for Down syndrome. *The Journal of Neuroscience, 30*, 8769–8779.

Bimonte-Nelson, H. A., Hunter, C. L., Nelson, M. E., & Granholm, A. C. E. (2003). Frontal cortex BDNF levels correlate with working memory in an animal model of Down syndrome. *Behavioural Brain Research, 139*, 47–57.

Blazek, J. D., Billingsley, C. N., Newbauer, A., & Roper, R. J. (2010). Embryonic and not maternal trisomy causes developmental attenuation in the Ts65Dn mouse model for Down syndrome. *Developmental Dynamics, 239*, 1645–1653.

Branchi, I., Bichler, Z., Minghetti, L., Delabar, J. M., Malchiodi-Albedi, F., Gonzalez, M. C., et al. (2004). Transgenic mouse in vivo library of human Down syndrome critical region 1: Association between DYRK1A overexpression, brain development abnormalities, and cell cycle protein alteration. *Journal of Neuropathology and Experimental Neurology, 63*, 429–440.

Braudeau, J., Delatour, B., Duchon, A., Lopes-Pereira, P., Dauphinot, L., de Chaumont, F., et al. (2011). Specific targeting of the GABA-A receptor {alpha}5 subtype by a selective inverse agonist restores cognitive deficits in Down syndrome mice. *Journal of Psychopharmacology, 25*(8), 1030–1042.

Brault, V., Besson, V., Magnol, L., Duchon, A., & Hérault, Y. (2007). Cre/loxP-mediated chromosome engineering of the mouse genome. *Handbook of Experimental Pharmacology, 189*, 1487–1495.

Brault, V., Pereira, P., Duchon, A., & Hérault, Y. (2006). Modeling chromosomes in mouse to explore the function of genes, genomic disorders, and chromosomal organization. *PLoS Genetics, 2*, e86.

Benavides-Piccione, R., Dierssen, M., Ballesteros-Yanez, I., Martinez de Lagran, M., Arbones, M. L., Fotaki, V., et al. (2005). Alterations in the phenotype of neocortical pyramidal cells in the Dyrk1A+/- mouse. *Neurobiology of Disease, 20*, 115–122.

Busciglio, J., Pelsman, A., Helguera, P., Ashur-Fabian, O., Pinhasov, A., Brenneman, D. E., et al. (2007). NAP and ADNF-9 protect normal and Down's syndrome cortical neurons from oxidative damage and apoptosis. *Current Pharmaceutical Design, 13*, 1091–1098.

Cardenas, A. M., Allen, D. D., Arriagada, C., Olivares, A., Bennett, L. B., Caviedes, R., et al. (2002a). Establishment and characterization of immortalized neuronal cell lines derived from the spinal cord of normal and trisomy 16 fetal mice, an animal model of Down syndrome. *Journal of Neuroscience Research*, *68*, 46–58.

Cardenas, A. M., Arriagada, C., Allen, D. D., Caviedes, R., Cortes, J. F., Martin, J., et al. (2002b). Cell lines derived from hippocampal neurons of the normal and trisomy 16 mouse fetus (a model for Down syndrome) exhibit neuronal markers, cholinergic function, and functional neurotransmitter receptors. *Experimental Neurology*, *177*, 159–170.

Cardenas, A. M., Rodriguez, M. P., Cortes, M. P., Alvarez, R. M., Wei, W. Z., Rapoport, S. I., et al. (1999). Calcium signals in cell lines derived from the cerebral cortex of normal and trisomy 16 mice. *Neuroreport*, *10*, 363–369.

Cataldo, A. M., Petanceska, S., Peterhoff, C. M., Terio, N. B., Epstein, C. J., Villar, A., et al. (2003). App gene dosage modulates endosomal abnormalities of Alzheimer's disease in a segmental trisomy 16 mouse model of Down syndrome. *The Journal of Neuroscience*, *23*, 6788–6792.

Caviedes, P., Caviedes, R., & Rapoport, S. I. (2006). Altered calcium currents in cultured sensory neurons of normal and trisomy 16 mouse fetuses, an animal model for human trisomy 21 (Down Syndrome). *Biological Research*, *39*, 471–481.

Chabert, C., Jamon, M., Cherfouh, A., Duquenne, V., Smith, D. J., Rubin, E., et al. (2004). Functional analysis of genes implicated in Down syndrome: 1. Cognitive abilities in mice transpolygenic for Down syndrome chromosomal region-1 (DCR-1). *Behavior Genetics*, *34*, 559–569.

Chakrabarti, L., Best, T. K., Cramer, N. P., Carney, R. S. E., Isaac, J. T. R., Galdzicki, Z., et al. (2010). Olig1 and Olig2 triplication causes developmental brain defects in Down syndrome. *Nature Neuroscience*, *13*, 927–934 U939.

Chakrabarti, L., Galdzicki, Z., & Haydar, T. F. (2007). Defects in embryonic neurogenesis and initial synapse formation in the forebrain of the Ts65Dn mouse model of Down syndrome. *The Journal of Neuroscience*, *27*, 11483–11495.

Chang, Q., & Gold, P. E. (2008). Age-related changes in memory and in acetylcholine functions in the hippocampus in the Ts65Dn mouse, a model of Down syndrome. *Neurobiology of Learning and Memory*, *89*, 167–177.

Chang, K. T., & Min, K. T. (2005). Drosophila melanogaster homolog of Down syndrome critical region 1 is critical for mitochondrial function. *Nature Neuroscience*, *8*, 1577–1585.

Chrast, R., Scott, H. S., Madani, R., Huber, L., Wolfer, D. P., Prinz, M., et al. (2000). Mice trisomic for a bacterial artificial chromosome with the single-minded 2 gene (Sim2) show phenotypes similar to some of those present in the partial trisomy 16 mouse models of Down syndrome. *Human Molecular Genetics*, *9*, 1853–1864.

Clark, S., Schwalbe, J., Stasko, M. R., Yarowsky, P. J., & Costa, A. C. S. (2006). Fluoxetine rescues deficient neurogenesis in hippocampus of the Ts65Dn mouse model for Down syndrome. *Experimental Neurology*, *200*, 256–261.

Contestabile, A., Fila, T., Ceccarelli, C., Bonasoni, P., Bonapace, L., Santini, D., et al. (2007). Cell cycle alteration and decreased cell proliferation in the hippocampal dentate gyrus and in the neocortical germinal matrix of fetuses with Down syndrome and in Ts65Dn mice. *Hippocampus*, *17*, 665–678.

Costa, A. C. S., & Grybko, M. J. (2005). Deficits in hippocampal CA1 LTP induced by TBS but not HFS in the Ts65Dn mouse: A model of Down syndrome. *Neuroscience Letters*, *382*, 317–322.

Costa, A. C. S., Scott-McKean, J. J., & Stasko, M. R. (2008). Acute injections of the NMDA receptor antagonist memantine rescue performance deficits of the Ts65Dn mouse model of Down syndrome on a fear conditioning test. *Neuropsychopharmacology*, *33*, 1624–1632.

Coussons-Read, M. E., & Crnic, L. S. (1996). Behavioral assessment of the Ts65Dn mouse, a model for Down syndrome: Altered behavior in the elevated plus maze and open field. *Behavior Genetics*, *26*, 7–13.

Dauphinot, L., Lyle, R., Rivals, I., Dang, M. T., Moldrich, R. X., Golfier, G., et al. (2005). The cerebellar transcriptome during postnatal development of the Ts1Cje mouse, a segmental trisomy model for Down syndrome. *Human Molecular Genetics*, *14*, 373–384.

Davisson, M. T., Schmidt, C., & Akeson, E. C. (1990). Segmental trisomy of murine chromosome 16: A new model system for studying Down syndrome. *Progress in Clinical and Biological Research*, *360*, 263–280.

Davisson, M. T., Schmidt, C., Reeves, R. H., Irving, N. G., Akeson, E. C., Harris, B. S., et al. (1993). Segmental trisomy as a mouse model for Down syndrome. *Progress in Clinical and Biological Research*, *384*, 117–133.

Deitz, S. L., & Roper, R. J. (2011). Trisomic and allelic differences influence phenotypic variability during development of Down syndrome mice. *Genetics*, *189*(4), 1487–1495.

Demas, G. E., Nelson, R. J., Krueger, B. K., & Yarowsky, O. J. (1996). Spatial memory deficits in segmental trisomic Ts65Dn mice. *Behavioural Brain Research*, *82*, 85–92.

Demas, G. E., Nelson, R. J., Krueger, B. K., & Yarowsky, P. J. (1998). Impaired spatial working and reference memory in segmental trisomy (Ts65Dn) mice. *Behavioural Brain Research*, *90*, 199–201.

Dierssen, M., Arque, G., McDonald, J., Andreu, N., Martinez-Cue, C., Florez, J., et al. (2011). Behavioral characterization of a mouse model overexpressing DSCR1/RCAN1. *PLoS One*, *6*(2), e17010.

Dierssen, M., Benavides-Piccione, R., Martinez-Cue, C., Estivill, X., Florez, J., Elston, G. N., et al. (2003). Alterations of neocortical pyramidal cell phenotype in the Ts65Dn mouse model of Down syndrome: Effects of environmental enrichment. *Cerebral Cortex*, *13*, 758–764.

Dierssen, M., Herault, Y., & Estivill, X. (2009). Aneuploidy: From a physiological mechanism of variance to Down syndrome. *Physiological Reviews*, *89*, 887–920.

Donato, R., Sorci, G., Riuzzi, F., Arcuri, C., Bianchi, R., Brozzi, F., et al. (2009). S100B's double life: Intracellular regulator and extracellular signal. *Biochimica et Biophysica Acta*, *1793*, 1008–1022.

Dorsey, S. G., Bambrick, L. L., Balice-Gordon, R. J., & Krueger, B. K. (2002). Failure of brain-derived neurotrophic factor-dependent neuron survival in mouse trisomy 16. *The Journal of Neuroscience*, *22*, 2571–2578.

Dorsey, S. G., Renn, C. L., Carim-Todd, L., Barrick, C. A., Bambrick, L., Krueger, B. K., et al. (2006). In vivo restoration of physiological levels of truncated TrkB.T1 receptor rescues neuronal cell death in a trisomic mouse model. *Neuron*, *51*, 21–28.

Duce, J. A., Tsatsanis, A., Cater, M. A., James, S. A., Robb, E., Wikhe, K., et al. (2010). Iron-export ferroxidase activity of beta-amyloid precursor protein is inhibited by zinc in Alzheimer's disease. *Cell*, *142*, 857–867.

Duchon, A., Pothion, S., Brault, V., Sharp, A. J., Tybulewicz, V. L. J., Fisher, E. M. C., et al. (2011a). The telomeric part of the human chromosome 21 from Cstb to Prmt2 is not necessary for the locomotor and short-term memory deficits observed in the Tc1 mouse model of Down syndrome. *Behavioural Brain Research*, *217*, 271–281.

Duchon, A., Raveau, M., Chevalier, C., Nalesso, V., Sharp, A. J., & Herault, Y. (2011b). Identification of the translocation breakpoints in the Ts65Dn and Ts1Cje mouse lines: Relevance for modeling Down syndrome. *Mammalian Genome*, *22*(11-12), 674–684.

Elson, A., Bernstein, Y., Degani, H., Levanon, D., Benhur, H., & Groner, Y. (1992). Gene dosage and Downs-syndrome—Metabolic and enzymatic changes in PC12 cells overexpressing transfected human liver-type phosphofructokinase. *Somatic Cell and Molecular Genetics*, *18*, 143–161.

Elson, A., Levanon, D., Weiss, Y., & Groner, Y. (1994). Overexpression of Liver-type phosphofructokinase (PFKL) in transgenic-PFKL Mice—Implication for gene dosage in Trisomy-21. *The Biochemical Journal*, *299*, 409–415.

Ema, M., Ikegami, S., Hosoya, T., Mimura, J., Ohtani, H., Nakao, K., et al. (1999). Mild impairment of learning and memory in mice overexpressing the mSim2 gene located on chromosome 16: An animal model of Down's syndrome. *Human Molecular Genetics*, *8*, 1409–1415.

Epstein, C. J., Avraham, K. B., Lovett, M., Smith, S., Elroystein, O., Rotman, G., et al. (1987). Transgenic mice with increased Cu/Zn-superoxide dismutase activity—Animal-model of dosage effects in Down-syndrome. *Proceedings of the National Academy of Sciences of the United States of America*, *84*, 8044–8048.

Ermak, G., Cheadle, C., Becker, K. G., Harris, C. D., & Davies, K. J. A. (2004). DSCR1 (Adapt78) modulates expression of SOD1. *The FASEB Journal*, *18*, 62–69.

Ermak, G., Harris, C. D., & Davies, K. J. A. (2002). The DSCR1 (Adapt78) isoform 1 protein calcipressin 1 inhibits calcineurin and protects against acute calcium-mediated stress damage, including transient oxidative stress. *The FASEB Journal*, *16*(8), 814–824.

Escorihuela, R. M., Vallina, I. F., Martínez-Cué, C., Baamonde, C., Dierssen, M., Tobeña, A., et al. (1998). Impaired short- and long-term memory in Ts65Dn mice, a model for Down syndrome. *Neuroscience Letters*, *247*, 171–174.

Faizi, M., Bader, P. L., Tun, C., Encarnacion, A., Kleschevnikov, A., Belichenko, P., et al. (2011). Comprehensive behavioral phenotyping of Ts65Dn mouse model of Down syndrome: Activation of pradrenergic receptor by xamoterol as a potential cognitive enhancer. *Neurobiology of Disease*, *43*, 397–413.

Fernandez, F., & Garner, C. C. (2007). Object recognition memory is conserved in Ts1Cje, a mouse model of Down syndrome. *Neuroscience Letters*, *421*, 137–141.

Fernandez, F., & Garner, C. C. (2008). Episodic-like memory in Ts65Dn, a mouse model of Down syndrome. *Behavioural Brain Research*, *188*, 233–237.

Fernandez, F., Morishita, W., Zuniga, E., Nguyen, J., Blank, M., Malenka, R. C., et al. (2007). Pharmacotherapy for cognitive impairment in a mouse model of Down syndrome. *Nature Neuroscience*, *10*, 411–413.

Fotaki, V., Dierssen, M., Alcantara, S., Martinez, S., Marti, E., Casas, C., et al. (2002). Dyrk1A haploinsufficiency affects viability and causes developmental delay and abnormal brain morphology in mice. *Molecular Cell Biology*, *22*, 6636–6647.

Fournier, R. E. K., & Ruddle, F. H. (1977). Microcell-mediated transfer of murine chromosomes into mouse, Chinese-hamster, and human somatic-cells. *Proceedings of the National Academy of Sciences of the United States of America*, *74*, 319–323.

Frederikse, P. H., & Ren, X. O. (2002). Lens defects and age-related fiber cell degeneration in a mouse model of increased APPP gene dosage in Down syndrome. *The American Journal of Pathology*, *161*, 1985–1990.

Friend, W. C., Clapoff, S., Landry, C., Becker, L. E., Ohanlon, D., Allore, R. J., et al. (1992). Cell-specific expression of high-levels of human S100-beta in transgenic mouse-brain is dependent on gene dosage. *The Journal of Neuroscience*, *12*, 4337–4346.

Fuentes, J. J., Genesca, L., Kingsbury, T. J., Cunningham, K. W., Perez-Riba, M., Estivill, X., et al. (2000). DSCR1, overexpressed in Down syndrome, is an inhibitor of calcineurin-mediated signaling pathways. *Human Molecular Genetics*, *9*, 1681–1690.

Fuentes, J. J., Pritchard, M. A., Planas, A. M., Bosch, A., Ferrer, I., & Estivill, X. (1995). A new human gene from the Down-syndrome critical region encodes a proline-rich protein highly expressed in fetal brain and heart. *Human Molecular Genetics*, *4*, 1935–1944.

Galante, M., Jani, H., Vanes, L., Daniel, H., Fisher, E. M. C., Tybulewicz, V. L. J., et al. (2009). Impairments in motor coordination without major changes in cerebellar plasticity in the Tc1 mouse model of Down syndrome. *Human Molecular Genetics*, *18*, 1449–1463.

Games, D., Adams, D., Alessandrini, R., Barbour, R., Berthelette, P., Blackwell, C., et al. (1995). Alzheimer-type neuropathology in transgenic mice overexpressing V717F beta-amyloid precursor protein. *Nature*, *373*, 523–527.

Gardiner, K., Fortna, A., Bechtel, L., & Davisson, M. T. (2003). Mouse models of Down syndrome: How useful can they be? Comparison of the gene content of human chromosome 21 with orthologous mouse genomic regions. *Gene*, *318*, 137–147.

Gerlai, R., & Roder, J. (1993). Female-specific hyperactivity in S100-beta transgenic mice does not habituate in open-field. *Behavioural Brain Research*, *59*, 119–124.

Gerlai, R., & Roder, J. (1996). Spatial and nonspatial learning in mice: Effects of S100 beta overexpression and age. *Neurobiology of Learning and Memory*, *66*, 143–154.

Granholm, A. C., Sanders, L. A., & Crnic, L. S. (2000). Loss of cholinergic phenotype in basal forebrain coincides with cognitive decline in a mouse model of Down's syndrome. *Experimental Neurology*, *161*, 647–663.

Groner, Y., Elroystein, O., Avraham, K. B., Yarom, R., Schickler, M., Knobler, H., et al. (1990). Down syndrome clinical symptoms are manifested in transfected cells and transgenic mice overexpressing the human Cu/Zn-superoxide dismutase gene. *Journal de Physiologie*, *84*, 53–77.

Gropp, A., Kolbus, U., & Giers, D. (1975). Systematic approach to study of trisomy in mouse. 2. *Cytogenetics and Cell Genetics*, *14*, 42–62.

Guedj, F., Sebrie, C., Rivals, I., Ledru, A., Paly, E., Bizot, J. C., et al. (2009). Green tea polyphenols rescue of brain defects induced by overexpression of DYRK1A. *PLoS One*, *4*(2), e4606.

Guimera, J., Casas, C., Pucharcos, C., Solans, A., Domenech, A., Planas, A. M., et al. (1996). A human homologue of Drosophila minibrain (MNB) is expressed in the neuronal regions affected in Down syndrome and maps to the critical region. *Human Molecular Genetics*, *5*, 1305–1310.

Gurney, M. E., Pu, H. F., Chiu, A. Y., Dalcanto, M. C., Polchow, C. Y., Alexander, D. D., et al. (1994). Motor-neuron degeneration in mice that express a human Cu, Zn superoxide-dismutase mutation. *Science*, *264*, 1772–1775.

Hammerle, B., Carnicero, A., Elizalde, C., Ceron, J., Martinez, S., & Tejedor, F. J. (2003). Expression patterns and subcellular localization of the Down syndrome candidate protein MNB/DYRK1A suggest a role in late neuronal differentiation. *The European Journal of Neuroscience*, *17*, 2277–2286.

Harris-Cerruti, C., Kamsler, A., Kaplan, B., Lamb, B., Segal, M., & Groner, Y. (2004). Functional and morphological alterations in compound transgenic mice overexpressing Cu/Zn superoxide dismutase and amyloid precursor protein. *The European Journal of Neuroscience*, *19*, 1174–1190 2913.

Haydar, T. F., Blue, M. E., Molliver, M. E., Krueger, B. K., & Yarowsky, P. J. (1996). Consequences of trisomy 16 for mouse brain development: Corticogenesis in a model of Down syndrome. *The Journal of Neuroscience*, *16*, 6175–6182.

Haydar, T. F., Nowakowski, R. S., Yarowsky, P. J., & Krueger, B. K. (2000). Role of founder cell deficit and delayed neuronogenesis in microencephaly of the trisomy 16 mouse. *The Journal of Neuroscience*, *20*, 4156–4164.

Herault, Y., Rassoulzadegan, M., Cuzin, F., & Duboule, D. (1998). Engineering chromosomes in mice through targeted meiotic recombination (TAMERE). *Nature Genetics*, *20*, 381–384.

Hewitt, C. A., Ling, K. H., Merson, T. D., Simpson, K. M., Ritchie, M. E., King, S. L., et al. (2010). Gene network disruptions and neurogenesis defects in the adult Ts1Cje mouse model of Down syndrome. *PLoS One*, *5*(7), e11561.

Higgins, L. S., Catalano, R., Quon, D., & Cordell, B. (1993). Transgenic mice expressing human beta-APP751, but not Mice expressing beta-APP695, display early Alzheimers disease-like histopathology. In R. M. Nitsch, J. H. Growdon, S. Corkin & R. J. Wurtman (Eds.), *Alzheimers disease: Amyloid precusor proteins, signal transduction, and neuronal transplantation.* (pp. 224–227). Annals of the New York Academy of Sciences.

Hill, C. A., Reeves, R. H., & Richtsmeier, J. T. (2007). Effects of aneuploidy on skull growth in a mouse model of Down syndrome. *Journal of Anatomy*, *210*, 394–405.

Hoeffer, C. A., Dey, A., Sachan, N., Wong, H., Patterson, R. J., Shelton, J. M., et al. (2007). The Down syndrome critical region protein RCAN1 regulates long-term potentiation and memory via inhibition of phosphatase signaling. *The Journal of Neuroscience*, *27*, 13161–13172.

Holtzman, D. M., Li, Y. W., Dearmond, S. J., McKinley, M. P., Gage, F. H., Epstein, C. J., et al. (1992). Mouse model of neurodegeneration—Atrophy of basal forebrain cholinergic neurons in trisomy-16 transplants. *Proceedings of the National Academy of Sciences of the United States of America*, *89*, 1383–1387.

Holtzman, D. M., Santucci, D., Kilbridge, J., ChuaCouzens, J., Fontana, D. J., Daniels, S. E., et al. (1996). Developmental abnormalities and age-related neurodegeneration in a mouse model of Down syndrome. *Proceedings of the National Academy of Sciences of the United States of America*, *93*, 13333–13338.

Hsiao, K. K., Borchelt, D. R., Olson, K., Johannsdottir, R., Kitt, C., Yunis, W., et al. (1995). Age-related CNS disorder and early death in transgenic FVB/N mice overexpressing Alzheimer amyloid precursor proteins. *Neuron*, *15*, 1203–1218.

Hsiao, K., Chapman, P., Nilsen, S., Eckman, C., Harigaya, Y., Younkin, S., et al. (1996). Correlative memory deficits, A beta elevation, and amyloid plaques in transgenic mice. *Science*, *274*, 99–102.

Hyde, L. A., Frisone, D. F., & Crnic, L. S. (2001). Ts65Dn mice, a model for Down syndrome, have deficits in context discrimination learning suggesting impaired hippocampal function. *Behavioural Brain Research*, *118*, 53–60.

Incerti, M., Toso, L., Vink, J., Roberson, R., Nold, C., Abebe, D., et al. (2011). Prevention of learning deficit in a Down syndrome model. *Obstetrics and Gynecology*, *117*, 354–361.

Insausti, A. M., Megías, M., Crespo, D., Cruz-Orive, L. M., Dierssen, M., Vallina, I. F., et al. (1998). Hippocampal volume and neuronal number in Ts65Dn mice: A murine model of Down syndrome. *Neuroscience Letters*, *253*, 175–178.

Ishihara, K., Amano, K., Takaki, E., Ebrahim, A. S., Shimohata, A., Shibazaki, N., et al. (2009). Increased lipid peroxidation in Down's syndrome mouse models. *Journal of Neurochemistry*, *110*, 1965–1976.

Ishihara, K., Amano, K., Takaki, E., Shimohata, A., Sago, H., Epstein, C. J., et al. (2010). Enlarged brain ventricles and impaired neurogenesis in the Ts1Cje and Ts2Cje mouse models of Down syndrome. *Cerebral Cortex*, *20*, 1131–1143.

Iwamoto, K., Bundo, M., Yamamoto, M., Ozawa, H., Saito, T., & Kato, T. (2004). Decreased expression of NEFH and PCP4/PEP19 in the prefrontal cortex of alcoholics. *Neuroscience Research*, *49*, 379–385.

Keating, D. J., Dubach, D., Zanin, M. P., Yu, Y., Martin, K., Zhao, Y. F., et al. (2008). DSCR1/RCAN1 regulates vesicle exocytosis and fusion pore kinetics: Implications for Down syndrome and Alzheimer's disease. *Human Molecular Genetics*, *17*, 1020–1030.

Kleschevnikov, A. M., Belichenko, P. V., Epstein, C. J., Villar, A. J., Malenka, R., & Mobley, W. C. (2004a). Reduced LTP and enhanced inhibition in dentate gyrus of Ts65Dn mice, a model for Down's syndrome. *Neurobiology of Aging*, *25*, S252 S252.

Kleschevnikov, A. M., Belichenko, P. V., Villar, A. J., Epstein, C. J., Malenka, R. C., & Mobley, W. C. (2004b). Hippocampal long-term potentiation suppressed by increased inhibition in the Ts65Dn mouse, a genetic model of Down syndrome. *The Journal of Neuroscience*, *24*, 8153–8160.

Knobler, H., Weiss, Y., Peled, M., & Groner, Y. (1997). Impaired glucose-induced insulin response in transgenic mice overexpressing the L-phosphofructokinase gene. *Diabetes*, *46*, 1414–1418.

Kuo, Y. M., Kokjohn, T. A., Beach, T. G., Sue, L. I., Brune, D., Lopez, J. C., et al. (2001). Comparative analysis of amyloid-beta chemical structure and amyloid plaque morphology of transgenic mouse and Alzheimer's disease brains. *The Journal of Biological Chemistry*, *276*, 12991–12998.

Kurt, M. A., Davies, D. C., Kidd, M., Dierssen, M., & Florez, J. (2000). Synaptic deficit in the temporal cortex of partial trisomy 16 (Ts65Dn) mice. *Brain Research*, *858*, 191–197.

Kurt, M. A., Kafa, M. I., Dierssen, M., & Davies, D. C. (2004). Deficits of neuronal density in CA1 and synaptic density in the dentate gyrus, CA3 and CA1, in a mouse model of Down syndrome. *Brain Research*, *1022*, 101–109.

Lamb, B. T., Sisodia, S. S., Lawler, A. M., Slunt, H. H., Kitt, C. A., Kearns, W. G., et al. (1993). Introduction and expression of the 400 kilobase precursor amyloid protein gene in transgenic mice. *Nature Genetics*, *5*, 22–29.

Le Pecheur, M., Bourdon, E., Paly, E., Farout, L., Friguet, B., & London, J. (2005). Oxidized SOD1 alters proteasome activities in vitro and in the cortex of SOD1 overexpressing mice. *FEBS Letters*, *579*, 3613–3618.

Levin, E. D., Christopher, N. C., Lateef, S., Elamir, B. M., Patel, M., Liang, L. P., et al. (2002). Extracellular superoxide dismutase overexpression protects against aging-induced cognitive impairment in mice. *Behavior Genetics*, *32*, 119–125.

Li, Z., Yu, T., Morishima, M., Pao, A., LaDuca, J., Conroy, J., et al. (2007). Duplication of the entire 22.9 Mb human chromosome 21 syntenic region on mouse chromosome 16 causes cardiovascular and gastrointestinal abnormalities. *Human Molecular Genetics*, *16*, 1359–1366.

Liu, F., Liang, Z. H., Wegiel, J., Hwang, Y. W., Iqbal, K., Grundke-Iqbal, I., et al. (2008). Overexpression of Dyrk1A contributes to neurofibrillary degeneration in Down syndrome. *The FASEB Journal*, *22*, 3224–3233.

Liu, C., Morishima, M., Yu, T., Matsui, S., Zhang, L., Fu, D., et al. (2011). Genetic analysis of Down syndrome-associated heart defects in mice. *Human Genetics*, *130*, 623–632.

Lopes Pereira, P., Magnol, L., Sahun, I., Brault, V., Duchon, A., Prandini, P., et al. (2009). A new mouse model for the trisomy of the Abcg1-U2af1 region reveals the complexity of the combinatorial genetic code of Down syndrome. *Human Molecular Genetics*, *18*, 4756–4769.

Lorenzi, H. A., & Reeves, R. H. (2006). Hippocampal hypocellularity in the Ts65Dn mouse originates early in development. *Brain Research*, *1104*, 153–159.

Mariggio, M. A., Fulle, S., Calissano, P., Nicoletti, I., & Fano, G. (1994). The brain protein S-100ab induces apoptosis in PC12 cells. *Neuroscience*, *60*, 29–35.

Marshak, D. R., Pesce, S. A., Stanley, L. C., & Griffin, W. S. T. (1992). Increased S100-beta neurotrophic activity in Alzheimers-disease temporal-lobe. *Neurobiology of Aging*, *13*, 1–7.

Martinez-Cue, C., Baamonde, C., Lumbreras, M., Paz, J., Davisson, M. T., Schmidt, C., et al. (2002). Differential effects of environmental enrichment on behavior and learning of male and female Ts65Dn mice, a model for Down syndrome. *Behavioural Brain Research*, *134*, 185–200.

Martinez-Cue, C., Baamonde, C., Lumbreras, M. A., Vallina, I. F., Dierssen, M., & Florez, J. (1999). A murine model for Down syndrome shows reduced responsiveness to pain. *Neuroreport*, *10*, 1119–1122.

Martinez-Cue, C., Rueda, N., Garcia, E., Davisson, M. T., Schmidt, C., & Florez, J. (2005). Behavioral, cognitive and biochemical responses to different environmental conditions in male Ts65Dn mice, a model of Down syndrome. *Behavioural Brain Research*, *163*, 174–185.

Moechars, D., Gilis, M., Kuiperi, C., Laenen, I., & Van Leuven, F. (1998). Aggressive behaviour in transgenic mice expressing APP is alleviated by serotonergic drugs. *Neuroreport, 9*, 3561–3564.

Moechars, D., Lorent, K., DeStrooper, B., Dewachter, I., & VanLeuven, F. (1996). Expression in brain of amyloid precursor protein mutated in the alpha-secretase site causes disturbed behavior, neuronal degeneration and premature death in transgenic mice. *The EMBO Journal, 15*, 1265–1274.

Moldrich, R. X., Dauphinot, L., Laffaire, J., Rossier, J., & Potier, M.-C. (2007). Down syndrome gene dosage imbalance on cerebellum development. *Progress in Neurobiology, 82*, 87–94.

Moldrich, R. X., Dauphinot, L., Laffaire, J., Vitalis, T., Herault, Y., Beart, P. M., et al. (2009). Proliferation deficits and gene expression dysregulation in Down's syndrome (Ts1Cje) neural progenitor cells cultured from neurospheres. *Journal of Neuroscience Research, 87*, 3143–3152.

Moon, J., Chen, M., Gandhy, S. U., Strawderman, M., Levitsky, D. A., Maclean, K. N., et al. (2010). Perinatal choline supplementation improves cognitive functioning and emotion regulation in the Ts65Dn mouse model of Down syndrome. *Behavioral Neuroscience, 124*, 346–361.

Moore, C. S. (2006). Postnatal lethality and cardiac anomalies in the Ts65Dn Down syndrome mouse model. *Mammalian Genome, 17*, 1005–1012.

Moran, P. M., Higgins, L. S., Cordell, B., & Moser, P. C. (1995). Age-related learning-deficits in transgenic mice expressing the 751-amino acid isoform of human beta-amyloid precursor protein. *Proceedings of the National Academy of Sciences of the United States of America, 92*, 5341–5345.

Morice, E., Andreae, L. C., Cooke, S. F., Vanes, L., Fisher, E. M. C., Tybulewicz, V. L. J., et al. (2008). Preservation of long-term memory and synaptic plasticity despite short-term impairments in the Tc1 mouse model of Down syndrome. *Learning & Memory, 15*, 492–500.

Mouton-Liger, F., Thomas, S., Rattenbach, R., Magnol, L., Larigaldie, V., Ledru, A., et al. (2011). PCP4 (PEP19) overexpression induces premature neuronal differentiation associated with Ca(2+)/calmodulin-dependent kinase II delta activation in mouse models of Down syndrome. *The Journal of Comparative Neurology, 519*(14), 2779–2802.

Muller, C. M., Akhavan, A. C., & Bette, M. (1993). Possible role of S-100 in glia neuronal signaling involved in activity-dependent plasticity in the developing mammalian cortex. *Journal of Chemical Neuroanatomy, 6*, 215–227.

Nagase, T., Ishikawa, K., Kikuno, R., Hirosawa, M., Nomura, N., & Ohara, O. (1999). Prediction of the coding sequences of unidentified human genes. XV. The complete sequences of 100 new cDNA clones from brain which code for large proteins in vitro. *DNA Research, 6*, 337–345.

Nambu, J. R., Franks, R. G., Hu, S., & Crews, S. T. (1990). The single-minded gene of drosophila is required for the expression of genes important for the development of CNS midline cells. *Cell, 63*, 63–75.

Nambu, J. R., Lewis, J. O., Wharton, K. A., & Crews, S. T. (1991). The drosophila single-minded gene encodes a helix-loop-helix protein that acts as a master regulator of CNS midline development. *Cell, 67*, 1157–1167.

Netzer, W. J., Powell, C., Nong, Y., Blundell, J., Wong, L., Duff, K., et al. (2010). Lowering beta-amyloid levels rescues learning and memory in a Down syndrome mouse model. *PLoS One, 5*, e10943.

Nishiyama, H., Knopfel, T., Endo, S., & Itohara, S. (2002). Glial protein S100B modulates long-term neuronal synaptic plasticity. *Proceedings of the National Academy of Sciences of the United States of America, 99*, 4037–4042.

O'Doherty, A., Ruf, S., Mulligan, C., Hildreth, V., Errington, M. L., Cooke, S., et al. (2005). An aneuploid mouse strain carrying human chromosome 21 with Down syndrome phenotypes. *Science, 309*, 2033–2037.

Oddo, S., Caccamo, A., Shepherd, J. D., Murphy, M. P., Golde, T. E., Kayed, R., et al. (2003). Triple-transgenic model of Alzheimer's disease with plaques and tangles: Intracellular a beta and synaptic dysfunction. *Neuron, 39*, 409–421.

Olson, L. E., Richtsmeier, J. T., Leszl, J., & Reeves, R. H. (2004a). A chromosome 21 critical region does not cause specific Down syndrome phenotypes. *Science, 306*, 687–690.

Olson, L. E., Roper, R. J., Baxter, L. L., Carlson, E. J., Epstein, C. J., & Reeves, R. H. (2004b). Down syndrome mouse models Ts65Dn, Ts1Cje, and Ms1Cje/Ts65Dn exhibit variable severity of cerebellar phenotypes. *Developmental Dynamics, 230*, 581–589.

Olson, L. E., Roper, R. J., Sengstaken, C. L., Peterson, E. A., Aquino, V., Galdzicki, Z., et al. (2007). Trisomy for the Down syndrome 'critical region' is necessary but not sufficient for brain phenotypes of trisomic mice. *Human Molecular Genetics, 16*, 774–782.

Oury, T. D., Ho, Y. S., Piantadosi, C. A., & Crapo, J. D. (1992). Extracellular superoxide dismutase, nitric oxide, and central nervous system O2 toxicity. *Proceedings of the National Academy of Sciences of the United States of America, 89*, 9715–9719.

Ortiz-Abalia, J., Sahún, I., Altafaj, X., Andreu, N., Estivill, X., Dierssen, M., et al. (2008). Targeting Dyrk1A with AAVshRNA attenuates motor alterations in TgDyrk1A, a mouse model of Down syndrome. *American Journal of Human Genetics, 83*, 479–488.

Peled-Kamar, M., Degani, H., Bendel, P., Margalit, R., & Groner, Y. (1998). Altered brain glucose metabolism in transgenic-PFKL mice with elevated L-phosphofructokinase: In vivo NMR studies. *Brain Research, 810*, 138–145.

Perez-Cremades, D., Hernandez, S., Blasco-Ibanez, J. M., Crespo, C., Nacher, J., & Varea, E. (2010). Alteration of inhibitory circuits in the somatosensory cortex of Ts65Dn mice, a model for Down's syndrome. *Journal of Neural Transmission, 117*, 445–455.

Popov, V. I., Kleschevnikov, A. M., Klimenko, O. A., Stewart, M. G., & Belichenko, P. V. (2011). Three-dimensional synaptic ultrastructure in the dentate gyrus and hippocampal area CA3 in the Ts65Dn mouse model of Down syndrome. *The Journal of Comparative Neurology, 519*, 1338–1354.

Pratico, D., Uryu, K., Leight, S., Trojanowswki, J. Q., & Lee, V. M. Y. (2001). Increased lipid peroxidation precedes amyloid plaque formation in an animal model of Alzheimer amyloidosis. *The Journal of Neuroscience, 21*, 4183–4187.

Priller, C., Bauer, T., Mitteregger, G., Krebs, B., Kretzschmar, H. A., & Herms, J. (2006). Synapse formation and function is modulated by the amyloid precursor protein. *The Journal of Neuroscience, 26*, 7212–7221.

Pruitt, K. D., Tatusova, T., & Maglott, D. R. (2003). NCBI reference sequence project: Update and current status. *Nucleic Acids Research, 31*, 34–37.

Quon, D., Wang, Y., Catalano, R., Scardina, J. M., Murakami, K., & Cordell, B. (1991). Formation of beta-amyloid protein deposits in brains of transgenic mice. *Nature, 352*, 239–241.

Rachidi, M., Lopes, C., Costantine, M., & Delabar, J. M. (2005). C21orf5, a new member of Dopey family involved in morphogenesis, could participate in neurological alterations and mental retardation in Down syndrome. *DNA Research, 12*, 203–210.

Rachidi, M., Lopes, C., VayssetteS, C., Smith, D. J., Rubin, E. M., & Delabar, J. M. (2007). New cerebellar phenotypes in YAC transgenic mouse in vivo library of human Down syndrome critical region-1. *Biochemical and Biophysical Research Communications, 364*, 488–494.

Ramirez-Solis, R., Liu, P., & Bradley, A. (1995). Chromosome engineering in mice. *Nature, 378*, 720–724.

Reeves, R. H., Irving, N. G., Moran, T. H., Wohn, A., Kitt, C., Sisodia, S. S., et al. (1995). A mouse model for Down syndrome exhibits learning and behaviour deficits. *Nature Genetics, 11*, 177–184.

Reeves, R. H., Yao, J. B., Crowley, M. R., Buck, S., Zhang, X. F., Yarowsky, P., et al. (1994). Astrocytosis and axonal proliferation in the hippocampus of S100b transgenic mice. *Proceedings of the National Academy of Sciences of the United States of America, 91*, 5359–5363.

Reinholdt, L. G., Ding, Y., Gilbert, G. T., Czechanskibr, A., Solzak, J. P., Roper, R. J., et al. (2011). Molecular characterization of the translocation breakpoints in the Down syndrome mouse model Ts65Dn. *Mammalian genome, 22*(11-12), 685–691.

Reynolds, L. E., Watson, A. R., Baker, M., Jones, T. A., D'Amico, G., Robinson, S. D., et al. (2010). Tumour angiogenesis is reduced in the Tc1 mouse model of Down's syndrome. *Nature, 465*, 813–817.

Richardson, J. A., & Burns, D. K. (2002). Mouse models of Alzheimer's disease: A quest for plaques and tangles. *ILAR Journal, 43*, 89–99.

Roper, R. J., Baxter, L. L., Saran, N. G., Klinedinst, D. K., Beachy, P. A., & Reeves, R. H. (2006). Defective cerebellar response to mitogenic Hedgehog signaling in Down's syndrome mice. *Proceedings of the National Academy of Sciences of the United States of America, 103*, 1452–1456.

Rothermundt, M., Peters, M., Prehn, J. H. M., & Arolt, V. (2003). S100B in brain damage and neurodegeneration. *Microscopy Research and Technique, 60*, 614–632.

Rueda, N., Florez, J., & Martinez-Cue, C. (2008). Chronic pentylenetetrazole but not donepezil treatment rescues spatial cognition in Ts65Dn mice, a model for Down syndrome. *Neuroscience Letters, 433*, 22–27.

Rueda, N., Llorens-Martin, M., Florez, J., Valdizan, E., Banerjee, P., Trejo, J. L., et al. (2010). Memantine normalizes several phenotypic features in the Ts65Dn mouse model of Down syndrome. *Journal of Alzheimer's Disease, 21*, 277–290.

Ryoo, S. R., Cho, H. J., Lee, H. W., Jeong, H. K., Radnaabazar, C., Kim, Y. S., et al. (2008). Dual-specificity tyrosine(Y)-phosphorylation regulated kinase 1A-mediated phosphorylation of amyloid precursor protein: Evidence for a functional link between Down syndrome and Alzheimer's disease. *Journal of Neurochemistry, 104*, 1333–1344.

Ryoo, S. R., Jeong, H. K., Radnaabazar, C., Yoo, J. J., Cho, H. J., Lee, H. W., et al. (2007). DYRK1A-mediated hyperphosphorylation of Tau—A functional link between Down syndrome and Alzheimer disease. *The Journal of Biological Chemistry, 282*, 34850–34857.

Sago, H., Carlson, E. J., Smith, D. J., Kilbridge, J., Rubin, E. M., Mobley, W. C., et al. (1998). Ts1Cje, a partial trisomy 16 mouse model for Down syndrome, exhibits learning and behavioral abnormalities. *Proceedings of the National Academy of Sciences of the United States of America, 95*, 6256–6261.

Sago, H., Carlson, E. J., Smith, D. J., Rubin, E. M., Crnic, L. S., Huang, T. T., et al. (2000). Genetic dissection of region associated with behavioral abnormalities in mouse models for Down syndrome. *Pediatric Research, 48*, 606–613.

Salehi, A., Delcroix, J. D., Belichenko, P. V., Zhan, K., Wu, C., Valletta, J. S., et al. (2006). Increased App expression in a mouse model of Down's syndrome disrupts NGF transport and causes cholinergic neuron degeneration. *Neuron, 51*, 29–42.

Salehi, A., Faizi, M., Colas, D., Valletta, J., Laguna, J., Takimoto-Kimura, R., et al. (2009). Restoration of norepinephrine-modulated contextual memory in a mouse model of Down syndrome. *Science Translational Medicine, 1*(7), 7ra17.

Sebrie, C., Chabert, C., Ledru, A., Guedj, F., Po, C., Smith, D. J., et al. (2008). Increased dosage of DYRK1A and brain volumetric alterations in a YAC model of partial trisomy 21. *The Anatomical Record: Advances in Integrative Anatomy and Evolutionary Biology, 291*, 254–262.

Seo, H., & Isacson, O. (2005). Abnormal APP, cholinergic and cognitive function in Ts65Dn Down's model mice. *Experimental Neurology, 193*, 469–480.

Shapiro, L. A., Marks, A., & Whitaker-Azmitia, P. M. (2004). Increased clusterin expression in old but not young adult S100B transgenic mice: Evidence of neuropathological aging in a model of Down Syndrome. *Brain Research, 1010*, 17–21.

Shapiro, L. A., & Whitaker-Azmitia, P. M. (2004). Expression levels of cytoskeletal proteins indicate pathological aging of S100B transgenic mice: An immunohistochemical study of MAP-2, drebrin and GAP-43. *Brain Research, 1019*, 39–46.

Shinohara, T., Tomizuka, K., Miyabara, S., Takehara, S., Kazuki, Y., Inoue, J., et al. (2001). Mice containing a human chromosome 21 model behavioral impairment and cardiac anomalies of Down's syndrome. *Human Molecular Genetics, 10*, 1163–1175.

Shukkur, E. A., Shimohata, A., Akagi, T., Yu, W. X., Yamaguchi, M., Murayama, M., et al. (2006). Mitochondrial dysfunction and tau hyperphosphorylation in Ts1Cje, a mouse model for Down syndrome. *Human Molecular Genetics, 15*, 2752–2762.

Siarey, R. J., Carlson, E. J., Epstein, C. J., Balbo, A., Rapoport, S. I., & Galdzicki, Z. (1999). Increased synaptic depression in the Ts65Dn mouse, a model for mental retardation in Down syndrome. *Neuropharmacology, 38*, 1917–1920.

Siarey, R. J., Stoll, J., Rapoport, S. I., & Galdzicki, Z. (1997). Altered long-term potentiation in the young and old Ts65Dn mouse, a model for Down Syndrome. *Neuropharmacology, 36*, 1549–1554.

Siarey, R. J., Villar, A. J., Epstein, C. J., & Galdzicki, Z. (2005). Abnormal synaptic plasticity in the Ts1Cje segmental trisomy 16 mouse model of Down syndrome. *Neuropharmacology, 49*, 122–128.

Skold, K., Svensson, M., Nilsson, A., Zhang, X. Q., Nydahl, K., Caprioli, R. M., et al. (2006). Decreased striatal levels of PEP-19 following MPTP lesion in the mouse. *Journal of Proteome Research, 5*, 262–269.

Slemmon, J. R., Morgan, J. I., Fullerton, S. M., Danho, W., Hilbush, B. S., & Wengenack, T. M. (1996). Camstatins are peptide antagonists of calmodulin based upon a conserved structural motif in PEP-19, neurogranin, and neuromodulin. *The Journal of Biological Chemistry, 271*, 15911–15917.

Smith, D. J., & Rubin, E. M. (1997). Functional screening and complex traits: Human 21q22.2 sequences affecting learning in mice. *Human Molecular Genetics, 6*, 1729–1733.

Smith, D. J., Stevens, M. E., Cheng, J. F., & Rubin, E. M. (1997a). Identification of genes affecting learning and memory in chromosome 21 YAC transgenic mice. *Cytogenetics and Cell Genetics, 79*, 49 49.

Smith, D. J., Stevens, M. E., Sudanagunta, S. P., Bronson, R. T., Makhinson, M., Watabe, A. M., et al. (1997b). Functional screening of 2 Mb of human chromosome 21q22.2 in transgenic mice implicates minibrain in learning defects associated with Down syndrome. *Nature Genetics, 16*, 28–36.

Smith, D. J., Zhu, Y. W., Zhang, J. L., Cheng, J. F., & Rubin, E. M. (1995). Construction of a panel of transgenic mice containing a contiguous 2-Mb set of YAC/P1 clones from human-chromosome 21q22.2. *Genomics, 27*, 425–434.

Song, W. J., Sternberg, L. R., KastenSportes, C., VanKeuren, M. L., Chung, S. H., Slack, A. C., et al. (1996). Isolation of human and murine homologues of the Drosophila minibrain gene: Human homologue maps to 21q22.2 in the Down syndrome "critical region" *Genomics, 38*, 331–339.

Stasko, M. R., & Costa, A. C. S. (2004). Experimental parameters affecting the Morris water maze performance of a mouse model of Down syndrome. *Behavioural Brain Research, 154*, 1–17.

Stewart, L. S., Persinger, M. A., Cortez, M. A., & Snead, O. C. (2007). Chronobiometry of behavioral activity in the Ts65Dn model of Down syndrome. *Behavior Genetics, 37*, 388–398.

Sumarsono, S. H., Wilson, T. J., Tymms, M. J., Venter, D. J., Corrick, C. M., Kola, R., et al. (1996). Down's syndrome-like skeletal abnormalities in Ets2 transgenic mice. *Nature, 379*, 534–537.

Sweeney, J. E., Hohmann, C. F., Ostergranite, M. L., & Coyle, J. T. (1989). Neurogenesis of the basal forebrain in euploid and trisomy-16 mice—An animal-model for developmental disorders in Down syndrome. *Neuroscience, 31*, 413–425.

Tecott, L. H. (2003). The genes and brains of mice and men. *The American Journal of Psychiatry, 160*, 646–656.

Tejedor, F., Zhu, X. R., Kaltenbach, E., Ackermann, A., Baumann, A., Canal, I., et al. (1995). Minibrain—A new-protein kinase family involved in postembryonic neurogenesis in drosophila. *Neuron, 14*, 287–301.

Trazzi, S., Mitrugno, V. M., Valli, E., Fuchs, C., Rizzi, S., Guidi, S., et al. (2011). APP-dependent up-regulation of Ptch1 underlies proliferation impairment of neural precursors in Down syndrome. *Human Molecular Genetics, 20*, 1560–1573.

Turner, P. R., O'Connor, K., Tate, W. P., & Abraham, W. C. (2003). Roles of amyloid precursor protein and its fragments in regulating neural activity, plasticity and memory. *Progress in Neurobiology, 70*, 1–32.

Utal, A. K., Stopka, A. L., Roy, M., & Coleman, P. D. (1998). PEP-19 immunohistochemistry defines the basal ganglia and associated structures in the adult human brain, and is dramatically reduced in Huntington's disease. *Neuroscience, 86*, 1055–1063.

Vacik, T., Ort, M., Gregorova, S., Strnad, P., Blatny, R., Conte, N., et al. (2005). Segmental trisomy of chromosome 17: A mouse model of human aneuploidy syndromes. *Proceedings of the National Academy of Sciences of the United States of America, 102*, 4500–4505.

Vankeuren, M., Drabkin, H., Hart, I., Harker, D., Patterson, D., & Vora, S. (1986). Regional assignment of human liver-type 6-phosphofructokinase to chromosome 21q22.3 by using somatic-cell hybrids and a monoclonal anti-l antibody. *Human Genetics*, *74*, 34–40.

Venter, J. C., Adams, M. D., Myers, E. W., Li, P. W., Mural, R. J., Sutton, G. G., et al. (2001). The sequence of the human genome. *Science*, *291*, 1304–1351.

Villar, A. J., Belichenko, P. V., Gillespie, A. M., Kozy, H. M., Mobley, W. C., & Epstein, C. J. (2005). Identification and characterization of a new Down syndrome model, Ts Rb (12.17(16)) 2Cje, resulting from a spontaneous Robertsonian fusion between T(17(16))65Dn and mouse Chromosome 12. *Mammalian Genome*, *16*, 79–90.

Vora, S., & Francke, U. (1981). Assignment of the human-gene for liver-type 6-phosphofructokinase isoenzyme (PFKL) to chromosome 21 by using somatic-cell hybrids and monoclonal anti-L antibody. *Proceedings of the National Academy of Sciences of the United States of America*, *78*, 3738–3742.

Wei, P., Blundon, J. A., Rong, Y., Zakharenko, S. S., & Morgan, J. I. (2011). Impaired locomotor learning and altered cerebellar synaptic plasticity in pep-19/PCP4-null mice. *Molecular and Cellular Biology*, *31*, 2838–2844.

Wenger, G. R., Schmidt, C., & Davisson, M. T. (2004). Operant conditioning in the Ts65Dn mouse: Learning. *Behavior Genetics*, *34*, 105–119.

Westmark, C. J., Westmark, P. R., & Malter, J. S. (2010). Alzheimer's disease and Down syndrome rodent models exhibit audiogenic seizures. *Journal of Alzheimer's Disease*, *20*, 1009–1013.

Whitaker-Azmitia, P. M., Wingate, M., Borella, A., Gerlai, R., Roder, J., & Azmitia, E. C. (1997). Transgenic mice overexpressing the neurotrophic factor S-100 beta show neuronal cytoskeletal and behavioral signs of altered aging processes: Implications for Alzheimer's disease and Down's syndrome. *Brain Research*, *776*, 51–60.

Winocur, G., Roder, J., & Lobaugh, N. (2001). Learning and memory in S100-beta transgenic mice: An analysis of impaired and preserved function. *Neurobiology of Learning and Memory*, *75*, 230–243.

Wolvetang, E. J., Bradfield, O. M., Hatzistavrou, T., Crack, P. J., Busciglio, J., Kola, I., et al. (2003). Overexpression of the chromosome 21 transcription factor Ets2 induces neuronal apoptosis. *Neurobiology of Disease*, *14*, 349–356.

Yan, W., Wilson, C. C., & Haring, J. H. (1997). 5-HT1A receptors mediate the neurotrophic effect of serotonin on developing dentate granule cells. *Brain Research. Developmental Brain Research*, *98*, 185–190.

Yarom, R., Sapoznikov, D., Havivi, Y., Avraham, K. B., Schickler, M., & Groner, Y. (1988). Premature aging changes in neuromuscular-junctions of transgenic mice with an extra human cuznsod gene—A model for tongue pathology in Downs-syndrome. *Journal of the Neurological Sciences*, *88*, 41–53.

Yu, T., Clapcote, S. J., Li, Z., Liu, C., Pao, A., Bechard, A. R., et al. (2010a). Deficiencies in the region syntenic to human 21q22.3 cause cognitive deficits in mice. *Mammalian Genome*, *21*, 258–267.

Yu, T., Li, Z., Jia, Z., Clapcote, S. J., Liu, C., Li, S., et al. (2010b). A mouse model of Down syndrome trisomic for all human chromosome 21 syntenic regions. *Human Molecular Genetics*, *19*, 2780–2791.

Yu, T., Liu, C. H., Belichenko, P., Clapcote, S. J., Li, S. M., Pao, A. N., et al. (2010c). Effects of individual segmental trisomies of human chromosome 21 syntenic regions on hippocampal long-term potentiation and cognitive behaviors in mice. *Brain Research*, *1366*, 162–171.

Yu, T., Liu, C. H., Belichenko, P., Clapcote, S. J., Li, S. M., Pao, A. N., et al. (2010d). Effects of individual segmental trisomies of human chromosome 21 syntenic regions on hippocampal long-term potentiation and cognitive behaviors in mice. *Brain Research*, *1366*, 162–171.

Zang, D. W., Yang, Q., Wang, H. X., Egan, G., Lopes, E. C., & Cheema, S. S. (2004). Magnetic resonance imaging reveals neuronal degeneration in the brainstem of the superoxide dismutase 1(G93A G1H) transgenic mouse model of amyotrophic lateral sclerosis. *European Journal of Neuroscience*, *20*, 1745–1751.

M. Dierssen and R. de la Torre (Eds.)
Progress in Brain Research, Vol. 197
ISSN: 0079-6123

CHAPTER 10

Discoveries in Down syndrome: moving basic science to clinical care

A. M. Kleschevnikov[†,‡,*], P. V. Belichenko[†,‡], A. Salehi[§,¶] and C. Wu[†,‡]

[†] *Department of Neurosciences, University of California, San Diego, La Jolla, CA, USA*
[‡] *Down Syndrome Center for Research and Treatment, University of California, San Diego, La Jolla, CA, USA*
[§] *VA Palo Alto Health Care System, Palo Alto, CA, USA*
[¶] *Department of Psychiatry and Behavioral Sciences, Stanford Medical School, Stanford, CA, USA*

Abstract: This review describes recent discoveries in neurobiology of Down syndrome (DS) achieved with use of mouse genetic models and provides an overview of experimental approaches aimed at development of pharmacological restoration of cognitive function in people with this developmental disorder. Changes in structure and function of synaptic connections within the hippocampal formation of DS model mice, as well as alterations in innervations of the hippocampus by noradrenergic and cholinergic neuromodulatory systems, provided important clues for potential pharmacological treatments of cognitive disabilities in DS. Possible molecular and cellular mechanisms underlying this genetic disorder have been addressed. We discuss novel mechanisms engaging misprocessing of amyloid precursor protein (App) and other proteins, through their affect on axonal transport and endosomal dysfunction, to "Alzheimer-type" neurodegenerative processes that affect cognition later in life. In conclusion, a number of therapeutic strategies have been defined that may restore cognitive function in mouse models of DS. In the juvenile and young animals, these strategists focus on restoration of synaptic plasticity, rate of adult neurogenesis, and functions of the neuromodulatory subcortical systems. Later in life, the major focus is on recuperation of misprocessed App and related proteins. It is hoped that the identification of an increasing number of potential targets for pharmacotherapy of cognitive deficits in DS will add to the momentum for creating and completing clinical trials.

Keywords: Down syndrome; cognition; synaptic structure; long-term potentiation; neurogenesis; Alzheimer's disease; axonal transport.

*Corresponding author.
Tel.: +1 (858) 534-1089; Fax: +1 (858) 534-4782
E-mail: akleschevnikov@yahoo.com

DOI: 10.1016/B978-0-444-54299-1.00010-8

Overview: An exciting time for research on Down syndrome (DS)

There has never been a more exciting time in research on Down syndrome (DS) nor one that promises with greater confidence, the development of new treatments for cognitive concerns in people with this condition. The past 10 years have seen significant advances in linking basic science discoveries to potential treatments. The many underpinnings for progress include (1) the sequencing of the human genome, in particular chromosome 21; (2) the development of advanced approaches for manipulation of the genome with respect to copy number, modulation of gene expression, functional regulation of synthesis and trafficking of mRNAs, and the discovery and exploration of the biological significance of non-coding DNA sequences and miRNAs; (3) development of increasingly sophisticated tools for exploring nervous system function; (4) increasing recognition and pursuit of collaborations that allow for multidisciplinary investigations; and (5) an ever-increasing attraction of this field of study to new and talented investigators. One could barely have hoped for a more fortunate concurrence of events in propelling studies on DS neurobiology. The question at hand is what we can hope for in the future—ideally the near future—for the translation of basic discoveries into clinically significant treatments for cognition in children and adults with DS. A signature of recent successes is that we can, as a field of investigation, ask this question with a sense of urgency. But it is important to note that the process involved in moving from discovery to clinical treatments is complex at best.

Today's landscape for translational medicine, for DS as well as essentially every other therapeutic target, is beset by a host of obstacles. In brief, the process that moves therapeutic possibilities to reality is severely compromised by fragmentation of expertise and effort, inefficiencies of practice, and lack of a systematic approach to defining and resourcing promising potential treatments. The result is that drug discovery, in particular, is challenged by increasing costs, decreased productivity, and product attrition (Tralau-Stewart et al., 2009). Pharmaceutical and large biotechnology companies, for example, increased R&D spending by 147% between 1993 and 2005 yet only saw an increase of new drug applications to the FDA of 38%. The report of the Clinical Research Roundtable of the Institute of Medicine pointed to the "disconnection between the promise of basic science and the delivery of better health" (Sung et al., 2003). Quoting further "the context of clinical research is increasingly encumbered by high costs, slow results, lack of funding, regulatory burdens, fragmented infrastructure, incompatible databases, and a shortage of qualified investigators and willing participants. These factors have contributed to two major obstacles, or translational blocks: impeding the translation of basic science discoveries into clinical studies and of clinical studies into medical practice and health decision making in systems of care." The situation today is essentially the same as that defined by this report, published in 2003. Clearly, extraordinary opportunities to heal will be lost if we do not define steps to relieve the existing translational blocks.

From genes to mechanisms: Gene dosage hypothesis

The ultimate goal of our studies has been to define treatments to enhance cognition in people with DS. Trisomy 21, the cause of DS, is the most frequent live-born aneuploidy, affecting ~1 in 700–800 births (CDC, 2006; Hook, 1983; Hook and Cross, 1982; Khoshnood et al., 2011; Parker et al., 2010). Approximately 95% of DS cases are due to the presence of an extra copy of the entire human chromosome 21 (HSA21); 5% are due to segmental trisomy for HSA21 or mosaic trisomy 21 (Galdzicki and Siarey, 2003; Lejeune et al., 1959; Pennington et al., 2003). DS exhibits a number of notable phenotypes affecting both general health and mental abilities. With the development of effective surgical treatments for congenital cardiac malformations, intellectual disability is the most important current burden—children fail to develop cognitive skills (Brown et al., 2003; Pennington et al., 2003), and all adults with DS show the pathology of Alzheimer's disease by age 40, with most also demonstrating severe cognitive decline beyond age 60 (Chapman and Hesketh, 2000; Lai and Williams, 1989; Wisniewski et al., 1985). Relatively little is known of the molecular mechanism(s) of cognitive disability in DS, and no treatments have yet been proven effective in humans.

Our studies, and those of others whose focus is nervous system function, have extensively used mouse models of DS. The mouse is currently the most important model organism for DS because of the syntenic conservation between the genomic segments on HSA21 and mouse chromosomes (MMU) 10, 16, and 17 (see Chapter 9). Utilizing existing mouse models, studies have attempted to define the genes and mechanisms that create unwanted cognitive phenotypes. Some of these studies test the gene dosage effect hypothesis for DS. This hypothesis states that all abnormal phenotypes in DS are caused by the triplication of one or more genes present on HSA21 (Epstein, 1990). To test the hypothesis, we strive to establish causal links between well-defined, quantitative brain phenotypes, and mouse genes homologous to those of HSA21. To decipher phenotype–genotype linkages, we have used the following strategy: (1) examine mice harboring an extra copy of the largest available set of mouse gene homologues to those on HSA21; (2) define and quantify phenotypes of interest—all of which are plausibly linked to changes in neuronal circuits that subserve cognitive pathways affected in people with DS; (3) through studies in mice harboring smaller trisomic segments, attempt to define an ever smaller set of gene candidates in whose presence there is conservation of the defined phenotypes; (4) to confirm or refute an individual candidate gene as responsible, studies are carried out in which just that gene is deleted in mice that, otherwise, carry all of the genes present in all of the larger trisomic segment(s); and (5) if deleting that gene significantly reduces or eliminates the phenotypes, add just this gene to the normal (i.e., euploid) mouse to ask if the phenotypes of interest are recreated. If deleting the candidate gene eliminates a phenotype, we can define the increased dose of the gene as necessary for creating a phenotype. If adding back just this gene recreates the phenotype, we interpret this as indicating that the increased dose of the gene is both necessary and sufficient for the phenotype. This so-called subtractive–additive approach is ideal not just for identifying a responsible gene(s) but also for defining the mechanistic pathways that may be deregulated as a result of increased gene dose.

Importantly, studies focused on defining causative genes are complemented and informed by studies to explore defined phenotypes using an approach in which the underlying mechanism(s) is examined (Belichenko et al., 2009a,b; Chakrabarti et al., 2010; Fernandez et al., 2007; Gardiner, 2010; Kleschevnikov et al., 2004; Salehi et al., 2009). These studies are supported by the burgeoning neuroscience literature which helps to suggest candidate mechanisms and provides the tools to explore the possibilities suggested. Using an approach in which gene discovery is informed by mechanistic studies allows for synergies that accelerate the pace of research.

As reviewed below, this combined approach is yielding new insights into the neurobiology of DS. Though there is much yet to discover, one can point at this time to no less than six therapeutic targets for people with DS and to increasing momentum in the movement toward clinical trials.

In this chapter, we focused mostly on the results recently elucidated in our laboratory. Many potentially effective approaches, for example, targeting mitochondria or reducing level of reactive oxygen species will not be discussed in this chapter. For those approaches, we refer the reader to several recent reviews (Arbuzova et al., 2002; Cramer et al., 2010; Das and Reeves, 2011; Galdzicki and Siarey, 2003; Gardiner, 2010; Gardiner et al., 2010; Kidd, 2005; Pallardo et al., 2010), as well as to other chapters of this book.

Synaptic structure: Defining the genetic basis of phenotypes

Mouse genetic models provide an indispensible tool for the exploration of possible links between genetic alterations and phenotypic features of subjects with DS (Barlow et al., 2001; Colas et al., 2008; Cramer et al., 2010; Das and Reeves, 2011; Davisson et al., 1993; Dierssen et al., 2001, 2011; Galdzicki and Siarey, 2003; Gardiner et al., 2003; Liu et al., 2011 and see Chapter 9). Our ultimate goal is to discover treatments enhancing cognition in people with DS. This requires solving four related tasks: (1) define critical changes in synaptic structure and function in cognitive circuits in mouse models of DS; (2) determine which genes are responsible for the observed phenotypes; (3) define the mechanisms underlying these changes and finally; (4) using these insights, accelerate development of effective treatments for people with DS. To understand genotype–phenotype relationships, in the past decade, we extensively explored several mouse models of DS (Belichenko et al., 2004, 2007, 2009a,b; Kleschevnikov et al., 2004, 2012; Liu et al., 2011; Popov et al., 2011; Salehi et al., 2006, 2007, 2009). So far, using the subtractive–additive genetic approach, we have examined phenotypes in Ts65Dn, Ts1Cje, and Ts1Rhr models (Fig. 1). More recently, we defined behavior and physiological phenotypes in a new triple-DS (TDS) mouse, in which most mouse genes homologous to genes on HSA21 are triplicated, the Dp(10)1Yey/+, Dp(16)1Yey/+, Dp(17)1Yey/+ (i.e., Ts1Yey) mouse model of DS (Liu et al., 2011; Yu et al., 2010a,b; Zhang et al., 2012).

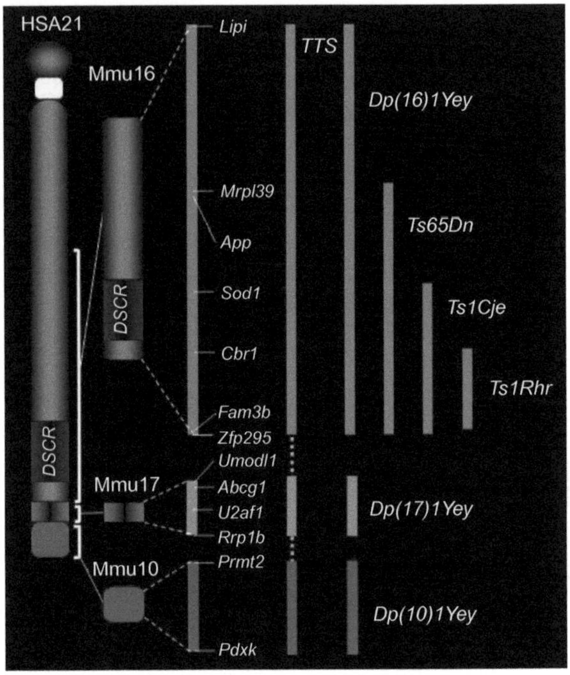

Fig. 1. Mouse models of DS and their relation to the human chromosome 21 (HSA21). (For color version of this figure, the reader is referred to the Web version of this chapter.)

Although intellectual disability and cognitive impairment in DS almost certainly result from changes in many regions of the brain, changes in the hippocampus, a region intimately involved in learning and memory, seem to be most severe (Belichenko et al., 2004, 2009b; Pennington et al., 2003). Consistent with these observations, hippocampus-dependent types of learning and memory are disproportionally affected in DS (Carlesimo et al., 1997; Pennington et al., 2003), as well as in mouse genetic models of DS (for review, see Galdzicki and Siarey, 2003). Thus, focusing on properties of synaptic pathways and subpopulations of neurons within the hippocampal formation is a rational choice for finding potential treatments to improve cognition in DS.

We showed that Ts65Dn, Ts1Cje, and Ts1Rhr mice had similar abnormalities in synaptic (Belichenko et al., 2004, 2009a,b), physiological (Belichenko et al., 2007, 2009a,b; Kleschevnikov et al., 2004), and behavioral phenotypes (Belichenko et al., 2007, 2009b; Faizi et al., 2011). The most striking synaptic phenotypic features in

DS mouse models were found in the dentate gyrus (DG) and included: (1) reduced spine density, (2) increased size of spine heads, and (3) decreased length of spine necks (Belichenko et al., 2004, 2007, 2009b). However, the last mentioned phenotype was not observed in the Ts1Rhr mouse model of DS (Belichenko et al., 2009a). An important conclusion from the study of synaptic phenotypes is conservation of changes in spine density and spine head size in Ts65Dn, Ts1Cje, and Ts1Rhr models.

An important question is how phenotypes revealed in mouse models of DS are related to phenotypes in people with DS. In a case study of a brain from a subject with DS (male, 25 years old), we performed Lucifer yellow microinjection into slices of neocortex and obtained images of dendrites using confocal microscopy. These observations documented important parallels between the status of neurons and synapses in the DS mouse models (Belichenko et al., 2004, 2007, 2009b; Popov et al., 2011) and individuals with DS. Thus, reduced spine density and enlarged spines were revealed in both mice and humans (Fig. 2). Because the phenotypes detected in mice appear to recapitulate those in people with DS in

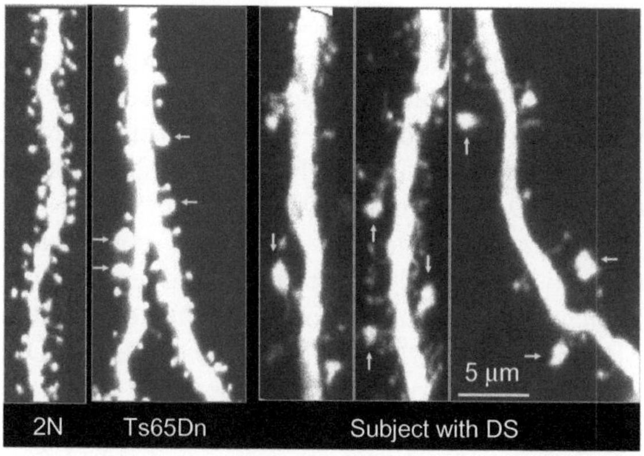

Fig. 2. Similarity of spine phenotypes in the Ts65Dn mouse model of DS and a subject with DS. *Note*: Both reduced density of spines and presence of enlarged spine heads (arrows) in Ts65Dn and the subject with DS as compared to wild-type (2N) mice. (For color version of this figure, the reader is referred to the Web version of this chapter.)

this and other studies (Marin-Padilla, 1976), histological observations of mouse models provide an effective means for evaluation of drug efficiency in the development of new treatments. Currently, this technique is effectively employed in our lab for evaluation of treatments directed at amelioration of neurodegenerative changes in mouse models of DS.

The causes of changes in spine morphology in Ts65Dn and other mouse models of DS have not been yet established with certainty. One possible explanation for the spine enlargement is a compensatory change due to an increased efficiency of the inhibitory system (see below). Abnormal expression of several synaptic proteins may contribute to these abnormalities. The proteins of interest include synaptojanin1 (*Synj1*), a polyphosphoinositide phosphatase involved in uncoating of clathrin-coated vesicles and in the downregulation of actin polymerization at endocytic zones (Slepnev and De Camilli, 2000); intersectin 1 (*Itsn1*), a scaffolding protein interacting with a number of proteins needed either for clathrin-mediated endocytosis or for actin assembly (O'bryan et al., 2001); T-lymphoma invasion and metastasis 1 protein (*Tiam1*), a guanine nucleotide exchange factor for Rac, a GTPase known to regulate endocytosis through effects on the actin cytoskeleton (Stamnes, 2002); Dual-specificity kinase 1a (*Dyrk1a*), phosphorylating a number of important targets including dynamin (Chen-Hwang et al., 2002); a protein critically linked to endocytosis (Schmid et al., 1998); and a number of other proteins.

Another striking set of phenotypic features in DS mouse models were both morphological and physiological changes in inhibitory inputs (Belichenko et al., 2004, 2009b; Kleschevnikov et al., 2004) and excessive inhibition in DG (Kleschevnikov et al., 2004). We characterized synapses, the distribution of receptors and several proteins essential for synaptic clustering of neurotransmitter receptors. The results demonstrate a significant alteration of inhibitory synapses in the DG of Ts65Dn mice (Belichenko et al., 2004, 2009b). Figure 3 summarizes our data on changes in the structure of pre- and postsynaptic elements and their innervation in Ts65Dn. Reorganization of inhibitory inputs, with a relative decrease in inputs to dendrite shafts and an increase in inputs to the neck of spines, was observed (Fig. 3).

It is essential to link changes in synaptic structure with changes in function and cognition. An important conclusion from the study of physiological phenotypes in Ts65Dn, Ts1Cje, and Ts1Rhr mouse models of DS was the absence of long-term potentiation (LTP) in the DG and restoration of LTP with picrotoxin (Belichenko et al., 2007, 2009a; Kleschevnikov et al., 2004). Conservation of changes was also observed in many behavioral tests including open field, T-maze, and novel object recognition in Ts65Dn, Ts1Cje, and Ts1Rhr mouse models of DS (Belichenko et al., 2007, 2009a; Faizi et al., 2011). To further examine the effect of triplicated genes on behavior and physiology, we recently tested a new TDS mouse model of DS (Yu et al., 2010a,b). We confirmed alterations in hippocampal (CA1 area) LTP and contextual fear conditioning test (Liu et al., 2011).

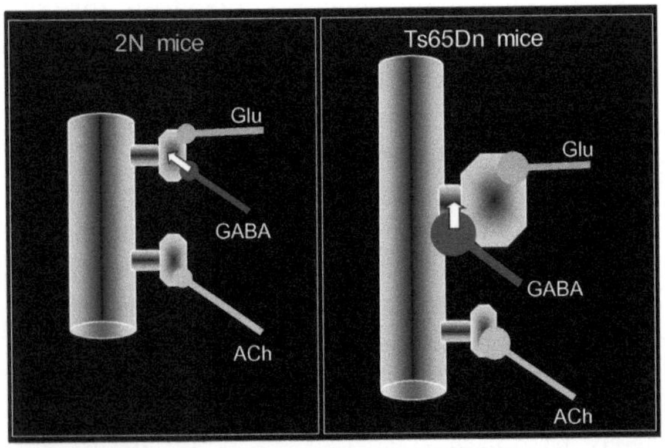

Fig. 3. Alteration of inhibitory inputs in the DG of Ts65Dn mouse model of DS as compared to wild type (2N) mice. *Note*: Both enlarged spines and redistribution of GABAergic input to the spine neck area in Ts65Dn mice. (For color version of this figure, the reader is referred to the Web version of this chapter.)

In conclusion, biologically significant synaptic, physiological, and behavioral phenotypes present in the Ts65Dn mouse are conserved in the Ts1Cje and Ts1Rhr mouse models of DS. This observation suggests that triplication of genes from the "DS critical region" of mouse chromosome 16 is sufficient to produce notable structural and functional phenotypes. In ongoing studies, we examine the mechanisms implicating these genes in reduced synaptic plasticity, changes in morphology, and impaired cognition in DS.

Studies of synaptic and neuronal function

Reduced synaptic plasticity

Synaptic plasticity is defined as the ability of synaptic connections to change in strength depending on afferent activity. This property of synaptic pathways is crucially important for learning and memory. It was found that synaptic plasticity and, in particular, LTP are severely impaired in Ts65Dn and other DS models (Costa and Grybko, 2005; Fernandez et al., 2007; Kleschevnikov et al., 2004; Siarey et al., 1997). In theory, pharmacological interventions aimed at restoration of LTP may improve impaired cognition. Properties of LTP in DS model mice have been rigorously explored in a number of laboratories (Belichenko et al., 2009a; Costa and Grybko, 2005; Fernandez et al., 2007; Kleschevnikov et al., 2004; Rueda et al., 2010; Salehi et al., 2009; Siarey et al., 1997).

LTP is characterized by three consecutive events: induction, maintenance, and expression. Which part(s) of this process is affected in DS? The induction of LTP in both the CA1 and DG requires activation of the *N*-methyl-D-aspartate (NMDA) receptors. We measured the NMDA receptor-mediated component of field EPSP (fEPSP) and found that it was significantly smaller in the DG of Ts65Dn versus 2N littermate mice (Kleschevnikov et al., 2004). Suppression of inhibitory neurotransmission with picrotoxin, the selective blocker of the GABAA receptor chloride channel, restored the NMDA receptor-mediated component of fEPSP. Importantly, restoration of the NMDA receptor-mediated fEPSP component was accompanied by restoration of LTP in slices from a Ts65Dn hippocampus (Fig. 4). Because properties of excitatory neurotransmission are largely spared in DS model mice (Costa and Grybko, 2005; Kleschevnikov et al., 2004; Siarey et al., 1997), restoration of LTP by antagonists of the GABAergic receptors suggests that the efficiency of inhibitory neurotransmission is increased, and that this change limits proper activation of the NMDA receptors during afferent activity. Indeed, many subsequent studies demonstrated changes in synaptic and neuronal properties in Ts65Dn and other DS models consistent with increased inhibition ("overinhibition") (Belichenko et al., 2009b; Best et al., 2007; Chakrabarti et al., 2010; Kleschevnikov et al., 2004; Perez-Cremades et al., 2010). In addition, we found that resting membrane potential is more negative in granule cells of Ts65Dn than 2N DG (Kleschevnikov et al., 2005). The change in the membrane potential may be attributable to increased expression of Kir3.2 protein (Harashima et al., 2006; Kleschevnikov et al., 2005), a subunit of G-protein-coupled inwardly rectifying potassium channels, whose gene (*Kcnj6*) is present in three copies in DS and Ts65Dn mice. Thus, induction of LTP in DS model mice could be negatively affected by "overinhibition" of hippocampal circuits and excessive hyperpolarization of neurons.

Whether or not LTP maintenance and expression are normal in DS model mice is an open question. Given that inhibitory neurotransmission is suppressed, both the magnitude and the time course of posttetanic changes are not altered in either Ts65Dn (Fig. 4) or other DS models (Belichenko et al., 2009a). If LTP maintenance and expression are not affected, efforts to treat pharmacologically should be directed at improving LTP induction. This could be done by increasing activity-dependent depolarization of neurons or greater activation of the NMDA receptors. However, it is still unknown whether or not later phases of LTP (lasting more than 1 h) are affected. Later LTP phases depend on synthesis of proteins and other energy-demanding events. Because subjects with DS have severe dysfunction of mitochondria (e.g., Arbuzova et al., 2002; Busciglio et al., 2002), it could be expected that the later LTP phases are abnormal and, hence, could be a target for additional pharmacotherapeutic approaches.

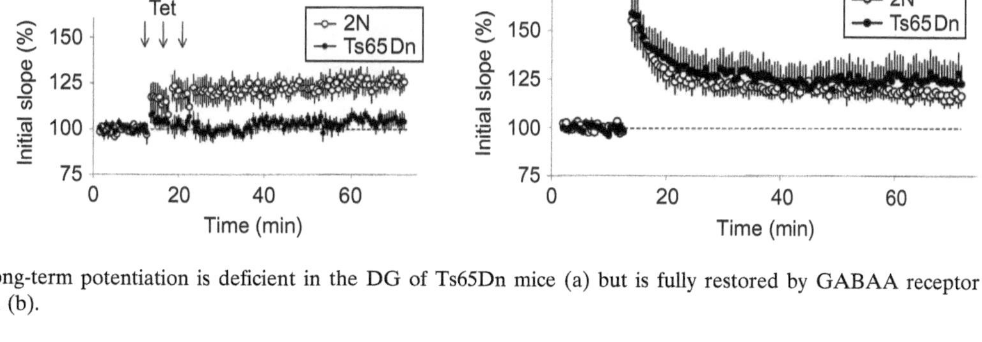

Fig. 4. Long-term potentiation is deficient in the DG of Ts65Dn mice (a) but is fully restored by GABAA receptor antagonist picrotoxin (b).

Reduced adult neurogenesis

Adult neurogenesis is another brain property important for learning and memory that is severely impaired in people with DS (Wisniewski et al., 1984) and DS model mice (Belichenko and Kleschevnikov, 2011; Bianchi et al., 2010; Chakrabarti et al., 2007; Clark et al., 2006; Llorens-Martin et al., 2010). It was shown that the rate of neurogenesis in Ts65Dn mice is reduced by 25–50% (Clark et al., 2006; Llorens-Martin et al., 2010). In the DG of Ts65Dn mice, this change is accompanied by a reduced number of cells per cluster (Fig. 5) (Belichenko and Kleschevnikov, 2011). Reduced neurogenesis in DS is observed early in development (Chakrabarti et al., 2007; Contestabile

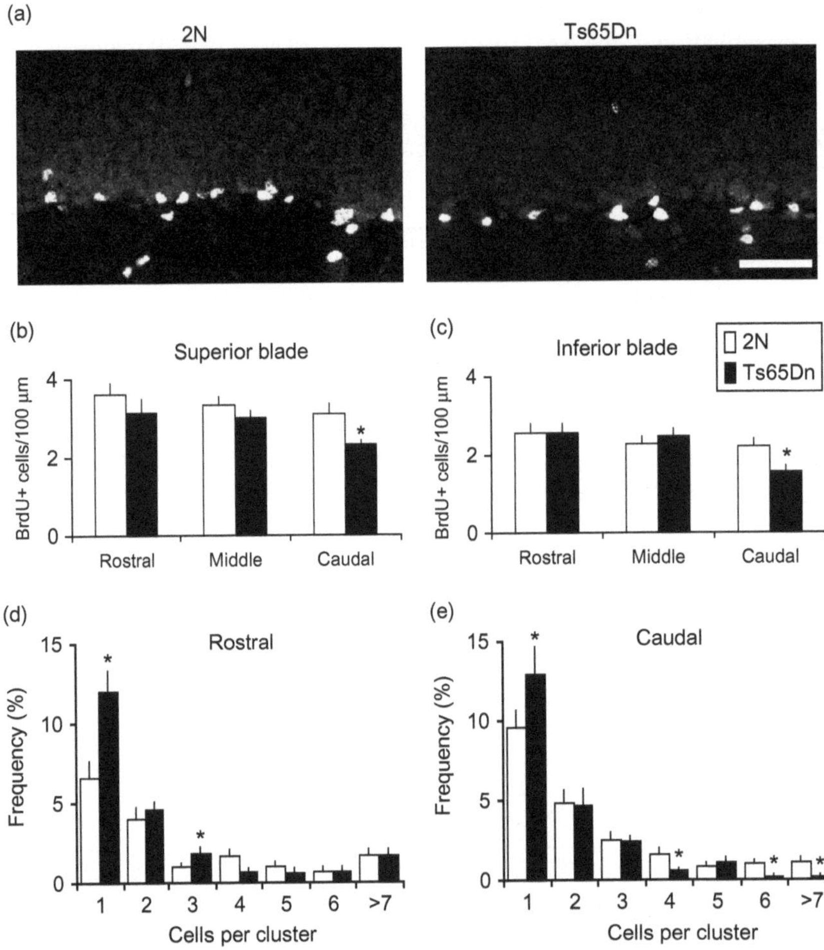

Fig. 5. Morphometry of BrdU+cells in 2N and Ts65Dn DG. (a) Confocal image of BrdU+cells in caudal part of dentate gyrus of 2N and Ts65Dn mice. BrdU+cell density in superior, (b) and inferior blades, (c) of DG. *Note*: Significant reduction of BrdU+cells in caudal DG in Ts65Dn mice. The number of cells per cluster in rostral (d) and caudal (e) part of DG. *$p < 0.05$ (Student's T-test). Scale bar=50 µm.

et al., 2007; Lorenzi and Reeves, 2006). Thus, at P6, Ts65Dn mice had about 20% fewer DG granule cells and 18% fewer mitotic cells versus 2N littermates (Lorenzi and Reeves, 2006). Interestingly, in P2, Ts65Dn pups precursor cell proliferation is impaired in all DG layers but became strictly localized to the subgranular proliferative layer by P18 (Contestabile et al., 2007). Reduced neurogenesis and an increased rate of apoptosis resulted in a reduced number of DG granule cells in adult Ts65Dn and other DS models (Contestabile et al., 2007; Insausti et al., 1998; Lorenzi and Reeves, 2006). The genetic basis for reduced neurogenesis is not fully defined. Increased expression of genes for oligodendrocyte transcription factor 1 and 2 (*Olig1* and *Olig2*) (Chakrabarti et al., 2010), amyloid precursor protein (*App*) (Trazzi et al., 2011), *Dyrk1A* (Dierssen and De Lagran, 2006), and other genes have been implicated (reviewed in Pallardo et al., 2010). In addition, it was shown that reduced neurogenesis in Ts65Dn mice is associated with reduced expression of BDNF (Bianchi et al., 2010). Because adult neurogenesis is required for synaptic plasticity, reduced neurogenesis likely contributes to reduced synaptic plasticity, as well as deficient cognition in DS model mice.

Potential treatments improving synaptic plasticity and cognition

Whatever the cause of changes in synaptic plasticity and neurogenesis, alterations in these properties should inevitably lead to poor cognition and, thus, these changes represent prominent targets for pharmacotherapy. Several potential approaches may be considered for improving cognition DS.

Suppression of inhibition and improvement of synaptic plasticity

Excitatory–inhibitory balance plays an important role in processing of information. Increased inhibitory efficiency could be an important factor in restricting synaptic plasticity and, hence, cognitive performance in DS model mice. It was hypothesized that partial suppression of the inhibitory circuits might allow for greater neuronal depolarization, better activation of the NMDA receptors, and, thus, improvement of synaptic plasticity (for review, see Rissman and Mobley, 2011). Indeed, we observed in acute experiments full recovery of LTP following suppression of the GABAA receptors with picrotoxin, the selective blocker of the GABAA receptor chloride channels (Kleschevnikov et al., 2004). However, an acute effect of a drug cannot predict whether or not its chronic administration is effective. The effect of chronic treatment was investigated with use of two selective antagonists of the GABAA receptors—picrotoxin and pentylenetetrazol (PTZ) (Fernandez et al., 2007; Rueda et al., 2008). Remarkably, it was observed that the chronic treatment improved LTP and cognition in Ts65Dn mice. The most exciting observation was that the changes persisted for several months after the treatment (Fernandez et al., 2007). Thus, chronic suppression of the GABAA receptors resulted in persistent changes, improving synaptic plasticity and cognition in DS models. The nature of those changes has not yet been established.

Signaling through metabotropic GABAB receptors is also considerably increased and likely contributes to the overinhibition of synaptic circuits in Ts65Dn and other DS models (Best et al., 2007; Kleschevnikov et al., 2012). Thus, suppression of the GABAB receptors could be used to increase depolarization of neurons during activity and, hence, to improve synaptic plasticity. This target may complement efforts directed at GABAA receptors. Interestingly, in contrast to antagonists of the GABAA receptor known for their proepileptic properties, antagonists of the GABAB receptors might have antiepileptic properties in DS subjects (Cortez et al., 2009).

Ionic channels that are effectors of the GABAergic receptors could also be regarded as targets for pharmacotherapy in DS. Postsynaptic

GABAB receptors use as effectors Kir3.2 subunit-containing K+ channels. GABAB/Kir3.2 signaling is selectively increased in Ts65Dn cells (Best et al., 2007; Kleschevnikov et al., 2012). Thus, blockers of Kir3.2-containing K+ channels could be considered as potential drugs. Interestingly, one of the compounds that effectively blocks ionic currents through these receptors is fluoxetine (Kobayashi et al., 2004). At physiological concentrations, it blocks up to 50% of the current. Thus, in addition to its effect on neurogenesis (see below), this drug may act to restore LTP in DS models through suppression of increased GABAB/Kir3.2 signaling.

Restoration of intracellular calcium homeostasis

A number of observations suggest that regulation of intracellular calcium is altered in DS, and that improper signaling through the NMDA receptors likely contributes to this abnormality. Thus, cultured hippocampal neurons from DS model mice with triplication of full chromosome 16 (Ts16) exhibited elevated basal levels of intracellular calcium, as well as prolonged recovery of the calcium concentration after brief glutamate application (Schuchmann et al., 1998). Ts65Dn mice showed exaggerated locomotor stimulatory effects upon administration of the high-affinity NMDA receptor channel blocker MK-801 (Costa et al., 2008). These and other observations suggest that restricting activation of the NMDA receptors may have beneficial effects on homeostasis of intracellular calcium and, possibly, cognition in DS model mice. Indeed, several recent studies described significant improvement of Ts65Dn mice in hippocampus-mediated memory tests following acute and/or chronic treatment with memantine, an uncompetitive partial antagonist of the NMDA receptors (Costa et al., 2008; Lockrow et al., 2011; Rueda et al., 2010). Memantine has been approved in the USA and Europe for moderate to severe dementia and was recently proposed for a clinical trial in people with DS (Costa, 2011).

Restoration of adult neurogenesis

Suppression of neurogenesis to the level seen in DS is believed to negatively impact synaptic plasticity as well as learning and memory (for review, see Koehl and Abrous, 2011). Thus, an increase in neurogenesis could lead to improvement of cognition in DS mouse models and people with DS. In line with this prediction, it was shown that fluoxetine, which positively affects neurogenesis in adult animals (Malberg et al., 2000), can improve neurogenesis in Ts65Dn mice to the level seen in 2N control animals (Clark et al., 2006). As mentioned, fluoxetine also suppresses currents through Kir3.2-containing K+ channels, thus reducing excessive hyperpolarization of neurons and, thereby, improving the conditions for induction of LTP. The effect of fluoxetine on neurogenesis and cognition in Ts65Dn DG was most profound when the treatment started early in development (Bianchi et al., 2010). Interestingly, restoration of neurogenesis and cognitive functions following treatment with fluoxetine was accompanied by normalization of the level of BDNF, suggesting that deficiency in the concentration of this factor in the brain of DS subjects contributes to abnormalities in neurogenesis and cognition. Whether or not changes in neurogenesis and BDNF are mechanistically linked is yet to be determined.

Synaptic maintenance and neurodegeneration

Defining in vivo *the genetic basis for neuronal degeneration*

Among multiple brain regions, the hippocampal formation and its cortical and subcortical inputs are the major target of degeneration in DS. Hippocampal networks receive extensive inputs from several neuronal populations that play a critical role in cognition. These include cholinergic neurons of the basal forebrain, noradrenergic (NA-ergic) neurons of the locus coeruleus (LC), serotonergic neurons of the raphe nuclei (RN), and neurons of the supramammillary area (SUMA). Degeneration in these neurons leads to significant loss of inputs to the hippocampus, that is, its "deafferentation." Understanding the mechanisms by which hippocampal inputs undergo degeneration in DS could be crucial in suggesting treatments for cognitive dysfunction in DS. For this matter, we examined in detail the status of cholinergic and NA-ergic systems in the Ts65Dn mouse model and its 2N controls.

Cholinergic system

Acetylcholine plays a significant role in regulating cerebral blood flow, sleep–wake cycle, attention, and cognitive functions (Schliebs and Arendt, 2006). Cholinergic neurons in the basal forebrain project to the entire hippocampus, particularly the DG. In DS, cholinergic neurons of the nucleus basalis of Meynert (NBM) undergo degeneration as shown by loss of these neurons in adults with DS. Furthermore, most adults with DS also show extensive cytoskeletal alterations (Mann et al., 1986) and senile plaques in this area (Mann et al., 1984). In fact, NBM degeneration in DS has been detected as early as 5.5 months of age (Mcgeer et al., 1985). Degeneration of basal forebrain cholinergic neurons (BFCN) in DS and relative success in treating cognitive dysfunction in AD has encouraged the use of cholinergic drugs to increase brain acetylcholine levels in DS. Donepezil, a second-generation reversible acetylcholinesterase inhibitor, has shown to improve expressive language in both children and adults with DS (Heller et al., 2003, 2004; Lott et al., 2002). Furthermore, in adults with DS, the use of donepezil has been shown to reduce the severity of deterioration in adaptive and global functioning (Prasher et al., 2003). An open label study using donepezil in children with DS led to a significant improvement in memory for names, narrative memory, and visual and sustained attention (Spiridigliozzi et al., 2007).

Using Ts65Dn mice, we found an age-dependent degeneration of BFCN as shown by a reduced number of neurons expressing p75 neurotrophin receptor ($p75^{NTR}$) (Cooper et al., 2001). Interestingly, degeneration of BFCNs in Ts65Dn mice was associated with a significant reduction in nerve growth factor (NGF) axonal transport in these neurons (Salehi et al., 2006). NGF plays an important role in the maintenance of structure and function of BFCNs (Sofroniew et al., 2001). Indeed, by infusing NGF directly into the lateral ventricles to reach directly BFCN cell bodies, thus bypassing the axonal transport, we were able to reverse advanced degenerative changes in the Ts65Dn mouse model of DS (Cooper et al., 2001).

To identify the gene(s) whose overexpression could be linked to failed NGF transport and BFCN degeneration in Ts65Dn mice, we used the Ts1Cje model, a mouse with a shorter fragment of MMU16 triplicated (see Fig. 1). Our investigations showed significant improvement in NGF retrograde transport in BFCNs and no apparent shrinkage of soma in BFCNs in Ts1Cje mice (Salehi et al., 2006). Among multiple genes not triplicated in Ts1Cje mice (Fig. 1), *App* has been the focus of intense study in DS. It has been shown that triplication of this gene is necessary for BFCN degeneration and extensive pathology in DS (Prasher et al., 1998). To test the effects of *App* in Ts65Dn mice, we compared Ts65Dn mice with either two or three copies of *App*. These mice were generated by crossing Ts65Dn mice with those lacking a copy of *App*. Deleting an extra copy of *App* in Ts65Dn mice led to a significant improvement in both NGF axonal transport and BFCNs morphology (Salehi et al., 2006). These data indicate that the cholinergic system, which is a major player in attention and cognitive functions, undergoes significant degeneration in Ts65Dn mice. Furthermore, among multiple genes triplicated in Ts65Dn mice, *App* overexpression plays a conspicuous role in BFCN degeneration and as a result hippocampal deafferentation in these mice (Salehi et al., 2006).

Noradrenergic system

A relatively small number of norepinephrine (NE)-expressing neurons in LC (50,000 in humans and around 2,500 in rodents) project to a majority of neurons in the cortex and hippocampus. In fact, very few regions (e.g., striatum) in the brain are spared from NA-ergic modulatory outputs. For this reason, LC neurons can induce powerful modulatory influence on a variety of brain functions, particularly cognition. LC neurons undergo degeneration in DS (Mann et al., 1986). Indeed, LC neurons show more severe degenerative alterations than cholinergic neurons of NBM in DS. For instance, Mann and colleagues found around a 24% loss of neuronal numbers in NBM in adults with DS compared to a 78% loss in LC neurons (Mann et al., 1984).

We investigated LC neuronal number and size, using an antibody against tyrosine hydroxylase (TH) to specifically label LC neurons. Both the number and size of TH-immunoreactive cells were significantly reduced in Ts65Dn mice at 6 and 18 months of age (Salehi et al., 2009). Interestingly, the reduction in LC neurons in young adult Ts65Dn mice was associated with a significant increase in the number of β1 adrenergic receptors-bearing neurons in the hippocampus, suggesting a compensatory effect of hippocampal deafferentation for this system.

To check the status of LC afferents to the hippocampus, we examined monoaminergic (MA)-terminals in the hippocampus by quantifying staining for vesicular monoamine transporter (VMAT2). Comparing Ts65Dn and 2N controls at 6 months of age, VMAT2 staining showed an overall decrease of this protein in Ts65Dn mice (Fig. 6). This finding again suggests a compensatory response that follows synaptic defects but precedes by months the loss of LC neurons. Degeneration of LC and loss of their hippocampal terminals in young adult Ts65Dn mice was associated with failed cognitive function in these mice. Indeed, we observed that contextual learning, a process in which spatial information is integrated with other salient features of the environment allowing the subject to produce appropriate responses to the environment, showed a significant deficit in the Ts65Dn model mice (Salehi et al., 2009).

To identify gene(s) whose overexpression can be linked to LC degeneration in Ts65Dn mice, we compared mouse models of DS harboring different fragments of mouse chromosome 16 (MMU16, Fig. 1). Ts1Cje mice with triplication of a shorter fragment of MMU16 showed no significant changes in size and number of LC neurons. To test whether LC degeneration is linked to increased dose for *App*, we examined these neurons in Ts65Dn mice bearing either two or

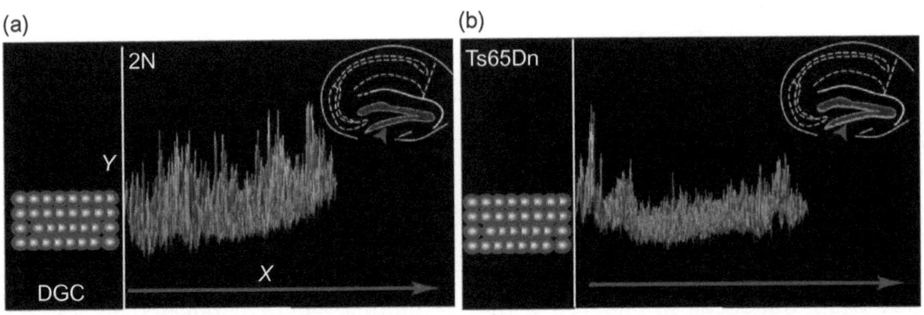

Fig. 6. Surface imaging of VMAT2 staining in the DG of young adult Ts65Dn mice and their 2N controls. Surface imaging was used to convert the optical density of staining to height. A significant reduction in VMAT2 optical density was found in the molecular layer of the DG in Ts65Dn mice. From Salehi et al. (2009). Reprinted with permission from AAAS. (For color version of this figure, the reader is referred to the Web version of this chapter.)

three copies of *App*. Deleting the extra copy of *App* in Ts65Dn mice alleviated the loss of LC neurons, suggesting that *App* overexpression is necessary for LC degeneration.

As mentioned, all adults with DS will eventually show significant AD pathology. Interestingly, similar to BFCNs, LC neurons also undergo significant degeneration in AD (Mann et al., 1984). Indeed, individuals with mild cognitive impairment also exhibit age-dependent LC structural changes, which negatively correlate with changes of mini-mental state examination scores (German et al., 1992). During the past 2 years, several new studies have shown that NE acts through nonsynaptic mechanisms, and perhaps on microglia, to reduce the severity of amyloid accumulation in mouse models of AD (Heneka et al., 2010). These studies suggest that the NE system plays a significant role in cognition and modulates aspects of AD pathology in DS.

Degeneration in DS: Lessons for therapy

To counter the effects of deafferentation of the hippocampus in DS, one would need to improve the function of affected systems projecting to the hippocampus. The use of drugs to enhance cholinergic function in DS has recently begun with modest success (Kondoh et al., 2011). However, since it has been shown that improving the NA system can also reduce Aβ pathology in mouse models of AD (Heneka et al., 2010), improving the NA-ergic system might be an effective method for therapy in adults with DS. We used two methods to improve β-adrenergic signaling in Ts65Dn mice:

(1) *Increasing NE levels.* L-threo-3,4-dihydroxyphenylserine (L-DOPS) is an NE pro-drug that readily crosses the blood–brain barrier (BBB). To evoke NE increases in only the central nervous system (CNS), L-DOPS (1 mg/g) was administered together with carbidopa (CD), a peripheral aromatic L-amino acid decarboxylase inhibitor that does not cross the BBB (Murchison et al., 2004). We also tested the effects of L-DOPS on fear conditioning. Treatment of 6-month-old Ts65Dn mice with L-DOPS and CD led to a significant improvement in nesting and contextual learning (Fig. 7). Recently, it has also been shown that the use of L-DOPS reduced the severity of Aβ accumulation in mouse models of AD (Heneka et al., 2010).
(2) *Improving β-adrenergic signaling.* It has been suggested that the beneficial effects of L-DOPS on cognition is through β1-adrenoceptors (Murchison et al., 2004). If that were also the case in Ts65Dn mice, treating these mice with a β1 adrenoceptor agonist would lead to similar beneficial effects on contextual learning. To test this idea, we treated Ts65Dn mice with xamoterol (Xam), a β1-adrenoceptor partial agonist, and found that it was also able to restore contextual learning (Salehi et al., 2009). This suggests that β1-adrenoreceptors play a significant role in cognition. Recently, using an extensive set of behavioral testing, Faizi and colleagues reported improvement in novel object recognition and contextual learning in Ts65Dn mice using Xam administration (Faizi et al., 2011). Studies indicating that the use of β-adrenergic drugs might also reduce the severity of pathology in mouse models of AD (Heneka et al., 2010) support that the use of β-adrenergic drugs in adults with DS with the possible additional benefit of reducing the severity of AD pathology in these individuals.

It is exciting that many therapeutic strategies have been defined that may restore cognitive function in mouse models of DS. Some strategies focused on restoring the function of affected systems, for example, GABAergic and NA-ergic, while others focused on reducing the expression of a specific triplicated gene. There is little doubt that a treatment for DS that reduces the expression of a responsible triplicated gene(s) would be ideal. However, at present using drugs to

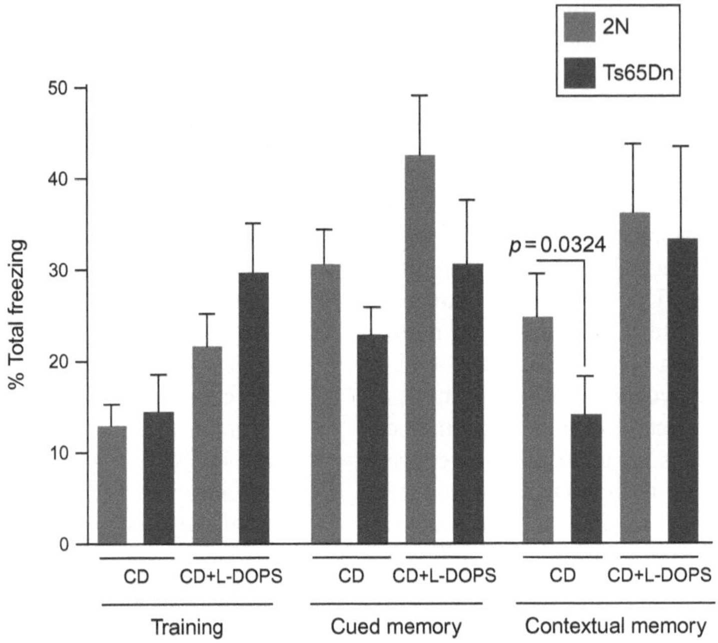

Fig. 7. The effects of L-DOPS on restoring contextual learning in Ts65Dn mice. Without treatment, there was a significant reduction in contextual learning in Ts65Dn mice compared with 2N controls. However, treating these mice with a combination of L-DOPS and CD led to a significant improvement in contextual learning. CD was used to prevent the peripheral effects of L-DOPS. From Salehi et al. (2009). Reprinted with permission from AAAS. (For color version of this figure, the reader is referred to the Web version of this chapter.)

restore the function of affected systems and neuronal networks is both attractive and more practical. Implication of pharmacotherapy to improve cognitive function in DS requires resolving a number of theoretical and practical issues. Chronic treatment with medicines suppressing or enhancing activity of neuronal receptors or channels often results in compensatory changes, reducing the sensitivity of the system and, therefore, negatively affecting the treatment efficiency. Investigation of acute versus chronic effects of the treatment is thus critical. Other potential problems include an increased risk of epilepsy and mood instabilities in people with DS. Indeed, many suggested treatments may affect the excitatory–inhibitory balance in the brain regions involved in learning and control of emotions. This may potentially increase the risk of epilepsy or affect general mental stability, especially during chronic treatments. These and other problems need to be addressed in the upcoming studies before pharmacotherapy of cognitive impairment in DS will become a reality.

Mechanisms of neurodegeneration and early insights into treatment

Full-length APP and APP cleavage products in DS

Triplication of the gene for App has been explicitly linked to degeneration of BFCN in Ts65Dn (Salehi et al., 2007). APP is processed by multiple secretases (α-, β-, and γ-secretases) to generate different carboxyl-terminal fragments (CTFs):

C99, C83, and APP-intracellular carboxyl domain (AICD), respectively. In addition, sequential cleavage of APP by β- and γ-secretases also gives rise to toxic Aβ amyloid peptides (Aβ40/42). It is conceivable that increased gene dose for APP leads to accumulation within nerve cells of C99, C83 AICD, and Aβ40/42, in addition to the full-length APP protein (Haass et al., 1993; Hare, 2010; Jin et al., 2004; Leblanc and Gambetti, 1994; Owen and Collins, 2010; Selkoe et al., 1996; Tang, 2009). Critical questions arising from these observations that need to be addressed are (1) what is the toxic species among APP or its C-terminal cleavage product(s) including Aβ40/42 and (2) what are the molecular and cellular events triggered by the species that lead to neurodegeneration.

Endosomal dysfunction in Ts65Dn mice

An important clue to the answers to these questions came from the observation that endocytic pathways are markedly altered in early stages of Alzheimer's disease and DS (Nixon, 2005). Nixon and colleagues observed enlarged early endosomes and increased immunoreactivity for markers of early endosomes (Rab5, early endosomal antigen 1, and Rabaptin5) most prominently in neurons of the basal forebrain in Ts65Dn mice (Cataldo et al., 2003). Interestingly, the abnormal endosomal phenotype was eliminated by selectively deleting one copy of *App* or a small portion of the chromosome 16 segment containing *App* from Ts65Dn mice (Cataldo et al., 2003). These results suggest that triplication of the *App* gene contributed to endosomal enlargement and dysfunction in Ts65Dn mice. How does increased level of App result in enlarged early endosome in Ts65Dn mice? Work from Neve's group uncovered that an increased level of App recruited the App binding protein APP-BP1 that, in turn, activated Rab5. Increased level of activated Rab5 caused enlargement of early endosomes and increased receptor-mediated endocytosis and apoptosis (Laifenfeld et al., 2007). These investigators further demonstrated that inhibition of Rab5 rescued neurons from apoptosis induced by either App (V642I) or APP-BP1, without affecting levels of intracellular or secreted Aβ peptides (Laifenfeld et al., 2007). These results raise the possibility that dis-regulation of Rab5 activity contributes prominently to neurodegeneration of BFCNs in Ts65Dn mice. However, it remains unclear how accumulation of excessive App, its cleavage products or Aβ40/42, leads to enhanced Rab5 activity and how enlarged early endosomes negatively impact neuronal function in Ts65Dn mice.

Impaired retrograde signaling of neurotrophic factors in Ts65Dn mice

A deficit in retrograde signaling by NGF has been implicated in causing degeneration of BFCNs in Ts65Dn mice (Cooper et al., 2001). The deficit was largely but not completely restored by the selective removal of the extra copy of the *App* gene in Ts65Dn mice (Salehi et al., 2007). One plausible explanation for these observations that links them to endosomal pathology is that the retrograde NGF trophic signal is carried in Rab5-early endosomes during transit from axonal terminals to the soma (Delcroix et al., 2003; Wu et al., 2007, 2009). In Ts65Dn mice, the Rab5-early endosomes are abnormally enlarged due to excessive accumulation of App, its cleavage products or Aβ40–42. The enlarged early endosomes may have retarded mobility within axons and deliver less NGF to the soma leading to a deficit in trophic support. It is also possible that increased Rab5 activity may trigger activation of Rab7, thus leading to premature trafficking of the NGF trophic signal from early endosome to late endosome/lysosome compartment. This would also cause a deficit in trophic support from axonal terminals. Therefore, targeting Rab5 activity may prove to be an attractive strategy for restoring axonal function and neuronal activity in DS. This could be achieved either by downregulating App, its cleavage product(s) or Aβ1–40/42, or compounds/molecules that can specifically inhibit Rab5 activity.

Targeting γ-secretase for treatment

Although there is a large body of supporting evidence that Aβ40/42 contributes significantly to neuronal degeneration in AD, the role of Aβ40/42 in impacting retrograde axonal signaling of NGF is much less clear. Yet Aβ40/42 has been found to be present in intraneuronal endosomes (Cataldo et al., 2004). Significant efforts have been invested in developing inhibitors that target the secretases that cleave App and give rise to Aβ40/42, that is, the β- or γ-secretase (Anderson et al., 2005; Evin et al., 2010; Kounnas et al., 2010; Pasternak et al., 2003; Sabo et al., 1999; Shinohara et al., 2010; Van Nostrand et al., 1992a,b; Vetrivel et al., 2004, 2011; Zhang et al., 2006). Yet the negative impact of inhibiting either the β- or the γ-secretase has not been fully appreciated until recently (Anderson et al., 2005; Campbell et al., 2003; Da Costa et al., 1999; Kim et al., 2005; Lee et al., 2002; Qin et al., 2003; Shizuka-Ikeda et al., 2002; Yagishita et al., 2006). For instance, drugs/compounds that specifically target and inhibit the γ-secretase cause unintended consequences. This is because the inhibition of γ-secretase affects not only App processing but also the cleavage of many other vital signaling molecules such as receptors (e.g., VEGFR-1, Notch, ErbB-4, IGFI-R) (Guardia-Laguarta et al., 2010; Haapasalo and Kovacs, 2011; Li, 2001; Selkoe and Kopan, 2003; Wolfe, 2008; Xia, 2008). For instance, repeated exposure to γ-secretase inhibitors often causes Notch-related changes in animals (e.g., intestinal proliferation of goblet cells). In addition, the use of γ-secretase inhibitors will most likely result in excessive accumulation of C99 and C83 within neurons. The negative impact of large amounts of these App products on cellular processes should be fully investigated.

To circumvent the negative consequences caused by the γ-secretase inhibitors, Wagner and colleagues screened over 1200 novel γ-secretase modulator (GSM) compounds and discovered a class of potent small molecules that were capable of further processing Aβ40–42 to Aβ37–38. These compounds did not inhibit epsilon-site cleavage of App and Notch and thus did not cause an increase in the level of the App C99 fragment and Notch intracellular domains (Kounnas et al., 2010). Impressively, these compounds were as effective as γ-secretase inhibitors in reducing both diffuse and neuritic plaques in Tg2576 transgenic AD mice (Kounnas et al., 2010). Therefore, GSMs could be of potential use in treating cellular dysfunction that results from increased expression of APP in DS.

Summary and a view to the future

The availability of an increasingly rich set of tools and concepts informed by recent successes promises to bring in an era of progress in not only understanding the neurobiology of DS but also the first successful attempts to enhance cognitive function. While the future promises success, there are still important limitations to be addressed. Essential next steps are to increase the effectiveness of work in the research community and to reach out to the larger therapeutic community. The objectives for effective translational medicine and the steps needed to be taken at this time, with special reference to what can be done in the academic community are as follows:

(1) Make basic discoveries, focused at least in part on those that are plausibly linked to enhancement of clinical care of people with DS. The most important limitation at this time is insufficient funding for discovery research. At a time when NIH funding is in peril, a greater emphasis on foundation-based funding is inevitable. The research community should continue to reach out to both

governmental and private funding sources to make the case that research on DS promises significant benefits to people with this condition as well as to others, especially those destined to develop Alzheimer's disease.

(2) Validate discoveries with potential for translation. Here, the challenge will be for the basic research community to create a mechanism in which discoveries can be validated quickly and effectively. Creating an infrastructure that defines this as its goal would be extremely useful. Conceivably, a variety of sources of funding could be tapped for this, including public, private, and industrial resources. A collaboration that brings academia together with industry would perhaps be most effective. One new model for such collaboration has been developed recently (Roth and Cuatrecasas, 2010). Evidence that the climate for engaging industry is changing are a recent increase in interest in clinical studies in DS by a number of pharmaceutical companies.

(3) Participate in creating the intellectual environment by which a new treatment is applied in man. Basic researchers should reach out to clinical colleagues, especially those with clinical trials expertise, to increase interest in the translational effort in DS, and to point to the need for collecting additional natural history data on the condition and for establishing the infrastructure for participating in clinical trials.

(4) Create tools that allow accurate estimates of efficacy in clinical trials. The need for tools and infrastructure is real and urgent. A tool to measure cognition in adults with DS was recently published (Edgin et al., 2010), encouraging the view that a collection of tools can be developed for studies across the age span in DS.

(5) Contribute energy and effort to ensure the timely transfer of information. The research community should find ways to more effectively communicate to its members, the results of recent experiments. Given current measures for acknowledging and rewarding performance in academia, investigators withhold disclosure of advances until a manuscript is accepted for publication. Unfortunately, this delays transmission of information important to the greater research community for many months or possibly a year or more. A compact is needed that allows for more rapid transfer of information. Equally useful would be the more ready transfer of research reagents, including the mouse models that allow for more targeted studies of candidate genes.

Addressing these limitations would allow the research, clinical, and industry communities to more rapidly and effectively build on recent advances to create a new era in which translation of discoveries to care becomes the norm for people with DS.

Acknowledgments

We thank Dr. William C Mobley for crucial contributions to the reviewed studies, as well as for critical reading of the chapter and providing helpful insights. This work was supported by grants from NIH Grants NS055371, NS066072, and EY016525, pilot P50 grant from ADRC of UCSD (to C. W.), Start-up grants from UCSD (to A. M. K., P. V. B., and C. W.), The Alzheimer's Association, The Larry L. Hillblom Foundation, the Thrasher Foundation, and The Down Syndrome Research and Treatment Foundation (DSRTF).

References

Anderson, J. J., Holtz, G., Baskin, P. P., Turner, M., Rowe, B., Wang, B., et al. (2005). Reductions in beta-amyloid concentrations in vivo by the gamma-secretase inhibitors BMS-289948 and BMS-299897. *Biochemical Pharmacology*, *69*, 689–698.

Arbuzova, S., Hutchin, T., & Cuckle, H. (2002). Mitochondrial dysfunction and Down's syndrome. *Bioessays*, *24*, 681–684.

Barlow, G. M., Micales, B., Lyons, G. E., & Korenberg, J. R. (2001). Down syndrome cell adhesion molecule is conserved in mouse and highly expressed in the adult mouse brain. *Cytogenetics and Cell Genetics*, *94*, 155–162.

Belichenko, N. P., Belichenko, P. V., Kleschevnikov, A. M., Salehi, A., Reeves, R. H., & Mobley, W. C. (2009a). The "Down syndrome critical region" is sufficient in the mouse model to confer behavioral, neurophysiological, and synaptic phenotypes characteristic of Down syndrome. *The Journal of Neuroscience*, *29*, 5938–5948.

Belichenko, P. V., & Kleschevnikov, A. M. (2011). Deficiency of adult neurogenesis in the Ts65Dn mouse model of Down syndrome. In S. Dey (Ed.), *Genetics and etiology of Down syndrome* (pp. 177–192). InTech., ISBN 978-953-307-631-7; www.intechopen.com/articles/show/title/deficiency-of-adult-neurogenesis-in-the-ts65dn-mouse-model-of-down-syndrome.

Belichenko, P. V., Kleschevnikov, A. M., Masliah, E., Wu, C., Takimoto-Kimura, R., Salehi, A., et al. (2009b). Excitatory-inhibitory relationship in the fascia dentata in the Ts65Dn mouse model of Down syndrome. *The Journal of Comparative Neurology*, *512*, 453–466.

Belichenko, P. V., Kleschevnikov, A. M., Salehi, A., Epstein, C. J., & Mobley, W. C. (2007). Synaptic and cognitive abnormalities in mouse models of Down syndrome: Exploring genotype-phenotype relationships. *The Journal of Comparative Neurology*, *504*, 329–345.

Belichenko, P. V., Masliah, E., Kleschevnikov, A. M., Villar, A. J., Epstein, C. J., Salehi, A., et al. (2004). Synaptic structural abnormalities in the Ts65Dn mouse model of Down Syndrome. *The Journal of Comparative Neurology*, *480*, 281–298.

Best, T. K., Siarey, R. J., & Galdzicki, Z. (2007). Ts65Dn, a mouse model of Down syndrome, exhibits increased GABAB-induced potassium current. *Journal of Neurophysiology*, *97*, 892–900.

Bianchi, P., Ciani, E., Guidi, S., Trazzi, S., Felice, D., Grossi, G., et al. (2010). Early pharmacotherapy restores neurogenesis and cognitive performance in the Ts65Dn mouse model for Down syndrome. *The Journal of Neuroscience*, *30*, 8769–8779.

Brown, J. H., Johnson, M. H., Paterson, S. J., Gilmore, R., Longhi, E., & Karmiloff-Smith, A. (2003). Spatial representation and attention in toddlers with Williams syndrome and Down syndrome. *Neuropsychologia*, *41*, 1037–1046.

Busciglio, J., Pelsman, A., Wong, C., Pigino, G., Yuan, M., Mori, H., et al. (2002). Altered metabolism of the amyloid beta precursor protein is associated with mitochondrial dysfunction in Down's syndrome. *Neuron*, *33*, 677–688.

Campbell, W. A., Reed, M. L., Strahle, J., Wolfe, M. S., & Xia, W. (2003). Presenilin endoproteolysis mediated by an aspartyl protease activity pharmacologically distinct from gamma-secretase. *Journal of Neurochemistry*, *85*, 1563–1574.

Carlesimo, G. A., Marotta, L., & Vicari, S. (1997). Long-term memory in mental retardation: Evidence for a specific impairment in subjects with Down's syndrome. *Neuropsychologia*, *35*, 71–79.

Cataldo, A. M., Petanceska, S., Peterhoff, C. M., Terio, N. B., Epstein, C. J., Villar, A., et al. (2003). App gene dosage modulates endosomal abnormalities of Alzheimer's disease in a segmental trisomy 16 mouse model of down syndrome. *The Journal of Neuroscience*, *23*, 6788–6792.

Cataldo, A. M., Petanceska, S., Terio, N. B., Peterhoff, C. M., Durham, R., Mercken, M., et al. (2004). Abeta localization in abnormal endosomes: Association with earliest Abeta elevations in AD and Down syndrome. *Neurobiology of Aging*, *25*, 1263–1272.

CDC. (2006). Improved national prevalence estimates for 18 selected major birth defects—United States, 1999–2001. *MMWR. Morbidity and Mortality Weekly Report*, *54*, 1301–1305.

Chakrabarti, L., Best, T. K., Cramer, N. P., Carney, R. S., Isaac, J. T., Galdzicki, Z., et al. (2010). Olig1 and Olig2 triplication causes developmental brain defects in Down syndrome. *Nature Neuroscience*, *13*, 927.

Chakrabarti, L., Galdzicki, Z., & Haydar, T. F. (2007). Defects in embryonic neurogenesis and initial synapse formation in the forebrain of the Ts65Dn mouse model of Down syndrome. *The Journal of Neuroscience*, *27*, 11483–11495.

Chapman, R. S., & Hesketh, L. J. (2000). Behavioral phenotype of individuals with Down syndrome. *Mental Retardation and Developmental Disabilities Research Reviews*, *6*, 84–95.

Chen-Hwang, M. C., Chen, H. R., Elzinga, M., & Hwang, Y. W. (2002). Dynamin is a minibrain kinase/dual specificity Yak1-related kinase 1A substrate. *Journal of Biological Chemistry*, *277*, 17597–17604.

Clark, S., Schwalbe, J., Stasko, M. R., Yarowsky, P. J., & Costa, A. C. (2006). Fluoxetine rescues deficient neurogenesis in hippocampus of the Ts65Dn mouse model for Down syndrome. *Experimental Neurology*, *200*, 256–261.

Colas, D., Valletta, J. S., Takimoto-Kimura, R., Nishino, S., Fujiki, N., Mobley, W. C., et al. (2008). Sleep and EEG features in genetic models of Down syndrome. *Neurobiology of Disease*, *30*, 1–7.

Contestabile, A., Fila, T., Ceccarelli, C., Bonasoni, P., Bonapace, L., Santini, D., et al. (2007). Cell cycle alteration and decreased cell proliferation in the hippocampal dentate gyrus and in the neocortical germinal matrix of fetuses with Down syndrome and in Ts65Dn mice. *Hippocampus*, *17*, 665–678.

Cooper, J. D., Salehi, A., Delcroix, J. D., Howe, C. L., Belichenko, P. V., Chua-Couzens, J., et al. (2001). Failed retrograde transport of NGF in a mouse model of Down's syndrome: Reversal of cholinergic neurodegenerative phenotypes following NGF infusion. *Proceedings of the National Academy of Sciences of the United States of America*, *98*, 10439–10444.

Cortez, M. A., Shen, L., Wu, Y., Aleem, I. S., Trepanier, C. H., Sadeghnia, H. R., et al. (2009). Infantile spasms and Down syndrome: A new animal model. *Pediatric Research, 65*, 499–503.

Costa, A. C. (2011). On the promise of pharmacotherapies targeted at cognitive and neurodegenerative components of Down syndrome. *Developmental Neuroscience, 33*(5), 414–427.

Costa, A. C., & Grybko, M. J. (2005). Deficits in hippocampal CA1 LTP induced by TBS but not HFS in the Ts65Dn mouse: A model of Down syndrome. *Neuroscience Letters, 382*, 317–322.

Costa, A. C., Scott-McKean, J. J., & Stasko, M. R. (2008). Acute injections of the NMDA receptor antagonist memantine rescue performance deficits of the Ts65Dn mouse model of Down syndrome on a fear conditioning test. *Neuropsychopharmacology, 33*, 1624–1632.

Cramer, N. P., Best, T. K., Stoffel, M., Siarey, R. J., & Galdzicki, Z. (2010). GABAB-GIRK2-mediated signaling in Down syndrome. *Advances in Pharmacology, 58*, 397–426.

da Costa, C. A., Ancolio, K., & Checler, F. (1999). C-terminal maturation fragments of presenilin 1 and 2 control secretion of APP alpha and A beta by human cells and are degraded by proteasome. *Molecular Medicine, 5*, 160–168.

Das, I., & Reeves, R. H. (2011). The use of mouse models to understand and improve cognitive deficits in Down syndrome. *Disease Models & Mechanisms, 4*(5), 596–606.

Davisson, M. T., Schmidt, C., Reeves, R. H., Irving, N. G., Akeson, E. C., Harris, B. S., et al. (1993). Segmental trisomy as a mouse model for Down syndrome. *Progress in Clinical and Biological Research, 384*, 117–133.

Delcroix, J. D., Valletta, J. S., Wu, C., Hunt, S. J., Kowal, A. S., & Mobley, W. C. (2003). NGF signaling in sensory neurons: Evidence that early endosomes carry NGF retrograde signals. *Neuron, 39*, 69–84.

Dierssen, M., Arque, G., McDonald, J., Andreu, N., Martinez-Cue, C., Florez, J., et al. (2011). Behavioral characterization of a mouse model overexpressing DSCR1/RCAN1. *PLoS One, 6*, e17010.

Dierssen, M., & de Lagran, M. M. (2006). DYRK1A (dual-specificity tyrosine-phosphorylated and -regulated kinase 1A): A gene with dosage effect during development and neurogenesis. *The Scientific World Journal, 6*, 1911–1922.

Dierssen, M., Fillat, C., Crnic, L., Arbones, M., Florez, J., & Estivill, X. (2001). Murine models for Down syndrome. *Physiology and Behavior, 73*, 859–871.

Edgin, J. O., Mason, G. M., Allman, M. J., Capone, G. T., Deleon, I., Maslen, C., et al. (2010). Development and validation of the Arizona Cognitive Test Battery for Down syndrome. *Journal of Neurodevelopmental Disorders, 2*, 149–164.

Epstein, C. J. (1990). The consequences of chromosome imbalance. *American Journal of Medical Genetics. Supplement, 7*, 31–37.

Evin, G., Barakat, A., & Masters, C. L. (2010). BACE: Therapeutic target and potential biomarker for Alzheimer's disease. *The International Journal of Biochemistry & Cell Biology, 42*, 1923–1926.

Faizi, M., Bader, P. L., Tun, C., Encarnacion, A., Kleschevnikov, A., Belichenko, P., et al. (2011). Comprehensive behavioral phenotyping of Ts65Dn mouse model of down syndrome: Activation of beta(1)-adrenergic receptor by xamoterol as a potential cognitive enhancer. *Neurobiology of Disease, 43*(2), 397–413.

Fernandez, F., Morishita, W., Zuniga, E., Nguyen, J., Blank, M., Malenka, R. C., et al. (2007). Pharmacotherapy for cognitive impairment in a mouse model of Down syndrome. *Nature Neuroscience, 10*, 411–413.

Galdzicki, Z., & Siarey, R. J. (2003). Understanding mental retardation in Down's syndrome using trisomy 16 mouse models. *Genes, Brain, and Behavior, 2*, 167–178.

Gardiner, K., Fortna, A., Bechtel, L., & Davisson, M. T. (2003). Mouse models of Down syndrome: How useful can they be? Comparison of the gene content of human chromosome 21 with orthologous mouse genomic regions. *Gene, 318*, 137–147.

Gardiner, K., Herault, Y., Lott, I. T., Antonarakis, S. E., Reeves, R. H., & Dierssen, M. (2010). Down syndrome: From understanding the neurobiology to therapy. *The Journal of Neuroscience, 30*, 14943–14945.

Gardiner, K. J. (2010). Molecular basis of pharmacotherapies for cognition in Down syndrome. *Trends in Pharmacological Sciences, 31*, 66–73.

German, D. C., Manaye, K. F., White, C. L., 3rd, Woodward, D. J., McIntire, D. D., Smith, W. K., et al. (1992). Disease-specific patterns of locus coeruleus cell loss. *Annals of Neurology, 32*, 667–676.

Guardia-Laguarta, C., Pera, M., & Lleo, A. (2010). gamma-Secretase as a therapeutic target in Alzheimer's disease. *Current Drug Targets, 11*, 506–517.

Haapasalo, A., & Kovacs, D. M. (2011). The many substrates of presenilin/gamma-secretase. *Journal of Alzheimer's Disease, 25*, 3–28.

Haass, C., Hung, A. Y., Schlossmacher, M. G., Oltersdorf, T., Teplow, D. B., & Selkoe, D. J. (1993). Normal cellular processing of the beta-amyloid precursor protein results in the secretion of the amyloid beta peptide and related molecules. *Annals of the New York Academy of Sciences, 695*, 109–116.

Harashima, C., Jacobowitz, D. M., Witta, J., Borke, R. C., Best, T. K., Siarey, R. J., et al. (2006). Abnormal expression of the G-protein-activated inwardly rectifying potassium channel 2 (GIRK2) in hippocampus, frontal cortex, and substantia nigra of Ts65Dn mouse: A model of Down syndrome. *The Journal of Comparative Neurology, 494*, 815–833.

Hare, J. (2010). Trafficking of amyloid beta-precursor protein products C83 and C99 on the endocytic pathway. *Biochemical and Biophysical Research Communications, 401*, 219–224.

Heller, J. H., Spiridigliozzi, G. A., Doraiswamy, P. M., Sullivan, J. A., Crissman, B. G., & Kishnani, P. S. (2004). Donepezil effects on language in children with Down syndrome: Results of the first 22-week pilot clinical trial. *American Journal of Medical Genetics Part A*, *130A*, 325–326.

Heller, J. H., Spiridigliozzi, G. A., Sullivan, J. A., Doraiswamy, P. M., Krishnan, R. R., & Kishnani, P. S. (2003). Donepezil for the treatment of language deficits in adults with Down syndrome: A preliminary 24-week open trial. *American Journal of Medical Genetics Part A*, *116A*, 111–116.

Heneka, M. T., Nadrigny, F., Regen, T., Martinez-Hernandez, A., Dumitrescu-Ozimek, L., Terwel, D., et al. (2010). Locus ceruleus controls Alzheimer's disease pathology by modulating microglial functions through norepinephrine. *Proceedings of the National Academy of Sciences of the United States of America*, *107*, 6058–6063.

Hook, E. B. (1983). Down syndrome rates and relaxed selection at older maternal ages. *American Journal of Human Genetics*, *35*, 1307–1313.

Hook, E. B., & Cross, P. K. (1982). Paternal age and Down's syndrome genotypes diagnosed prenatally: No association in New York state data. *Human Genetics*, *62*, 167–174.

Insausti, A. M., Megias, M., Crespo, D., Cruz-Orive, L. M., Dierssen, M., Vallina, I. F., et al. (1998). Hippocampal volume and neuronal number in Ts65Dn mice: A murine model of Down syndrome. *Neuroscience Letters*, *253*, 175–178.

Jin, L. W., Shie, F. S., Maezawa, I., Vincent, I., & Bird, T. (2004). Intracellular accumulation of amyloidogenic fragments of amyloid-beta precursor protein in neurons with Niemann-Pick type C defects is associated with endosomal abnormalities. *American Journal of Pathology*, *164*, 975–985.

Khoshnood, B., Greenlees, R., Loane, M., & Dolk, H. (2011). Paper 2: EUROCAT public health indicators for congenital anomalies in Europe. *Birth Defects Research. Part A, Clinical, and Molecular Teratology*, *91*(Suppl. 1), S16–S22.

Kidd, P. M. (2005). Neurodegeneration from mitochondrial insufficiency: Nutrients, stem cells, growth factors, and prospects for brain rebuilding using integrative management. *Alternative Medicine Review*, *10*, 268–293.

Kim, D. Y., Ingano, L. A., Carey, B. W., Pettingell, W. H., & Kovacs, D. M. (2005). Presenilin/gamma-secretase-mediated cleavage of the voltage-gated sodium channel beta2-subunit regulates cell adhesion and migration. *Journal of Biological Chemistry*, *280*, 23251–23261.

Kleschevnikov, A. M., Belichenko, P. V., Gall, J., George, L., Nosheny, R., Maloney, M. T., et al. (2012). Increased efficiency of the GABAA and GABAB receptor-mediated neurotransmission in the Ts65Dn mouse model of Down syndrome. *Neurobiology of Disease*, *45*(2), 683–691.

Kleschevnikov, A. M., Belichenko, P. V., Villar, A. J., Epstein, C. J., Malenka, R. C., & Mobley, W. C. (2004). Hippocampal long-term potentiation suppressed by increased inhibition in the Ts65Dn mouse, a genetic model of Down syndrome. *The Journal of Neuroscience*, *24*, 8153–8160.

Kleschevnikov, A. M., Van Volkinburg, J., Zhan, K., & Mobley, W. C. (2005). *Expression of GIRK2 and related proteins after chronic administrations of Prozac in Ts65Dn mice, a genetic model of Down syndrome.* Washington, DC: Society for Neuroscience Program No. 115.7. Abstract Viewer/Itinerary Planner.

Kobayashi, T., Washiyama, K., & Ikeda, K. (2004). Inhibition of G protein-activated inwardly rectifying K+ channels by various antidepressant drugs. *Neuropsychopharmacology*, *29*, 1841–1851.

Koehl, M., & Abrous, D. N. (2011). A new chapter in the field of memory: Adult hippocampal neurogenesis. *European Journal of Neuroscience*, *33*, 1101–1114.

Kondoh, T., Kanno, A., Itoh, H., Nakashima, M., Honda, R., Kojima, M., et al. (2011). Donepezil significantly improves abilities in daily lives of female Down syndrome patients with severe cognitive impairment: A 24-week randomized, double-blind, placebo-controlled trial. *International Journal of Psychiatry in Medicine*, *41*, 71–89.

Kounnas, M. Z., Danks, A. M., Cheng, S., Tyree, C., Ackerman, E., Zhang, X., et al. (2010). Modulation of gamma-secretase reduces beta-amyloid deposition in a transgenic mouse model of Alzheimer's disease. *Neuron*, *67*, 769–780.

Lai, F., & Williams, R. S. (1989). A prospective study of Alzheimer disease in Down syndrome. *Archives of Neurology*, *46*, 849–853.

Laifenfeld, D., Patzek, L. J., McPhie, D. L., Chen, Y., Levites, Y., Cataldo, A. M., et al. (2007). Rab5 mediates an amyloid precursor protein signaling pathway that leads to apoptosis. *The Journal of Neuroscience*, *27*, 7141–7153.

LeBlanc, A. C., & Gambetti, P. (1994). Production of Alzheimer 4kDa beta-amyloid peptide requires the C-terminal cytosolic domain of the amyloid precursor protein. *Biochemical and Biophysical Research Communications*, *204*, 1371–1380.

Lee, H. J., Jung, K. M., Huang, Y. Z., Bennett, L. B., Lee, J. S., Mei, L., et al. (2002). Presenilin-dependent gamma-secretase-like intramembrane cleavage of ErbB4. *Journal of Biological Chemistry*, *277*, 6318–6323.

Lejeune, J., Gautier, M., & Turpin, R. (1959). Study of somatic chromosomes from 9 mongoloid children. *Comptes Rendus Hebdomadaires des Séances de l'Académie des Sciences*, *248*, 1721–1722.

Li, Y. M. (2001). Gamma-secretase: A catalyst of Alzheimer disease and signal transduction. *Molecular Interventions*, *1*, 198–207.

Liu, C., Belichenko, P. V., Zhang, L., Fu, D., Kleschevnikov, A. M., Baldini, A., et al. (2011). Mouse models for Down syndrome-associated developmental cognitive disabilities. *Developmental Neuroscience*, *33*(5), 404–413.

Llorens-Martin, M. V., Rueda, N., Tejeda, G. S., Florez, J., Trejo, J. L., & Martinez-Cue, C. (2010). Effects of voluntary physical exercise on adult hippocampal neurogenesis and behavior of Ts65Dn mice, a model of Down syndrome. *Neuroscience*, *171*, 1228–1240.

Lockrow, J., Boger, H., Bimonte-Nelson, H., & Granholm, A. C. (2011). Effects of long-term memantine on memory and neuropathology in Ts65Dn mice, a model for Down syndrome. *Behavioural Brain Research*, *221*, 610–622.

Lorenzi, H. A., & Reeves, R. H. (2006). Hippocampal hypocellularity in the Ts65Dn mouse originates early in development. *Brain Research*, *1104*, 153–159.

Lott, I. T., Osann, K., Doran, E., & Nelson, L. (2002). Down syndrome and Alzheimer disease: Response to donepezil. *Archives of Neurology*, *59*, 1133–1136.

Malberg, J. E., Eisch, A. J., Nestler, E. J., & Duman, R. S. (2000). Chronic antidepressant treatment increases neurogenesis in adult rat hippocampus. *The Journal of Neuroscience*, *20*, 9104–9110.

Mann, D. M., Yates, P. O., & Marcyniuk, B. (1984). Alzheimer's presenile dementia, senile dementia of Alzheimer type and Down's syndrome in middle age form an age related continuum of pathological changes. *Neuropathology and Applied Neurobiology*, *10*, 185–207.

Mann, D. M., Yates, P. O., Marcyniuk, B., & Ravindra, C. R. (1986). The topography of plaques and tangles in Down's syndrome patients of different ages. *Neuropathology and Applied Neurobiology*, *12*, 447–457.

Marin-Padilla, M. (1976). Pyramidal cell abnormalities in the motor cortex of a child with Down's syndrome. A Golgi study. *The Journal of Comparative Neurology*, *167*, 63–81.

McGeer, E. G., Norman, M., Boyes, B., O'Kusky, J., Suzuki, J., & McGeer, P. L. (1985). Acetylcholine and aromatic amine systems in postmortem brain of an infant with Down's syndrome. *Experimental Neurology*, *87*, 557–570.

Murchison, C. F., Zhang, X. Y., Zhang, W. P., Ouyang, M., Lee, A., & Thomas, S. A. (2004). A distinct role for norepinephrine in memory retrieval. *Cell*, *117*, 131–143.

Nixon, R. A. (2005). Endosome function and dysfunction in Alzheimer's disease and other neurodegenerative diseases. *Neurobiology of Aging*, *26*, 373–382.

O'Bryan, J. P., Mohney, R. P., & Oldham, C. E. (2001). Mitogenesis and endocytosis: What's at the INTERSECTIoN? *Oncogene*, *20*, 6300–6308.

Owen, D. J., & Collins, B. M. (2010). Vesicle transport: A new player in APP trafficking. *Current Biology*, *20*, R413–R415.

Pallardo, F. V., Lloret, A., Lebel, M., d'Ischia, M., Cogger, V. C., Le Couteur, D. G., et al. (2010). Mitochondrial dysfunction in some oxidative stress-related genetic diseases: Ataxia-Telangiectasia, Down syndrome, Fanconi Anaemia and Werner syndrome. *Biogerontology*, *11*, 401–419.

Parker, S. E., Mai, C. T., Canfield, M. A., Rickard, R., Wang, Y., Meyer, R. E., et al. (2010). Updated National Birth Prevalence estimates for selected birth defects in the United States, 2004–2006. *Birth Defects Research. Part A, Clinical and Molecular Teratology*, *88*, 1008–1016.

Pasternak, S. H., Bagshaw, R. D., Guiral, M., Zhang, S., Ackerley, C. A., Pak, B. J., et al. (2003). Presenilin-1, nicastrin, amyloid precursor protein, and gamma-secretase activity are co-localized in the lysosomal membrane. *Journal of Biological Chemistry*, *278*, 26687–26694.

Pennington, B. F., Moon, J., Edgin, J., Stedron, J., & Nadel, L. (2003). The neuropsychology of Down syndrome: Evidence for hippocampal dysfunction. *Child Development*, *74*, 75–93.

Perez-Cremades, D., Hernandez, S., Blasco-Ibanez, J. M., Crespo, C., Nacher, J., & Varea, E. (2010). Alteration of inhibitory circuits in the somatosensory cortex of Ts65Dn mice, a model for Down's syndrome. *Journal of Neural Transmission*, *117*, 445–455.

Popov, V. I., Kleschevnikov, A. M., Klimenko, O. A., Stewart, M. G., & Belichenko, P. V. (2011). Three-dimensional synaptic ultrastructure in the dentate gyrus and hippocampal area CA3 in the Ts65Dn mouse model of Down syndrome. *The Journal of Comparative Neurology*, *519*, 1338–1354.

Prasher, V. P., Adams, C., & Holder, R. (2003). Long term safety and efficacy of donepezil in the treatment of dementia in Alzheimer's disease in adults with Down syndrome: Open label study. *International Journal of Geriatric Psychiatry*, *18*, 549–551.

Prasher, V. P., Farrer, M. J., Kessling, A. M., Fisher, E. M., West, R. J., Barber, P. C., et al. (1998). Molecular mapping of Alzheimer-type dementia in Down's syndrome. *Annals of Neurology*, *43*, 380–383.

Qin, W., Ho, L., Pompl, P. N., Peng, Y., Zhao, Z., Xiang, Z., et al. (2003). Cyclooxygenase (COX)-2 and COX-1 potentiate beta-amyloid peptide generation through mechanisms that involve gamma-secretase activity. *Journal of Biological Chemistry*, *278*, 50970–50977.

Rissman, R. A., & Mobley, W. C. (2011). Implications for treatment: GABAA receptors in aging, Down syndrome and Alzheimer's disease. *Journal of Neurochemistry*, *117*, 613–622.

Roth, D., & Cuatrecasas, P. (2010). *The distributed partnering model for drug discovery and development.* Ewing Marion Kauffman Foundation, January 26, 2010, 9pp.

Rueda, N., Florez, J., & Martinez-Cue, C. (2008). Chronic pentylenetetrazole but not donepezil treatment rescues spatial cognition in Ts65Dn mice, a model for Down syndrome. *Neuroscience Letters*, *433*, 22–27.

Rueda, N., Llorens-Martin, M., Florez, J., Valdizan, E., Banerjee, P., Trejo, J. L., et al. (2010). Memantine normalizes several phenotypic features in the Ts65Dn mouse model of Down syndrome. *Journal of Alzheimer's Disease*, *21*, 277–290.

Sabo, S. L., Lanier, L. M., Ikin, A. F., Khorkova, O., Sahasrabudhe, S., Greengard, P., et al. (1999). Regulation of beta-amyloid secretion by FE65, an amyloid protein precursor-binding protein. *Journal of Biological Chemistry*, *274*, 7952–7957.

Salehi, A., Delcroix, J. D., Belichenko, P. V., Zhan, K., Wu, C., Valletta, J. S., et al. (2006). Increased App expression in a mouse model of Down's syndrome disrupts NGF transport and causes cholinergic neuron degeneration. *Neuron*, *51*, 29–42.

Salehi, A., Faizi, M., Belichenko, P. V., & Mobley, W. C. (2007). Using mouse models to explore genotype-phenotype relationship in Down syndrome. *Mental Retardation and Developmental Disabilities Research Reviews*, *13*, 207–214.

Salehi, A., Faizi, M., Colas, D., Valletta, J. S., Laguna, J., Takimoto-Kimura, R., et al. (2009). Restoration of norepinephrine-modulated contextual memory in a mouse model of Down syndrome. *Science Translational Medicine*, *1*, 7–17.

Schliebs, R., & Arendt, T. (2006). The significance of the cholinergic system in the brain during aging and in Alzheimer's disease. *Journal of Neural Transmission*, *113*, 1625–1644.

Schmid, S. L., McNiven, M. A., & De Camilli, P. (1998). Dynamin and its partners: A progress report. *Current Opinion in Cell Biology*, *10*, 504–512.

Schuchmann, S., Muller, W., & Heinemann, U. (1998). Altered Ca2+ signaling and mitochondrial deficiencies in hippocampal neurons of trisomy 16 mice: A model of Down's syndrome. *The Journal of Neuroscience*, *18*, 7216–7231.

Selkoe, D., & Kopan, R. (2003). Notch and presenilin: Regulated intramembrane proteolysis links development and degeneration. *Annual Review of Neuroscience*, *26*, 565–597.

Selkoe, D. J., Yamazaki, T., Citron, M., Podlisny, M. B., Koo, E. H., Teplow, D. B., et al. (1996). The role of APP processing and trafficking pathways in the formation of amyloid beta-protein. *Annals of the New York Academy of Sciences*, *777*, 57–64.

Shinohara, M., Sato, N., Kurinami, H., Takeuchi, D., Takeda, S., Shimamura, M., et al. (2010). Reduction of brain beta-amyloid (Abeta) by fluvastatin, a hydroxymethylglutaryl-CoA reductase inhibitor, through increase in degradation of amyloid precursor protein C-terminal fragments (APP-CTFs) and Abeta clearance. *Journal of Biological Chemistry*, *285*, 22091–22102.

Shizuka-Ikeda, M., Matsubara, E., Ikeda, M., Kanai, M., Tomidokoro, Y., Ikeda, Y., et al. (2002). Generation of amyloid beta protein from a presenilin-1 and betaAPP complex. *Biochemical and Biophysical Research Communications*, *292*, 571–578.

Siarey, R. J., Stoll, J., Rapoport, S. I., & Galdzicki, Z. (1997). Altered long-term potentiation in the young and old Ts65Dn mouse, a model for Down syndrome. *Neuropharmacology*, *36*, 1549–1554.

Slepnev, V. I., & De Camilli, P. (2000). Accessory factors in clathrin-dependent synaptic vesicle endocytosis. *Nature Reviews. Neuroscience*, *1*, 161–172.

Sofroniew, M. V., Howe, C. L., & Mobley, W. C. (2001). Nerve growth factor signaling, neuroprotection, and neural repair. *Annual Review of Neuroscience*, *24*, 1217–1281.

Spiridigliozzi, G. A., Heller, J. H., Crissman, B. G., Sullivan-Saarela, J. A., Eells, R., Dawson, D., et al. (2007). Preliminary study of the safety and efficacy of donepezil hydrochloride in children with Down syndrome: A clinical report series. *American Journal of Medical Genetics. Part A*, *143A*, 1408–1413.

Stamnes, M. (2002). Regulating the actin cytoskeleton during vesicular transport. *Current Opinion in Cell Biology*, *14*, 428–433.

Sung, N. S., Crowley, W. F., Jr., Genel, M., Salber, P., Sandy, L., Sherwood, L. M., et al. (2003). Central challenges facing the national clinical research enterprise. *JAMA: The Journal of the American Medical Association*, *289*, 1278–1287.

Tang, B. L. (2009). Neuronal protein trafficking associated with Alzheimer disease: From APP and BACE1 to glutamate receptors. *Cell Adhesion & Migration*, *3*, 118–128.

Tralau-Stewart, C. J., Wyatt, C. A., Kleyn, D. E., & Ayad, A. (2009). Drug discovery: New models for industry-academic partnerships. *Drug Discovery Today*, *14*, 95–101.

Trazzi, S., Mitrugno, V. M., Valli, E., Fuchs, C., Rizzi, S., Guidi, S., et al. (2011). APP-dependent up-regulation of Ptch1 underlies proliferation impairment of neural precursors in Down syndrome. *Human Molecular Genetics*, *20*, 1560–1573.

Van Nostrand, W. E., Schmaier, A. H., & Wagner, S. L. (1992a). Potential role of protease nexin-2/amyloid beta-protein precursor as a cerebral anticoagulant. *Annals of the New York Academy of Sciences*, *674*, 243–252.

Van Nostrand, W. E., Wagner, S. L., Shankle, W. R., Farrow, J. S., Dick, M., Rozemuller, J. M., et al. (1992b). Decreased levels of soluble amyloid beta-protein precursor in cerebrospinal fluid of live Alzheimer disease patients. *Proceedings of the National Academy of Sciences of the United States of America*, *89*, 2551–2555.

Vetrivel, K. S., Barman, A., Chen, Y., Nguyen, P. D., Wagner, S. L., Prabhakar, R., et al. (2011). Loss of cleavage at {beta}'-site contributes to apparent increase in {beta}-amyloid peptide (A{beta}) secretion by {beta}-secretase (BACE1)-glycosylphosphatidylinositol (GPI) processing of amyloid precursor protein. *Journal of Biological Chemistry*, *286*, 26166–26177.

Vetrivel, K. S., Cheng, H., Lin, W., Sakurai, T., Li, T., Nukina, N., et al. (2004). Association of gamma-secretase with lipid rafts in post-Golgi and endosome membranes. *Journal of Biological Chemistry*, *279*, 44945–44954.

Wisniewski, K. E., Laure-Kamionowska, M., & Wisniewski, H. M. (1984). Evidence of arrest of neurogenesis and synaptogenesis in brains of patients with Down's syndrome. *The New England Journal of Medicine*, *311*, 1187–1188.

Wisniewski, K. E., Wisniewski, H. M., & Wen, G. Y. (1985). Occurrence of neuropathological changes and dementia of Alzheimer's disease in Down's syndrome. *Annals of Neurology*, *17*, 278–282.

Wolfe, M. S. (2008). Gamma-secretase inhibition and modulation for Alzheimer's disease. *Current Alzheimer Research*, *5*, 158–164.

Wu, C., Cui, B., He, L., Chen, L., & Mobley, W. C. (2009). The coming of age of axonal neurotrophin signaling endosomes. *Journal of Proteomics*, *72*, 46–55.

Wu, C., Ramirez, A., Cui, B., Ding, J., Delcroix, J. D., Valletta, J. S., et al. (2007). A functional dynein-microtubule network is required for NGF signaling through the Rap1/MAPK pathway. *Traffic*, *8*, 1503–1520.

Xia, W. (2008). From presenilinase to gamma-secretase, cleave to capacitate. *Current Alzheimer Research*, *5*, 172–178.

Yagishita, S., Morishima-Kawashima, M., Tanimura, Y., Ishiura, S., & Ihara, Y. (2006). DAPT-induced intracellular accumulations of longer amyloid beta-proteins: Further implications for the mechanism of intramembrane cleavage by gamma-secretase. *Biochemistry*, *45*, 3952–3960.

Yu, T., Li, Z., Jia, Z., Clapcote, S. J., Liu, C., Li, S., et al. (2010a). A mouse model of Down syndrome trisomic for all human chromosome 21 syntenic regions. *Human Molecular Genetics*, *19*, 2780–2791.

Yu, T., Liu, C., Belichenko, P., Clapcote, S. J., Li, S., Pao, A., et al. (2010b). Effects of individual segmental trisomies of human chromosome 21 syntenic regions on hippocampal long-term potentiation and cognitive behaviors in mice. *Brain Research*, *1366*, 162–171.

Zhang, L., Fu, D., Belichenko, P. V., Liu, C., Kleschevnikov, A. M., Pao, A., et al. (2012). Genetic analysis of Down syndrome facilitated by mouse chromosome engineering. *Bioengineered Bugs*, *45*(2), 683–691 (in press).

Zhang, M., Haapasalo, A., Kim, D. Y., Ingano, L. A., Pettingell, W. H., & Kovacs, D. M. (2006). Presenilin/gamma-secretase activity regulates protein clearance from the endocytic recycling compartment. *The FASEB Journal*, *20*, 1176–1178.

M. Dierssen and R. de la Torre (Eds.)
Progress in Brain Research, Vol. 197
ISSN: 0079-6123

CHAPTER 11

A Sonic hedgehog (Shh) response deficit in trisomic cells may be a common denominator for multiple features of Down syndrome

Duane G. Currier, Renita C. Polk and Roger H. Reeves*

Department of Physiology, The McKusick-Nathans Institute for Genetic Medicine, Johns Hopkins University School of Medicine, Baltimore, MA, USA

Abstract: The hedgehog (HH) family of growth factors is involved in many aspects of growth and development, from the establishment of left–right axes at gastrulation to the patterning and formation of multiple structures in essentially every tissue, to the maintenance and regulation of stem cell populations in adults. Sonic hedgehog (Shh) in particular acts as a mitogen, regulating proliferation of target cells, a growth factor that triggers differentiation in target populations, and a morphogen causing cells to respond differently based on their positions along a spatial and temporal concentration gradient. Given its very broad range of effects in development, it is not surprising that many of the structures affected by a disruption in Shh signaling are also affected in Down syndrome (DS). However, recent studies have shown that trisomic cerebellar granule cell precursors have a deficit, compared to their euploid counterparts, in their response to the mitogenic effects of Shh. This deficit substantially contributes to the hypocellular cerebellum in mouse models that parallels the human DS phenotype and can be corrected in early development by a single exposure to a small-molecule agonist of the Shh pathway.

Here, we consider how an attenuated Shh response might affect several aspects of development to produce multiple phenotypic outcomes observed in DS.

Keywords: SHH signaling; Down syndrome; brain development; common denominators of Down syndrome; neural crest; cerebellum.

Therapeutic approaches in Down syndrome

Trisomy for human chromosome 21 (Hsa21) results in Down syndrome (DS) which is among the most complex genetic perturbations compatible with survival past term. While trisomy affects development of every tissue, reduced cognitive ability in DS is among the most limiting features, and DS is one of the leading genetic causes of intellectual disability. The development and characterization of mouse models of DS, especially Ts65Dn, demonstrates that orthologous gene dosage effects produce comparable outcomes for

*Corresponding author.
E-mail: rreeves@jhmi.edu

DOI: 10.1016/B978-0-444-54299-1.00011-X

some phenotypes, including cognitive impairment (Fernandez et al., 2007; Hanson et al., 2007; Kleschevnikov et al., 2004; Reeves et al., 1995). As detailed elsewhere in this volume (de la Torre and Dierssen, 2012; Fillat and Altafaj, 2012), several drugs with the potential to ameliorate cognitive deficits in DS are making their way to clinical trials.

Studies of mice have played an important role in understanding the brain regions that are especially affected in DS (Lott and Dierssen, 2010). Functional outcomes as well as anatomical and physiological studies demonstrate three regions among those with the largest effects: prefrontal cortex, a contributor to executive function; hippocampus, a crucial site for learning and memory; and cerebellum, which shows a dramatic reduction in size and cellularity. It is notable that initial studies of learning and memory deficits in trisomic mice, which showed affects on hippocampus, informed the development of the first cognitive tests focused on deficits associated with the hippocampus in DS (Pennington et al., 2003; Reeves et al., 1995). That effort has been carried forward, resulting recently in the Arizona Cognitive Test Battery for DS (ACTB) (Edgin et al., 2010). The ACTB is a sensitive set of tests focused on brain regions affected in DS (see Edgin et al., 2012). Clinical trials with the goal of ameliorating cognitive deficits in DS have begun; many proposed efforts with this goal will utilize the ACTB tests as part of their assessments (for updated information, see http://clinicaltrials.gov/).

Currently, approaches to therapy in DS may be thought of in three very broad areas. First, people with DS frequently exhibit early onset of geriatric diseases. The histopathology of Alzheimer disease (AD) is present in all persons with DS along with the sequellae of the disease, including dementia in a substantial fraction of the DS population, and is certainly related at least in part to overexpression of the amyloid precursor protein gene, APP (Salehi et al., 2006). Age-related loss of afferents to the hippocampus from the locus coeruleus of neurons that use norepinephrine as a neurotransmitter and degeneration of basal forebrain cholinergic neurons are also hallmarks of DS shared with AD (Salehi et al., 2006, 2009).

A second general area for DS therapy involves correction of perturbed neuronal function in older children or adults. For example, restoration of an imbalance of inhibitory and excitatory inputs to the hippocampus forms the basis for major clinical trials going forward (Braudeau et al., 2011; Fernandez et al., 2007). This approach is based on the observation that downregulation of the GABAergic inhibitory PV neurons in Ts65Dn mice restores the balance of inhibitory:excitatory inputs and normalizes performance in hippocampal-based tasks such as the Novel Object Recognition Task and the Morris Water Maze (see Reeves and Garner, 2007; Rueda et al., 2008; Salehi et al., 2007). Several other efforts that have been carried out in trisomic mice and in some cases piloted in human studies look at a variety of hippocampal pathways (Lott and Dierssen, 2010).

A third potential area for therapy that is further downstream in the drug development pipeline addresses the initial basis of cognitive deficits, that is, antenatal brain development (Haydar and Reeves, 2012). Anatomical and morphological changes in the developing trisomic brain are being studied in detail in animal models, while imaging techniques are increasingly providing information about development of the DS brain. One approach of this type has been shown to normalize early deficits in postnatal development of the cerebellum, which is markedly hypocellular in DS and mouse models (Roper et al., 2006b); this example involves Shh signaling and is considered in detail here.

Shh signaling

Canonical Shh pathway

The Shh ligand is produced in cells distinct and often separated from those receiving the signal. The precursor protein is substantially modified by a cleavage that involves addition of a cholesterol moiety followed by palmitoylation (Mann and Beachy, 2004). Fully processed Shh (Shh-Np) is secreted from the producing cell and likely assembles into multimers (Zeng et al., 2001). Extracellular Shh-Np is sensed by the receiving cell via interactions with the 12-pass transmembrane protein, Patched (Ptch) (Marigo et al., 1996; Stone et al., 1996). In the pathway-off state (Fig. 1a), Ptch catalyzes the production of an unidentified repressor of Smoothened (Smo), a seven-pass transmembrane protein with possible G-protein-coupled receptor activity (Ayers and Therond, 2010; Chen et al., 2002). When Smo is repressed (pathway-off), the transcription factors Gli2 and Gli3 are targeted to the proteasome for processing to produce their transcriptional repressor forms ($Gli2^R$, $Gli3^R$) (Asai et al., 2006; Wang et al., 2000).

Another pathway element, Suppressor of Fused (SuFu), is found in both the cytoplasm and the nucleus and interacts with Gli1 and Gli2 proteins to further suppress pathway activity (Barnfield et al., 2005; Kogerman et al., 1999). SuFu/Gli complexes are exported from the nucleus and tethered in a SuFu-dependent manner in the cytoplasm. Further, SuFu inhibits Gli-mediated transcriptional activation by binding and inhibiting DNA-bound Gli1 or Gli2. The pathway is activated when Shh binds to Ptch, inhibiting the catalytic activity of the latter, thereby reversing the repression on Smo (Fig. 1b). This results in degradation of SuFu and Gli phosphorylation to produce activator Gli proteins that move to the nucleus and promote transcription (Chen et al., 2002; Humke et al., 2010; Yue et al., 2009; Zhang et al., 2004).

Noncanonical Shh signaling

Jenkins (2009) broadly defines several mechanisms for pathway activation outside of the canonical derepression of Gli transcription factors following Shh binding to Ptch. For example, Ptch can interact directly with Cyclin B1 to affect cell cycle

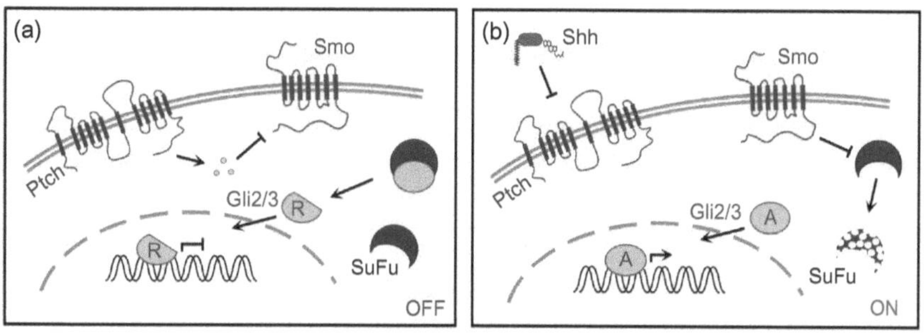

Fig. 1. The Shh pathway. (a) In the pathway-off state, Patched (Ptch) catalytically inhibits Smoothened (Smo) activity through an unidentified intermediate. Suppressor of Fused (SuFu) mediates cleavage of Gli2 and Gli3 into their repressor forms, lacking transactivation domains. The Gli2/3 repressors translocate to the nucleus where they repress transcription by binding target gene promoter sequences. (b) Shh binding to Ptch inhibits its catalytic repression of Smo, resulting in activation of Smo and degradation of SuFu. In the absence of SuFu, Gli2/3 are phosphorylated to become the activator forms. Gli2/3 activators translocate to the nucleus where they promote transcription by recruiting other transcriptional activators to target gene promoters. (For color version of this figure, the reader is referred to the Web version of this chapter.)

progression (Barnes et al., 2001) and can initiate apoptosis in neuroepithelial cells until it is blocked by Shh binding (Thibert et al., 2003). Although Ptch is the primary receptor for Shh, several other membrane bound proteins compete for Shh and are capable of enhancing or inhibiting pathway activity. Cell Adhesion Molecule-Related/Downregulated by Oncogenes (*CDO*) and Brother of CDO (*BOC*) both bind Shh though Fn3 domains. Growth Arrest Specific 1 (*Gas1*) (Martinelli and Fan, 2007) and Hedgehog-Interacting Protein (*Hhip*) (Bosanac et al., 2009) interact with Shh, but not through Fn3 domains. Of these, expression of CDO, BOC, or Gas1 increases Shh pathway activity, while Hhip negatively regulates the pathway (Beachy et al., 2010). SCUBE2, a secreted SCUBE protein family member, interacts with both Shh and Ptch and enhances Shh signaling (Tsai et al., 2009). The precise relationships between these receptor molecules and the Shh/Ptch interaction have yet to be described in detail.

The Gli transcription factors can also be regulated outside of the canonical Shh pathway. Borycki et al. demonstrated that Wnt1 and Wnt4 can induce *Gli2* expression and repress *Gli3* expression in a quail segmental plate mesoderm explant culture system (Borycki et al., 2000). Others have suggested that Gli1 protein may be regulated independently of Shh through the MAPK pathway (Seto et al., 2009). This would raise some interesting possibilities for activating Gli-regulated genes in the absence of Shh as well as for synergizing with Shh to superactivate the pathway.

While precise definitions of noncanonical pathways are lacking, results of multiple perturbations of the Shh pathway support the involvement of many of its components in multiple signaling paradigms. Indeed, it would be surprising if the enormous range of Shh effects as mitogen and morphogen in essentially every tissue could be reduced to one relatively simple pathway with three transcription factor effectors. Elaboration of these additional pathways for Shh signaling and their roles in specific processes will be a rich source for potential targets of therapeutic molecules that are fine-tuned to specific effects that are perturbed in disease states.

Phenotypes of Shh pathway mutants

Hedgehog signaling is a fundamental pathway involved in many aspects of prenatal development. Varied roles have been described from a number of studies in model organisms using constitutive and targeted gene "knockouts" in mice and chick/quail chimeras (Table 1). The first report of a constitutive Shh knockout demonstrated cephalic neural tube defects as early as day 8.5 of gestation (E8.5) that becomes more severe a day later, resulting in a markedly hypomorphic central nervous system (Chiang et al., 1996). Mutant embryos die around the time of birth and exhibit defects in development of heart, lung, kidney, and foregut in addition to the forming CNS. Many of the same phenotypes have been observed in embryos lacking Smo, an intermediate member of the Shh pathway that functions to positively regulate pathway activation, and embryos lacking Dispatched A (*mDispA*), a factor that is essential for efficient Shh release from cells producing it. Embryos lacking *Smo* or *mDispA* have somewhat more severe phenotypes resembling *Shh* and Indian hedgehog (*Ihh*) double knockout. *mDispA* and *Smo* are required for Ihh as well as for Shh signaling.

Similar results are seen after exposure to an inhibitor of Shh signaling, cyclopamine, the effective agent in *Veratrum californicum* that induces cyclopia in fetuses of pregnant sheep. In addition to cyclopia, malformations of the nose and skull, notably the premaxilla, are also present (Binns et al., 1963). From these studies, it is clear that successful embryo development requires restriction of Shh activation to specific levels in both a temporal and spatial manner.

Table 1. Phenotypes caused by alterations and interruptions in *Shh* signaling that may relate to deficits in DS

Perturbation	Affected system	Phenotypes	Age	Notes	References
Shh$^{-/-}$	Cardiac	Pharyngeal arch artery defects, ASD, VSD, Tetralogy of Fallot-like, persistent truncus arteriosus (PTA)	E10.5–15.5		Hildreth et al. (2009) and Smoak et al. (2005)
	Brain	Midline fusion of anterior lips of cephalic neural plate, incomplete separation of primitive optic vesicles[a]	E8.25		Abdelwahid et al. (2002)
		Loss of ventral structures, growth deficit in forebrain	E11.5		Chiang et al. (1996)
	Branchial arches	Reduced mandibular component[a]	E9.5		Abdelwahid et al. (2002)
	Craniofacial bones	Trace	E15.5		Chiang et al. (1996)
Nkx2.5$^{Cre/+}$, *Shh*$^{flox/-}$	Cardiac	Pharyngeal arch artery defects, PTA, AVSD	E10.5	*Shh* ablated in *Nkx2.5*-expressing cells	Goddeeris et al. (2007)
Mef2c-AHF-Cre, Smo$^{flox/-}$	Cardiac	PTA, ASD, VSD, AVSD, rounded and short AV valves	E14.5–18.5	*Smo* ablated in anterior heart field	Goddeeris et al. (2008)
Wnt1-Cre, Smo$^{flox/-}$	Cardiac	PTA	E10.5	*Smo* ablated in neural crest	Goddeeris et al. (2007)
Smo$^{Gli1\text{-}CreERT2}$	Cardiac	ASD and AVSD	E13.5	Floxed *Smo* allele under the control of inducible *Gli1:Cre*	Hoffmann et al. (2009)
Shh$^{Nkx2.1\text{-}Cre}$	Cardiac	ASD	E13.5	*Shh* ablated in *Nkx2.1*-expressing cells	Hoffmann et al. (2009)
Shh^{c}/*Shh*^{n}, *Pax2-Cre*	Cerebellum	Absence of EGL, disorganized PL, fewer lobes	P5	*Shh* ablated in Pax2 expressing precursors to Purkinje cells	Lewis et al. (2004)
Shh^{c}/*Shh*^{n}, *L7-Cre*	Cerebellum	Absence of EGL, disorganized PL, fewer lobes	P5	*Shh* ablated in precursors to Purkinje cells	Lewis et al. (2004)
SuFu$^{-/loxP}$, *Hoxb.7-Cre*	Cerebellum	Hypoplastic vermis, lack of foliation, disorganized cell layers	P21	*SuFu* ablated in precursors to CGNPs	Kim et al. (2011)
5E1 Hybridoma	Neural crest/ branchial arches	Hypomorphic branchial arches, developmental delay	Stages 9–11 +24h	Hybridoma cells injected lateral mesenchyme of developing chick	Ahlgren and Bronner-Fraser (1999)
Kif3a$^{fl/fl}$, *hGFAP-Cre*	Cerebellum	Atrophic cerebellum, fused folia, thicker PL	P25	*Kif3a* ablated in GFAP expressing CGNPS, Bergman glia, and Radial glia	Spassky et al. (2008)
Ptc$^{-/-}$	Embryo	Failure of neural tube closure, lethality	E10.5	Constitutive Ptc knockout	Goodrich et al. (1997)
Gli2$^{flox/flox}$, *En1-Cre*	Cerebellum	Reduced foliation, hypocellular EGL	P5/P8	Gli2 is ablated by E9.0 in En1-expressing cells which give rise to the cerebellum	Blaess et al. (2006)

[a]Similar phenotypes were seen in Smo −/− and mDispA −/− knockout mice.

Shh response deficit as a "common denominator" of DS phenotypes

Trisomy and Shh in cerebellar development

The first direct demonstration of Shh response perturbation due to trisomy came from analysis of cerebellar development in the Ts65Dn "Down syndrome" mouse (Baxter et al., 2000). Mouse models play a critical role in the study of gene dosage mechanisms that produce the features of DS as reviewed in detail in this volume and elsewhere (Das and Reeves, 2011; de la Torre and Dierssen, 2012; Herault et al., 2012; Mobley, in this volume; Moore and Roper, 2007; O'Doherty et al., 2005). Ts65Dn mice, like people with trisomy 21, have a smaller cerebellum and show specific deficits of Purkinje cells and of the granule cell neurons that make up the internal granule layer (IGL) of the cerebellum. Further, the reduced density of GC in the IGL of Ts65Dn mice was shown to occur in people with DS, as well (Baxter et al., 2000).

The IGL is not present at birth in mice nor is it fully formed in newborn humans. Rather, granule cell precursors (GCPs) form the external germinal layer on the surface layer of the cerebellum. Over the first 3weeks of life in mice (2–3years in human beings), Purkinje cells produce Shh which stimulates GCPs to divide and migrate inward to form the IGL (Dahmane and Ruiz i Altaba, 1999; Wallace, 1999; Wechsler-Reya and Scott, 1999). The granule cell neuron deficit in Ts65Dn is already detectable from 1week after birth (Roper et al., 2006b). On the day of birth, the number of GCPs in the external germinal layer is the same in Ts65Dn and euploid mice; however, the frequency of mitosis is significantly reduced in Ts65Dn. This reduced mitotic rate is a major contributor to the deficit in granule cell generation in trisomic mice (Roper et al., 2006b) and in DS (Guidi et al., 2011). Similarly, deleting a floxed Shh gene in late gestation by driving Cre expression with either the *Pax2* or *L7* promoters results in reduced cerebellar volume, hypocellularity, and disorganization of GCPs in the EGL (Table 1; Lewis et al., 2004).

When GCPs were isolated from trisomic and euploid cerebella and cultured in the presence of increasing amounts of Shh, two important things were observed (Roper et al., 2006b). First, trisomic GCP responded less to the mitogenic effects of Shh at every concentration (Fig. 2a). Second, the trisomic cells did exhibit a dosage response, suggesting that stimulation of Shh signaling *in vivo* might overcome some of the mitogenic deficit in trisomic cells that was observed *in vitro*.

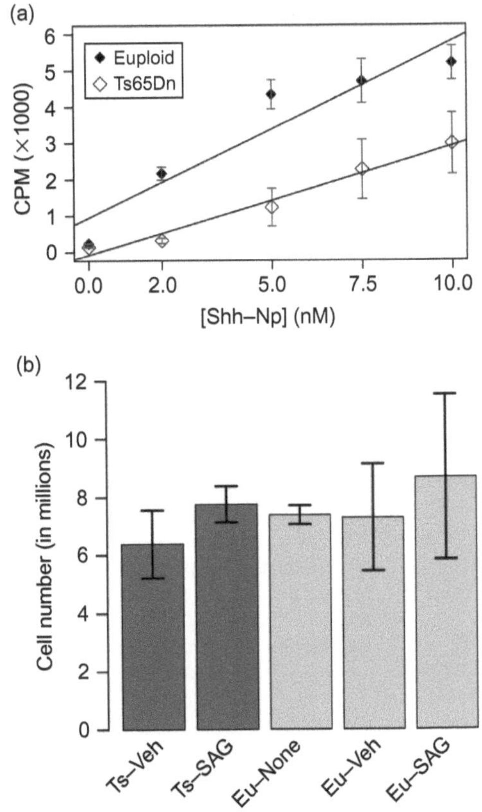

Fig. 2. SAG corrects Ts65Dn cerebellar dysmorphology. (a) Ts65Dn GCPs display an attenuated response to Shh treatment over a range of concentrations. (b) A single dose of SAG given on the day of birth restores GCP number at P6 to similar levels as Euploid littermates. (This figure is derived from Roper et al., 2006b.)

This was indeed the case. Trisomic mice that received a single dose of SAG on the day of birth had the same number of GCPs and of mitotic GCPs one week later, whereas vehicle-treated trisomic mice already showed a significant deficit of these cells (Fig. 2b).

The Shh hypothesis for DS

These results raise the question, "is the attenuated response to Shh in trisomic mice restricted to GCP, or do all Shh-responsive cells in a trisomic individual show a reduced reaction to Shh stimulation?" If the latter is the case, could stimulation of those developing populations at the appropriate stages of development represent a common approach to ameliorate diverse structural deficits in a wide range of cells and tissues that are affected to produce the DS phenotype? Based on the demonstration that trisomy results in a reduced response to the mitogenic effects of Shh in cerebellum, we consider here the possible effects of attenuated Shh response in three additional systems that are frequently or always affected during development in DS: craniofacial skeleton, heart, and the enteric nervous system. Observations of parallel effects of Shh disruption and of trisomy suggest that this mechanism may contribute to multiple DS phenotypes. Effects in development of the face and enteric nervous system further suggest that Shh signaling effects may impinge on neural crest cells (NCCs) which contribute to each of these structures.

Craniofacial development

The appearance of the DS face is very characteristic of this syndrome and is due substantially to hypoplasia of the midface skeleton. This is reflected in the Ts65Dn mouse and other models in an absolute correspondence between affected bones across the two species (Richtsmeier et al., 2000, 2002). In particular, the midface and mandible are significantly smaller and dysmorphic due to trisomy. These bones arise from an embryonic precursor, Meckel's cartilage, which is itself a product of the NCCs that contribute substantially to the first pharyngeal arch (PA1). To identify the earliest changes that lead to midface hypoplasia, we studied the formation of PA1 in Ts65Dn mice and their euploid littermates at embryonic day 9.5 (E9.5) in crosses with mice that express lacZ under control of the *Wnt1* promoter, marking NCCs (Roper et al., 2009).

Development of both trisomic and euploid embryos was highly variable at E9.5 with somite numbers ranging from 7 to 43, but no difference in developmental stage was observed between trisomic and euploid embryos (Roper et al., 2006a). When embryos at the 20–24 somite stage were considered, the trisomic PA1 was smaller, contained fewer neural crest-derived cells and these cells had a lower mitotic index than did their euploid counterparts. The number of migrating NCCs (lacZ + cells between the neural tube and PA1) was not significantly different at this stage; however, fewer migrating NCCs were present in slightly less mature, 17–19 somite embryos (Roper et al., 2009). Earlier experiments in both chick and mice show that Shh from endoderm of the ventral foregut is required to maintain migrating NCCs and to promote proliferation in PA1 (Ahlgren and Bronner-Fraser, 1999; Brito et al., 2006; Jeong et al., 2004). Note that if trisomic NCCs, like GCP, respond less to Shh than their euploid counterparts, some of these migrating cells might differentiate since they would "see" less Shh signal at the same concentration.

We then dissected the neural tubes from trisomic or euploid embryos and cultured them *ex vivo* to determine whether the delamination of NCCs from the tube is affected. Twenty-four hours after being placed in culture, Ts65Dn explants showed fewer cells migrating from the neural tube and those trisomic cells that did delaminate migrated for a shorter distance. Finally, we isolated cells from PA1 of euploid or trisomic embryos and cultured them to measure proliferation. Trisomic cells showed lower proliferation than did euploid.

However, addition of Shh to the cultures increased cell division, bringing the rate in trisomic cells to that seen in euploid cultures (Roper et al., 2006a).

Thus, the earliest trisomy-related deficits leading to midface skeletal hypoplasia arise from reduced delamination and migration of NCCs and from reduced proliferation of these cells in PA1, which provides the anlage for the cartilaginous model from which the mandible and midfacial bones will form. The known effects of Shh as well as our observations are consistent with the hypothesis that an important contribution to this deficit is the reduced responsiveness to Shh in cranial neural crest from trisomic mice, with direct phenotypic consequences.

Trisomy and Shh in cardiovascular development

Nearly half of all children born with DS have a congenital heart defect (CHD) (Ferencz et al., 1989). Atrioventricular septal defects (AVSDs) are the most common followed by ventricular septal defects (VSDs) and atrial septal defects (ASDs). Several mouse models of DS show similar patterns of CHD, indicating conservation of the effects of trisomic genes during mammalian heart development (Liu et al., 2011; O'Doherty et al., 2005; Williams et al., 2008). Nearly half of Tc1 transchromosomic mice, which carry a freely segregating copy of Hsa21, present with heart defects. VSDs are most common in trisomic mice, while AVSD and persistent truncus arteriousus (PTA) are also observed. About 15% of newborn Ts65Dn mice have cardiac defects, including ASD, VSD, PTA, and various errors of branching of the pulmonary and outflow tracts. Mice that carry a duplication of the Hsa21 region of conserved synteny on mouse chromosome 16 (Mmu16) show cardiovascular defects reminiscent of those seen in individuals with DS. They display ASD, VSD, and a tetralogy of Fallot-like phenotype (Li et al., 2007). Mice that are trisomic for all regions conserved with Hsa21 on mouse chromosomes 10, 16, and 17 have cardiovascular defects at a similar frequency (Yu et al., 2010).

Shh is secreted from cells in both the pulmonary endoderm, where it is required for proper atrial septation, and in the pharyngeal endoderm, where it is necessary for proper outflow tract septation (Goddeeris et al., 2008). Shh signaling marks cells within the second heart field (SHF) as progenitors of the atrial septum and outflow tract. Labeling of hedgehog-responsive cells early in heart development demonstrates that those cells migrate from the SHF and contribute to the primary atrial septum, dorsal mesenchymal protrusion (DMP), endocardial cushions, and pulmonary trunk (Hoffmann et al., 2009). The atrial septum, DMP, and endocardial cushions all combine to form the mesenchymal complex of the atrioventricular septum (Snarr et al., 2007). The appearance of this complex is necessary to complete AV septation and to anchor AV valves.

NCCs contribute to heart development by migrating into the outflow tract of the heart, contributing to septation and alignment. *Smo* is necessary for Shh pathway activation, and the loss of this gene in NCCs resulted in errors in septation and alignment of the aorta and pulmonary trunk, as well as defects in pharyngeal arch arteries (Goddeeris et al., 2008). A Shh response deficit could thus contribute to heart defects through direct effects in SHF, or because of an impaired response of trisomic neural crest. As noted, several steps in NCC delamination, migration, and proliferation require Shh signaling.

In support of this idea, several mouse models with impaired Shh signaling also display errors in septation (Table 1). A knockout of *Shh* in which exon 2 and its flanking introns are removed displays AVSD and other structural defects (Hildreth et al., 2009; Smoak et al., 2005). Similarly, when Shh signaling is blocked by cyclopamine at HH stage 14 chick embryos, they exhibit PTA, VSD, and pulmonary atresia secondary to reduced proliferation in the SHF (Dyer and Kirby, 2009). Similar outcomes occur when other components of the pathway are altered. Conditional knockouts of *Smo* and *Shh* result in AVSD and PTA in mouse embryos. Deletion of a floxed *Shh* allele in all cells expressing either *Nkx2.5* or *Gli1* results in AVSD (Goddeeris et al., 2007; Hoffmann et al., 2009). Thus Shh signaling mutants present AVSDs, VSDs, and ASDs, structural defects that are common in DS (Ferencz et al., 1989).

Septal defects were attributed primarily to errors in the endocardial cushions for many years, but evidence has emerged recently that points to a critical role for DMP as a contributing factor, especially to AVSD and secundum ASD (Goddeeris et al., 2008; Hoffmann et al., 2009). In this light, it is relevant that Shh signaling is not required for endocardial cushion contributions to septation but is necessary for proper contributions to DMP from the SHF. When Shh signaling is disrupted in the SHF, the DMP is hypoplastic or does not form and an AVSD results (Goddeeris et al., 2008; Hoffmann et al., 2009). Hypoplastic DMP has also been described in human fetuses with DS and in mice trisomic for all of Mmu16 (Blom et al., 2003; Snarr et al., 2007; Webb et al., 1999). Thus there is an important role for Shh signaling in formation of the DMP, and for DMP involvement in AVSDs; DS is a major risk factor for AVSD (Ferencz et al., 1989). Overall, there are substantial similarities between heart phenotypes caused by trisomy and those seen in Shh signaling mutants. These results do not prove causation but they are consistent with the effects expected from reduced response to Shh signaling in the developing heart.

Enteric nervous system

The small and large intestines are innervated by vagal NCCs that migrate along the primitive gut from the rostral toward the caudal end in response to glial derived neurotrophic factor (GDNF) (Young et al., 2001). In humans, these enteric neuron precursors (ENPs) colonize the gut beginning week 7 of gestation, with the primitive enteric ganglia reaching the rectum in week 12 (Kenny et al., 2010). Failure of the ENPs to reach the caudal end of the colon results in a condition known as aganglionic megacolon, or Hirschsprung's disease (HSCR) (Kenny et al., 2010). Though still rare, the incidence of HSCR in conjunction with DS is significantly increased over the rate in the population at large (Arnold et al., 2009). Mutations in the mouse *Ret* gene, a receptor tyrosine kinase that is activated by GDNF, cause NCCs colonizing the gut to migrate less efficiently and these mutants phenocopy HSCR (Asai et al., 2006). Human *RET* gene mutations contribute to susceptibility to the development of HSCR in people (Amiel et al., 2008; Angrist et al., 1995).

Shh is expressed by epithelial cells on the inner membrane of the gut and signals via BMP4 to inhibit differentiation of ENPs that are located in the central mesenchyme but are not close to the (outer) surface mesenchyme (Sukegawa et al., 2000). Inhibition of ENP differentiation could result in HSCR and, given the increased incidence in DS, it appears plausible that dosage-sensitive genes located on Hsa21 may contribute to the aganglionic phenotype. Decreased responsiveness to Shh could result in the expansion of the pro-differentiation environment to a point deeper in the gut mesenchyme than normal. Early differentiation of these ENPs could then deplete the migratory pool of cells before the entire length of colon has been colonized.

Hsa21 genes and Shh signaling

None of the genes encoding canonical Shh signaling pathway components are encoded on Hsa21. However, upregulation of *Ptch* (resulting in downregulation of the Shh pathway) has been reported in Ts65Dn mice for a specific, small group of stem cells in the subventricular zone (SVZ), the origin of granule cells in the dentate gyrus (Trazzi et al., 2011). In cultured neurospheres developed from the SVZ region, a C-terminal fragment of the APP protein, AICD, can contribute to the upregulation of *Ptch* transcription (Trazzi et al., 2011). Since the *APP* gene is found on Hsa21 and thus is chronically upregulated in DS (and also in Ts65Dn mice), this provides a possible explanation for the attenuated mitogenic response to Shh by trisomic cells. At the phenotypic level, the number of cells in dentate gyrus is reduced by about 20% in Ts65Dn mice compared to euploid (Insausti et al., 1998; Lorenzi and Reeves, 2006). Drugs developed for AD that modulate APP cleavage to reduce C-terminal fragments might thus have an additional ameliorative benefit in DS.

Molecular pathway analysis has implicated several additional Hsa21 genes whose expression may impinge on Shh signaling directly or indirectly, especially on the regulation of *Gli1, 2,* and/or *3* (see review by Sturgeon et al., 2012). To date, however, there is no direct demonstration of a dosage-sensitive trisomic gene disrupting Shh signaling in the developing cerebellum, heart, or the cranial or vagal neural crest. Trisomic mouse models of DS provide a sensitized genetic background for dissection of these mechanisms.

Discussion

Trisomy for Hsa21 results in increased dosage for more than 300 genes, and numerous studies of gene expression in DS and in animal models suggest that most of these will be upregulated by ~50% whenever and wherever they are normally expressed. Viewed from this perspective, the challenge of finding "cures" based on the modulation of individual gene function is daunting. The availability of segmental trisomies in animal models that recreate the dosage imbalances seen in DS and the demonstration that this produces features analogous to those in DS (Reeves et al., 1995) have led to a productive phenotype-based approach to the development of therapies (Reeves and Garner, 2007; Reeves et al., 1995; Salehi et al., 2007)

The phenotype-based approach suggests the possibility that multiple effects of trisomy in different tissues may result from perturbations in the same developmental pathways and regulatory processes, as we posit here for Shh. A deficit in response to the mitogenic effects of Shh has been demonstrated in trisomic cerebellar GCP. Trisomic NCC-derived cells in PA1 also appear to respond less to Shh than do their euploid counterparts. A similar response deficit in other trisomic cell types could affect development of the face, heart, enteric nervous system, and perhaps other tissues affected in DS. The cerebellar GCP response deficit to Shh is amenable to amelioration through the application of a small-molecule agonist of the Shh pathway (Chen et al., 2002; Roper et al., 2006b). Might a similar positive effect be possible in other cells and tissues that develop abnormally in DS if the Shh pathway could be stimulated to an appropriate degree at the appropriate time and place?

The Shh pathway is utilized in so many aspects of development that suggesting it as a therapeutic target seems highly improbable at first glance. Development is substantially disrupted in mice that are engineered to over- or underexpress Shh. In some case, the effects are concentration dependent, as when Shh acts as a morphogen to program cell response based on temporal and spatial gradients in anterior–posterior patterning of the limb (Harfe et al., 2004). Indeed, delivery of any molecule that stimulates or inhibits this pathway would likely need to be strictly limited in space and time to avoid deleterious side effects.

However, Shh should have no effect on cells that do not possess appropriate receptor and signal transduction pathways. We argue here that many if not all trisomic cells that are Shh responsive might show the attenuated response seen in cerebellar GCP. To the degree that this is the case, off-target effects would be reduced and could possibly have a beneficial effect.

The basic tenets of this model are testable in cell and mouse model systems. While it is clear that there is a substantial amount to learn about Shh signaling in all situations where it occurs, models of DS can play an important part in understanding these pathways. If this single molecular mechanism does prove to be a "common denominator" of multiple trisomic phenotypes, there are attendant prospects that a single kind of pharmaceutical treatment might ameliorate multiple features of DS.

Acknowledgments

The authors thank Phil Beachy for stimulating discussion about possible roles of Shh in DS. Critical support for this work was provided by the Down Syndrome Research and Treatment Foundation and Research Down Syndrome. This work was also supported by PHS awards 1R01 HD038384 from the National Institute of Child Health and Human Development and R01 HL083300 from the National Heart, Lung and Blood Institute. The content is solely the responsibility of the authors and does not necessarily represent the official views of the National Institutes of Health.

Abbreviations

ACTB	Arizona Cognitive Test Battery
AD	Alzheimer disease
ASD	atrial septal defect
AVSD	atrioventricular septal defect
DMP	dorsal mesenchymal protrusion
DS	Down syndrome
EGL	external germinal layer of the cerebellum
ENP	enteric neuron precursors
GCP	granule cell precursor
HSCR	Hirschsprung's disease
IGL	internal granule layer
NCCs	neural crest cells
PA1	first pharyngeal arch
PTA	persistent truncus arteriosus
SAG	Sonic agonist
SHF	second heart field
Shh	Sonic hedgehog
SVZ	subventricular zone
VSD	ventricular septal defect

References

Abdelwahid, E., Pelliniemi, L. J., & Jokinen, E. (2002). Cell death and differentiation in the development of the endocardial cushion of the embryonic heart. *Microscopy Research and Technique*, *58*, 395–403.

Ahlgren, S. C., & Bronner-Fraser, M. (1999). Inhibition of sonic hedgehog signaling in vivo results in craniofacial neural crest cell death. *Current Biology*, *9*, 1304–1314.

Amiel, J., Sproat-Emison, E., Garcia-Barcelo, M., Lantieri, F., Burzynski, G., Borrego, S., et al. (2008). Hirschsprung disease, associated syndromes and genetics: A review. *Journal of Medical Genetics*, *45*, 1–14.

Angrist, M., Bolk, S., Thiel, B., Puffenberger, E. G., Hofstra, R. M., Buys, C. H., et al. (1995). Mutation analysis of the RET receptor tyrosine kinase in Hirschsprung disease. *Human Molecular Genetics*, *4*, 821–830.

Arnold, S., Pelet, A., Amiel, J., Borrego, S., Hofstra, R., Tam, P., et al. (2009). Interaction between a chromosome 10 RET enhancer and chromosome 21 in the Down syndrome-Hirschsprung disease association. *Human Mutation*, *30*, 771–775.

Asai, N., Fukuda, T., Wu, Z., Enomoto, A., Pachnis, V., Takahashi, M., et al. (2006). Targeted mutation of serine 697 in the Ret tyrosine kinase causes migration defect of enteric neural crest cells. *Development (Cambridge, England)*, *133*, 4507–4516.

Ayers, K. L., & Therond, P. P. (2010). Evaluating Smoothened as a G-protein-coupled receptor for Hedgehog signalling. *Trends in Cell Biology*, *20*, 287–298.

Barnes, E. A., Kong, M., Ollendorff, V., & Donoghue, D. J. (2001). Patched1 interacts with cyclin B1 to regulate cell cycle progression. *The EMBO Journal*, *20*, 2214–2223.

Barnfield, P. C., Zhang, X., Thanabalasingham, V., Yoshida, M., & Hui, C. C. (2005). Negative regulation of Gli1 and Gli2 activator function by Suppressor of fused through multiple mechanisms. *Differentiation*, *73*, 397–405.

Baxter, L. L., Moran, T. H., Richtsmeier, J. T., Troncoso, J., & Reeves, R. H. (2000). Discovery and genetic localization of Down syndrome cerebellar phenotypes using the Ts65Dn mouse. *Human Molecular Genetics*, *9*, 195–202.

Beachy, P. A., Hymowitz, S. G., Lazarus, R. A., Leahy, D. J., & Siebold, C. (2010). Interactions between Hedgehog proteins and their binding partners come into view. *Genes & Development*, *24*, 2001–2012.

Binns, W., James, L. F., Shupe, J. L., & Everett, G. (1963). A congenital cyclopian-type malformation in lambs induced by maternal ingestion of a range plant, Veratrum californicum. *American Journal of Veterinary Research*, *24*, 1164–1175.

Blaess, S., Corrales, J. D., & Joyner, A. L. (2006). Sonic hedgehog regulates Gli activator and repressor functions with spatial and temporal precision in the mid/hindbrain region. *Development (Cambridge, England)*, *133*, 1799–1809.

Blom, N. A., Ottenkamp, J., Wenink, A. G., & Gittenberger-de Groot, A. C. (2003). Deficiency of the vestibular spine in atrioventricular septal defects in human fetuses with down syndrome. *The American Journal of Cardiology*, *91*, 180–184.

Borycki, A., Brown, A. M., & Emerson, C. P. Jr. (2000). Shh and Wnt signaling pathways converge to control Gli gene activation in avian somites. *Development (Cambridge, England)*, *127*, 2075–2087.

Bosanac, I., Maun, H. R., Scales, S. J., Wen, X., Lingel, A., Bazan, J. F., et al. (2009). The structure of SHH in complex with HHIP reveals a recognition role for the Shh pseudo active site in signaling. *Nature Structural and Molecular Biology*, *16*, 691–697.

Braudeau, J., Delatour, B., Duchon, A., Lopes-Pereira, P., Dauphinot, L., de Chaumont, F., et al. (2011). Specific targeting of the GABA-A receptor [alpha]5 subtype by a selective inverse agonist restores cognitive deficits in Down syndrome mice. *Journal of Psychopharmacology*, *25*, 1030–1042.

Brito, J. M., Teillet, M. A., & Le Douarin, N. M. (2006). An early role for sonic hedgehog from foregut endoderm in jaw development: Ensuring neural crest cell survival. *Proceedings of the National Academy of Sciences of the United States of America*, *103*, 11607–11612.

Chen, J. K., Taipale, J., Young, K. E., Maiti, T., & Beachy, P. A. (2002). Small molecule modulation of Smoothened activity. *Proceedings of the National Academy of Sciences of the United States of America*, *99*, 14071–14076.

Chiang, C., Litingtung, Y., Lee, E., Young, K. E., Corden, J. L., Westphal, H., et al. (1996). Cyclopia and defective axial patterning in mice lacking Sonic hedgehog gene function. *Nature*, *383*, 407–413.

Dahmane, N., & Ruiz i Altaba, A. (1999). Sonic hedgehog regulates the growth and patterning of the cerebellum. *Development (Cambridge, England)*, *126*, 3089–3100.

Das, I., & Reeves, R. H. (2011). The use of mouse models to understand and improve cognitive deficits in Down syndrome. *Disease Models & Mechanisms*, *4*, 596–606.

de la Torre, R., & Dierssen, M. (2012). Therapeutic approaches in the improvement of cognitive performance in Down syndrome: Past, present and future. *Progress in Brain Research*, *197*, 1–14.

Dyer, L. A., & Kirby, M. L. (2009). The role of secondary heart field in cardiac development. *Developmental Biology*, *336*, 137–144.

Edgin, J. O., Mason, G. M., Allman, M. J., Capone, G. T., Deleon, I., Maslen, C., et al. (2010). Development and validation of the Arizona Cognitive Test Battery for Down syndrome. *Journal of Neurodevelopmental Disorders*, *2*, 149–164.

Edgin, J. O., Mason, G. M., Spanò, G., Fernández, A., & Nadel, L. (2012). Human and mouse model cognitive phenotypes in Down syndrome: Implications for assessment. *Progress in Brain Research*, *197*, 123–151.

Ferencz, C., Neill, C. A., Boughman, J. A., Rubin, J. D., Brenner, J. I., & Perry, L. W. (1989). Congenital cardiovascular malformations associated with chromosome abnormalities: An epidemiologic study. *The Journal of Pediatrics*, *114*, 79–86.

Fernandez, F., Morishita, W., Zuniga, E., Nguyen, J., Blank, M., Malenka, R. C., et al. (2007). Pharmacotherapy for cognitive impairment in a mouse model of Down syndrome. *Nature Neuroscience*, *10*, 411–413.

Fillat, C., & Altafaj, X. (2012). Gene therapy for Down syndrome. *Progress in Brain Research*, *197*, 237–247.

Goddeeris, M. M., Rho, S., Petiet, A., Davenport, C. L., Johnson, G. A., Meyers, E. N., et al. (2008). Intracardiac septation requires hedgehog-dependent cellular contributions from outside the heart. *Development (Cambridge, England)*, *135*, 1887–1895.

Goddeeris, M. M., Schwartz, R., Klingensmith, J., & Meyers, E. N. (2007). Independent requirements for Hedgehog signaling by both the anterior heart field and neural crest cells for outflow tract development. *Development (Cambridge, England)*, *134*, 1593–1604.

Goodrich, L. V., Milenković, L., Higgins, K. M., & Scott, M. P. (1997). Altered neural cell fates and medulloblastoma in mouse patched mutants. *Science (New York, NY)*, *277*, 1109–1113.

Guidi, S., Ciani, E., Bonasoni, P., Santini, D., & Bartesaghi, R. (2011). Widespread proliferation impairment and hypocellularity in the cerebellum of fetuses with Down syndrome. *Brain Pathology*, *21*, 361–373.

Hanson, J. E., Blank, M., Valenzuela, R. A., Garner, C. C., & Madison, D. V. (2007). The functional nature of synaptic circuitry is altered in area CA3 of the hippocampus in a mouse model of Down's syndrome. *The Journal of Physiology*, *579*, 53–67.

Harfe, B. D., Scherz, P. J., Nissim, S., Tian, H., McMahon, A. P., & Tabin, C. J. (2004). Evidence for an expansion-based temporal Shh gradient in specifying vertebrate digit identities. *Cell*, *118*, 517–528.

Haydar, T., & Reeves, R. H. (2012). Direct and indirect effects of trisomy on early brain development. *Trends in Neurosciences*, *35*, 81–91.

Herault, Y., Duchon, A., Velot, E., Maréchal, D., & Brault, V. (2012). The in vivo Down syndrome genomic library in mouse. *Progress in Brain Research*, *197*, 169–197.

Hildreth, V., Webb, S., Chaudhry, B., Peat, J. D., Phillips, H. M., Brown, N., et al. (2009). Left cardiac isomerism in the Sonic hedgehog null mouse. *Journal of Anatomy*, *214*, 894–904.

Hoffmann, A. D., Peterson, M. A., Friedland-Little, J. M., Anderson, S. A., & Moskowitz, I. P. (2009). Sonic hedgehog is required in pulmonary endoderm for atrial septation. *Development (Cambridge, England)*, *136*, 1761–1770.

Humke, E. W., Dorn, K. V., Milenkovic, L., Scott, M. P., & Rohatgi, R. (2010). The output of Hedgehog signaling is controlled by the dynamic association between Suppressor of Fused and the Gli proteins. *Genes & Development, 24*, 670–682.

Insausti, A. M., Megias, M., Crespo, D., Cruz-Orive, L. M., Dierssen, M., Vallina, I. F., et al. (1998). Hippocampal volume and neuronal number in Ts65Dn mice: A murine model of Down syndrome. *Neuroscience Letters, 253*, 175–178.

Jenkins, D. (2009). Hedgehog signalling: Emerging evidence for non-canonical pathways. *Cellular Signalling, 21*, 1023–1034.

Jeong, J., Mao, J., Tenzen, T., Kottmann, A. H., & McMahon, A. P. (2004). Hedgehog signaling in the neural crest cells regulates the patterning and growth of facial primordia. *Genes & Development, 18*, 937–951.

Kenny, S. E., Tam, P. K., & Garcia-Barcelo, M. (2010). Hirschsprung's disease. *Seminars in Pediatric Surgery, 19*, 194–200.

Kim, J. J., Gill, P. S., Rotin, L., van Eede, M., Henkelman, R. M., Hui, C.-c., et al. (2011). Suppressor of fused controls mid-hindbrain patterning and cerebellar morphogenesis via GLI3 repressor. *The Journal of Neuroscience, 31*, 1825–1836.

Kleschevnikov, A. M., Belichenko, P. V., Villar, A. J., Epstein, C. J., Malenka, R. C., & Mobley, W. C. (2004). Hippocampal long-term potentiation suppressed by increased inhibition in the Ts65Dn mouse, a genetic model of Down syndrome. *The Journal of Neuroscience, 24*, 8153–8160.

Kogerman, P., Grimm, T., Kogerman, L., Krause, D., Unden, A. B., Sandstedt, B., et al. (1999). Mammalian suppressor-of-fused modulates nuclear-cytoplasmic shuttling of Gli-1. *Nature Cell Biology, 1*, 312–319.

Lewis, P. M., Gritli-Linde, A., Smeyne, R., Kottmann, A., & McMahon, A. P. (2004). Sonic hedgehog signaling is required for expansion of granule neuron precursors and patterning of the mouse cerebellum. *Developmental Biology, 270*, 393–410.

Li, Z., Yu, T., Morishima, M., Pao, A., LaDuca, J., Conroy, J., et al. (2007). Duplication of the entire 22.9 Mb human chromosome 21 syntenic region on mouse chromosome 16 causes cardiovascular and gastrointestinal abnormalities. *Human Molecular Genetics, 16*, 1359–1366.

Liu, C., Morishima, M., Yu, T., Matsui, S. I., Zhang, L., Fu, D., et al. (2011). Genetic analysis of Down syndrome-associated heart defects in mice. *Human Genetics, 130*, 623–632.

Lorenzi, H. A., & Reeves, R. H. (2006). Hippocampal hypocellularity in the Ts65Dn mouse originates early in development. *Brain Research, 1104*, 153–159.

Lott, I. T., & Dierssen, M. (2010). Cognitive deficits and associated neurological complications in individuals with Down's syndrome. *Lancet Neurology, 9*, 623–633.

Mann, R. K., & Beachy, P. A. (2004). Novel lipid modifications of secreted protein signals. *Annual Review of Biochemistry, 73*, 891–923.

Marigo, V., Davey, R. A., Zuo, Y., Cunningham, J. M., & Tabin, C. J. (1996). Biochemical evidence that patched is the Hedgehog receptor. *Nature, 384*, 176–179.

Martinelli, D. C., & Fan, C. M. (2007). Gas1 extends the range of Hedgehog action by facilitating its signaling. *Genes & Development, 21*, 1231–1243.

Moore, C. S., & Roper, R. J. (2007). The power of comparative and developmental studies for mouse models of Down syndrome. *Mammalian Genome, 18*, 431–443.

O'Doherty, A., Ruf, S., Mulligan, C., Hildreth, V., Errington, M. L., Cooke, S., et al. (2005). An aneuploid mouse strain carrying human chromosome 21 with down syndrome phenotypes. *Science, 309*, 2033–2037.

Pennington, B. F., Moon, J., Edgin, J., Stedron, J., & Nadel, L. (2003). The neuropsychology of Down syndrome: Evidence for hippocampal dysfunction. *Child Development, 74*, 75–93.

Reeves, R. H., & Garner, C. C. (2007). A year of unprecedented progress in Down syndrome basic research. *Mental Retardation and Developmental Disabilities Research Reviews, 13*, 215–220.

Reeves, R., Irving, N., Moran, T., Wohn, A., Kitt, C., Sisodia, S., et al. (1995). A mouse model for Down syndrome exhibits learning and behaviour deficits. *Nature Genetics, 11*, 177–183.

Richtsmeier, J., Baxter, L., & Reeves, R. (2000). Parallels of craniofacial maldevelopment in Down Syndrome and Ts65Dn mice. *Developmental Dynamics, 217*, 137–145.

Richtsmeier, J. T., Zumwalt, A., Carlson, E. J., Epstein, C. J., & Reeves, R. H. (2002). Craniofacial phenotypes in segmentally trisomic mouse models for Down syndrome. *American Journal of Medical Genetics, 107*, 317–324.

Roper, R. J., Baxter, L. L., Saran, N. G., Klinedinst, D. K., Beachy, P. A., & Reeves, R. H. (2006). Defective cerebellar response to mitogenic Hedgehog signaling in Down [corrected] syndrome mice. *Proceedings of the National Academy of Sciences of the United States of America, 103*, 1452–1456.

Roper, R., St. John, H., Philip, J., Lawler, A., & Reeves, R. (2006). Perinatal loss of Ts65Dn mice, a model of Down syndrome. *Genetics, 172*, 437–443.

Roper, R. J., VanHorn, J. F., Cain, C. C., & Reeves, R. H. (2009). A neural crest deficit in Down syndrome mice is associated with deficient mitotic response to Sonic hedgehog. *Mechanisms of Development, 126*, 212–219.

Rueda, N., Florez, J., & Martinez-Cue, C. (2008). Chronic pentylenetetrazole but not donepezil treatment rescues spatial cognition in Ts65Dn mice, a model for Down syndrome. *Neuroscience Letters, 433*, 22–27.

Salehi, A., Delcroix, J. D., Belichenko, P. V., Zhan, K., Wu, C., Valletta, J. S., et al. (2006). Increased App expression in a mouse model of Down's syndrome disrupts NGF transport and causes cholinergic neuron degeneration. *Neuron, 51*, 29–42.

Salehi, A., Faizi, M., Belichenko, P. V., & Mobley, W. C. (2007). Using mouse models to explore genotype-phenotype relationship in Down syndrome. *Mental Retardation and Developmental Disabilities Research Reviews*, *13*, 207–214.

Salehi, A., Faizi, M., Colas, D., Valletta, J., Laguna, J., Takimoto-Kimura, R., et al. (2009). Restoration of norepinephrine-modulated contextual memory in a mouse model of Down syndrome. *Science Translational Medicine*, *1*, 7ra17.

Seto, M., Ohta, M., Asaoka, Y., Ikenoue, T., Tada, M., Miyabayashi, K., et al. (2009). Regulation of the hedgehog signaling by the mitogen-activated protein kinase cascade in gastric cancer. *Molecular Carcinogenesis*, *48*, 703–712.

Smoak, I. W., Byrd, N. A., Abu-Issa, R., Goddeeris, M. M., Anderson, R., Morris, J., et al. (2005). Sonic hedgehog is required for cardiac outflow tract and neural crest cell development. *Developmental Biology*, *283*, 357–372.

Snarr, B. S., O'Neal, J. L., Chintalapudi, M. R., Wirrig, E. E., Phelps, A. L., Kubalak, S. W., et al. (2007). Isl1 expression at the venous pole identifies a novel role for the second heart field in cardiac development. *Circulation Research*, *101*, 971–974.

Spassky, N., Han, Y.-G., Aguilar, A., Strehl, L., Besse, L., Laclef, C., et al. (2008). Primary cilia are required for cerebellar development and Shh-dependent expansion of progenitor pool. *Developmental Biology*, *317*, 246–259.

Stone, D. M., Hynes, M., Armanini, M., Swanson, T. A., Gu, Q., Johnson, R. L., et al. (1996). The tumour-suppressor gene patched encodes a candidate receptor for Sonic hedgehog. *Nature*, *384*, 129–134.

Sturgeon, X., Le, T., Ahmed, M. M., & Gardiner, K. J. (2012). Pathways to cognitive deficits in Down syndrome. *Progress in Brain Research*, *197*, 73–100.

Sukegawa, A., Narita, T., Kameda, T., Saitoh, K., Nohno, T., Iba, H., et al. (2000). The concentric structure of the developing gut is regulated by Sonic hedgehog derived from endodermal epithelium. *Development (Cambridge, England)*, *127*, 1971–1980.

Thibert, C., Teillet, M. A., Lapointe, F., Mazelin, L., Le Douarin, N. M., & Mehlen, P. (2003). Inhibition of neuroepithelial patched-induced apoptosis by sonic hedgehog. *Science*, *301*, 843–846.

Trazzi, S., Mitrugno, V. M., Valli, E., Fuchs, C., Rizzi, S., Guidi, S., et al. (2011). APP-dependent up-regulation of Ptch1 underlies proliferation impairment of neural precursors in Down syndrome. *Human Molecular Genetics*, *20*, 1560–1573.

Tsai, M. T., Cheng, C. J., Lin, Y. C., Chen, C. C., Wu, A. R., Wu, M. T., et al. (2009). Isolation and characterization of a secreted, cell-surface glycoprotein SCUBE2 from humans. *The Biochemical Journal*, *422*, 119–128.

Wallace, V. A. (1999). Purkinje-cell-derived Sonic hedgehog regulates granule neuron precursor cell proliferation in the developing mouse cerebellum. *Current Biology*, *9*, 445–448.

Wang, B., Fallon, J. F., & Beachy, P. A. (2000). Hedgehog-regulated processing of Gli3 produces an anterior/posterior repressor gradient in the developing vertebrate limb. *Cell*, *100*, 423–434.

Webb, S., Anderson, R. H., Lamers, W. H., & Brown, N. A. (1999). Mechanisms of deficient cardiac septation in the mouse with trisomy 16. *Circulation Research*, *84*, 897–905.

Wechsler-Reya, R. J., & Scott, M. P. (1999). Control of neuronal precursor proliferation in the cerebellum by Sonic Hedgehog. *Neuron*, *22*, 103–114.

Williams, A. D., Mjaatvedt, C. H., & Moore, C. S. (2008). Characterization of the cardiac phenotype in neonatal Ts65Dn mice. *Developmental Dynamics*, *237*, 426–435.

Young, H. M., Hearn, C. J., Farlie, P. G., Canty, A. J., Thomas, P. Q., & Newgreen, D. F. (2001). GDNF is a chemoattractant for enteric neural cells. *Developmental Biology*, *229*, 503–516.

Yu, T., Li, Z., Jia, Z., Clapcote, S. J., Liu, C., Li, S., et al. (2010). A mouse model of Down syndrome trisomic for all human chromosome 21 syntenic regions. *Human Molecular Genetics*, *19*, 2780–2791.

Yue, S., Chen, Y., & Cheng, S. Y. (2009). Hedgehog signaling promotes the degradation of tumor suppressor Sufu through the ubiquitin-proteasome pathway. *Oncogene*, *28*, 492–499.

Zeng, X., Goetz, J. A., Suber, L. M., Scott, W. J., Jr., Schreiner, C. M., & Robbins, D. J. (2001). A freely diffusible form of Sonic hedgehog mediates long-range signalling. *Nature*, *411*, 716–720.

Zhang, C., Williams, E. H., Guo, Y., Lum, L., & Beachy, P. A. (2004). Extensive phosphorylation of Smoothened in Hedgehog pathway activation. *Proceedings of the National Academy of Sciences of the United States of America*, *101*, 17900–17907.

M. Dierssen and R. de la Torre (Eds.)
Progress in Brain Research, Vol. 197
ISSN: 0079-6123

CHAPTER 12

Gene therapy for Down syndrome

Cristina Fillat[†,‡,*] and Xavier Altafaj[‡,§]

[†] *Institut d'Investigacions Biomèdiques August Pi i Sunyer (IDIBAPS), Barcelona, Spain*
[‡] *Centro de Investigación Biomédica en Red de Enfermedades Raras (CIBERER), Barcelona, Spain*
[§] *Institute of Neuropathology, IDIBELL, L'Hospitalet de Llobregat, Barcelona, Spain*

Abstract: The presence of an additional copy of HSA21 chromosome in Down syndrome (DS) individuals leads to the overexpression of 30–50% of HSA21 genes. This upregulation can, in turn, trigger a deregulation on the expression of non-HSA21 genes. Moreover, the overdose of HSA21 microRNAs (miRNAs) may result in the downregulation of its target genes. Additional complexity can also arise from epigenetic changes modulating gene expression. Thus, a myriad of transcriptional and posttranscriptional alterations participate to produce abnormal phenotypes in almost all tissues and organs of DS individuals. The study of the physiological roles of genes dysregulated in DS, as well as their characterization in murine models with gene(s) dosage imbalance, pointed out several genes, and functional noncoding elements to be particularly critical in the etiology of DS. Recent findings indicate that gene therapy strategies—based on the introduction of genetic elements by means of delivery vectors—toward the correction of phenotypic abnormalities in DS are also very promising tool to identify HSA21 and non-HSA21 gene candidates, contributing to DS phenotype. In this chapter, we focus on the impact of normalizing the expression levels of up or downregulated genes to rescue particular phenotypes of DS. Attempts toward gene-based treatment approaches in mouse models will be discussed as new opportunities to ameliorate DS alterations.

Keywords: Down syndrome; gene therapy; microRNAs; HSA21 genes; viral vectors; adeno-associated virus; lentivirus; central nervous system.

Introduction

Down syndrome (DS) is most likely a multigenic condition, in which the extradose of a subset of genes in the trisomy 21 plays a major contributing role. Interactions within HSA21 genes and between genes located in other genomic regions (e.g., in disomy) may also exert an effect. Moreover, the overexpression of functional noncoding elements in the HSA21 might result in the deregulation of specific proteins, that theoretically may further contribute to the DS phenotype.

*Corresponding author.
Tel.: +34 93 227 54 00 Ext. 4579; Fax: +34 93 227 9240
E-mail: cfillat@clinic.ub.es

DOI: 10.1016/B978-0-444-54299-1.00012-1

Functional analysis of HSA21-encoded proteins has revealed their involvement in a plethora of biological processes (up to 87), with 81 different functions and a localization pattern in 26 different cellular compartments (Antonarakis et al., 2004). Also, five microRNAs (miRNAs) have been identified on 21q, although limited knowledge is available on their potential involvement in DS (Kuhn et al., 2008; Fig. 1).

As our understanding of the molecular contributors to DS phenotypes draws a more detailed picture, the possibility to normalize the excess or defective expression of altered gene expression by gene therapy emerges as a potential therapeutic strategy. This approach is only at its infancy and requires of many proof-of-principle studies to envision all its potential. DS involves alterations in many different tissues and in different regions or cell types in a given tissue. Thus, there is not a universal approach, and specific designed strategies to target identified phenotypes will be required. The application of these strategies must take into account numerous complexities, such as the adequate control of the introduced target sequence to finely regulate gene expression up to normalization and devoid of toxicity. It is also fundamental to identify the optimal time of intervention (prenatal, postnatal, or adult) for proper phenotype rescue. The selection of an optimal delivery system and the definition of the best routes of administration are also critical issues that might be adapted to the selected target gene(s) (Fig. 2).

Despite a lot of preclinical research needs to be developed, promising advances have been recently done. Thus, gene therapy for DS can be envisioned, together with pharmacological approaches, as an attempt to ameliorate the quality of life of affected individuals.

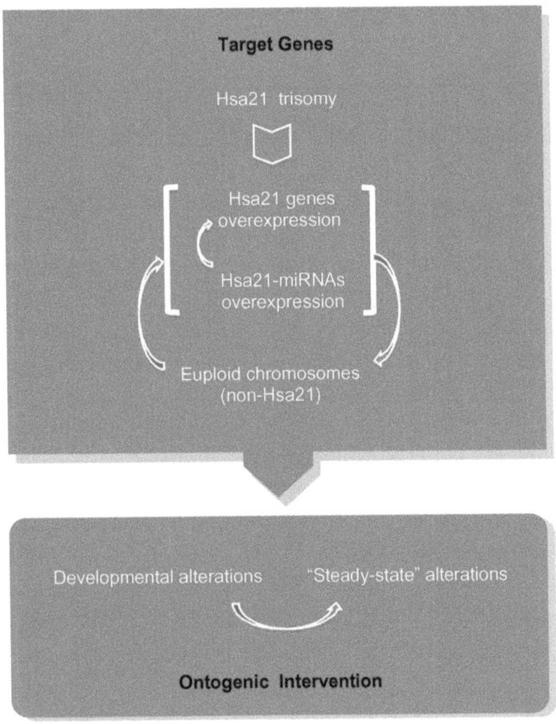

Fig. 1. Complex and cross talk alterations in DS transcriptome. The scheme represents the transcriptomic alterations derived from HSA21 trisomy. The additional dose of HSA21 triggers an overexpression of HSA21 genes and HSA21-encoded miRNAs. These products alter the expression of euploid and trisomic genes, which in turn can modify the transcriptome. Considering this complex network, as well as the developmental stages in which the alterations occur (lower box), it is crucial for gene therapy intervention.

Therapeutic targets to modulate DS phenotypes

HSA21 genes

The development of potential therapies for DS requires first the assignment of candidate genes to the different DS phenotypes. In this line, mouse models are extremely useful to associate abnormal phenotypes to specific genes. A direct approach to assess HSA21 gene function consisted to generate single-gene transgenic mice overexpressing an individual HSA21-orthologous gene in euploid mice. Several transgenic mice have been generated following this strategy, and the association of a particular gene to DS phenotypes has been assessed (Fillat et al., 2010; Seregaza et al., 2006). For example, evidence has been found of elevated oxidative stress in

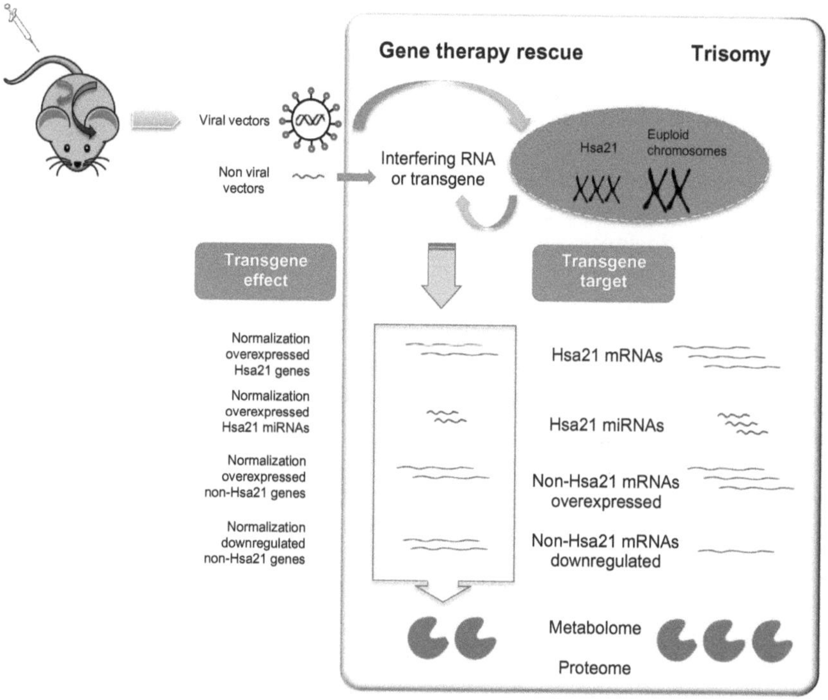

Fig. 2. Current and putative gene strategies toward the normalization of altered expression profiles in DS murine models. The cartoon depicts the administration of the transgene directly to the target organ or indirectly using a vector with a high tropism for the target tissue. Once the transgenes penetrate into the cell, their targets (central column) are putatively corrected, leading to a normalization of the proteome and the metabolome of the treated tissue, allowing a phenotypic rescue. (For color version of this figure, the reader is referred to the Web version of this chapter.)

individuals with DS and experiments with mouse models have identified *SOD1* and *RCAN1* HSA21 genes as being important in this process (Le Pecheur et al., 2005; Porta et al., 2007). Studies in animal models have also assigned a contributing role of *DYRK1A* in cognition (Altafaj et al., 2001; Arque et al., 2009). These studies strongly contributed to identify specific HSA21 dosage-sensitive genes as targets to develop therapies based on pharmacological or on gene therapy approaches.

An alternative strategy to identify the contribution of specific genes to a phenotype has been the generation of mouse models that are disomic for one gene in the context of a trisomy. These mice are obtained by breeding trisomic mice with HSA21 heterozygous mouse mutants. One example of this strategy has been reported to highlight the relevance of *the Synj1* HSA21gene in cognition. (Voronov et al., 2008). Restoring *Synj1* to disomy by breeding Ts65Dn trisomic mice and Synj1$^{+/-}$ mice rescued the metabolism of phosphatidylinositol-4,5-biphosphate (PtdIns(4,5)P_2) phospholipid that is altered in Ts65Dn mice. Normal PtdIns(4,5)P_2 balance is required at synapses for proper neurotransmission. Neurotransmitters release requires the balance between vesicle-mediated exocytosis of neurotransmitter and synaptic vesicle recycling. In this line, clathrin-dependent endocytosis of synaptic vesicles depends, among others, on the action of gene products of HSA21 genes *SYNJ1* (Synaptojanin 1) and *ITSN1* (Intersectin 1) (Voronov et al., 2008; Yu et al., 2008). The rescue of Inositol in these experiments strongly supports the role of Synj1 in neurotransmitter turnover and defines this gene as a putative target for further gene therapy studies. In view of these results *ITSN1* could also be postulated as a target in cognitive defects, although it remains to be demonstrated. *ITSN1*, has been shown that when overexpressed, inhibits clathrin-mediated endocytosis (Pucharcos et al., 2000). Thus, it can be hypothesized that *ITSN1* overexpression during DS fetal life (Pucharcos et al., 1999) might critically impair synaptic vesicles endocytosis affecting the neurodevelopment of DS individuals. Consequently, normalizing *ITSN1* during early embryonic stages might complement developmental and adult alterations.

An alternative approach to study the impact of an HSA21-overexpressed gene to DS-specific phenotypes consists of modulating gene overexpression with RNA interference (RNAi). RNAi is a nearly ubiquitous cellular pathway that modulates gene expression by posttranscriptional mechanisms. Basically, this mechanism consists on the ability of a subset of small RNA molecules to interfere with the translation of target mRNAs by an initial step of binding to the complementary sequences (in the 3′ untranslated region, 3′UTR), followed by either the cleavage of the target mRNA or its translational arrest (Dorsett and Tuschl, 2004). Due to their high specificity and easy handling, RNAi technologies are attractive therapeutic approaches to silence or downregulate the expression of candidate genes in a variety of pathologies. The potential interest of RNAi in DS is based on the fact that the overexpression of candidate genes contributes to specific phenotypes; thus, normalizing gene expression in particular genes might modify the course of the disease.

Importantly, many aspects must be considered in the design and the development of RNAi strategies for DS. Off-target activity of siRNAs can lead to unanticipated phenotypes and hurdle the interpretation of siRNA effects. A way to reduce off-target effects consists on pooling multiple siRNAs against the same target. As all members of the pool recognize the intended target but have unique off-target signatures, there is a selective reduction in the off-target effects (Kittler et al., 2007). Chemical modifications of single positions in the siRNA guide have also shown to reduce off-target silencing (Fedorov et al., 2006). These and other strategies toward the mitigation of the off-target effects of this technology are an active area of research (Jackson and Linsley, 2010).

Regarding shRNAs, their efficacy as well as off-target effects do not necessarily follow the same design rules than their synthetic siRNAs counterparts. This difference could be related to their specific cellular processing. Many studies in mouse models have demonstrated the efficacy of shRNA delivery for the correction of impaired functions in murine models. However, hints of potential toxicity have also been reported. Most of the toxicity has been related to highly expressed shRNAs or to the specific sequences. The use of promoters with lower potency or the embedding of DNA sequences in a miRNA backbone has been proposed as an additional molecular "brake" to mitigate off-target effects.

Normalizing HSA21-overexpressed genes

The application of RNAi approaches to ameliorate DS phenotypes is particularly tricky because the silencing rate must be very precise: the goal is to normalize gene dosage and not to fully silence gene expression. Thus, the reduction in target gene expression has to be very tightly controlled. Genetic engineering strategies to regulate gene expression of shRNA approaches as well as dose-delivery adjustment are key factors in defining such strategies. In the work of Ortiz-abalia et al., downregulation of the HSA21 gene *Dyrk1A* in the striatum of transgenic mice (TgDyrk1A mice) by the delivery of an shRNA led to a normalization of *Dyrk1A* gene expression with the rescue of the motor phenotype (Ortiz-Abalia et al., 2008). In this study, the use of a serotype 2 adeno-associated virus (AAV2) which is a mildly efficient delivery vector, in combination with the viral dose injected, was a key factor to enable the accuracy on *Dyrk1A* gene expression normalization. It will be interesting to investigate the impact of this normalization in a trisomic mouse model, in which many other genes are found to be overexpressed.

HSA21-encoded miRNAs

In addition to the extensively reported contribution of dosage-sensitive genes in DS phenotypic alterations, overrepresentation of functional nonprotein-coding elements might also be involved in some DS phenotypic abnormalities. Bioinformatic analysis using TargetScan software predicted up to 1695 genes with putative HSA21 miRNA target sites (Elton et al., 2010). HSA21 chromosome harbors five different miRNAs, named miR-99a, let-7c, miR-125b-2, miR-155, and miR-802. All these five miRNAs have been found to be overexpressed in human fetal hippocampus and heart samples from individuals with DS. This observation suggests that the overexpression of HSA21-derived miRNAs might result in the decreased expression of specific target proteins and contribute to DS phenotype (Kuhn et al., 2008). As mentioned above, miRNAs are small nonprotein-coding RNAs involved in posttranscriptional regulation. Their activity consists on base pairing in the 3′UTR of specific mRNAs, leading to translational repression or mRNA cleavage of target transcripts. It is estimated that 30–90% of human genes are regulated by miRNAs. Another estimation indicates that each miRNA can interact with 100 target mRNAs, which could potentially regulate a large number of protein-coding genes (Brennecke et al., 2005). Therefore, the overexpression of HSA21-encoded miRNAs might putatively exert a pleiotropic effect on the entire genome of DS individuals.

Up to now, there is very little knowledge on the functionality of HSA21-encoded miRNAs. miR-99a, let-7c, and miR-125b-2 map into the intronic region of *C21orf34*. This genomic region has been recently associated with a frequent loss of heterozygosity in lung cancer and is downregulated in human cancer cell lines as well as in lung cancer tissues, suggesting a potential tumor suppressor role (Yamada et al., 2008). Human miR-155 has been found to downregulate one allelic form of *AGTR1* associated with hypertension, but not a nonassociated allele. A single SNP in the 3′UTR sequence was sufficient to discriminate the activity of miR-155. The overexpression of miR-155 leading to allele specific underexpression of *AGTR1* was proposed to contribute to the reduction in blood pressure observed in trisomy 21 individuals (Sethupathy et al., 2007). The methyl-CpG-binding protein (MeCP2) has been validated as a target for miR-155 and miR-802. Interestingly, this protein is underexpressed in DS brain specimens (Kuhn et al., 2010). Since MeCP2 has been implicated in neurogenesis (Tsujimura et al., 2009), the proposed HSA21 miRNAs-driven downregulation of MeCP2 expression could result in the dysregulation of neuronal circuits contributing to cognitive defects observed in DS.

Normalizing HSA21-overexpressed miRNAs

Inhibition of miRNA function is envisioned as a strategy to rescue gene expression in target genes. One approach to inhibit miRNA function consists on the use of antisense oligonucelotides complementary to the miRNA guide strand. Oligonucelotide chemistry has been applied to develop optimal inhibitors. *In vivo*, miRNA inhibition has been successfully achieved using antagomirs. These molecules are single stranded 2′-*O*-methyl oligonucleotides with two and four phosphorotioate linkages at 5′- and 3′-ends, respectively, linked at the 3′-end with a cholesterol moiety to reduce renal clearance and enhance uptake. Intracerebroventricular (ICV) injections of antagomir-155 and antagomir-802 in Ts65Dn mice that overexpress such miRNAs resulted in the attenuation of 30–40% of endogenous miRNA-155 and 802 and increased MeCP2 target gene expression (65–170%) 7 days after administration (Kuhn et al., 2010). These data show the possibility to reduce miRNA using antagomirs, which could have a potential therapeutic value. However, to our knowledge, the impact of this approach on the phenotypic rescue of trisomic mice has not been already evaluated.

The inhibition of specific miRNAs in cells can also be achieved using the so-called "miRNA sponges" or "decoy miRNAs." This strategy is based on the use of transcripts containing miRNA target sites, that are designed to accumulate in the cell and act as a decoy or sponge for the cognate miRNA, interfering with its ability to regulate its natural targets. The group of Naldini has been the first to explore this strategy, using lentiviral vectors to regulate transgene expression according to tissue, lineage, and differentiation cell state (Brown and Naldini, 2009; Brown et al., 2007). A key point to achieve miRNA inhibition is to create suitable conditions, so the artificial decoy is a good competitor of the natural targets. Despite micromanaging miRNAs still requires optimization, it is envisioned as a very promising strategy to control the impact of HSA21 overdose miRNAs in DS phenotypes.

Non-HSA21 therapeutic targets in DS

An alternative approach to ameliorate DS alterations could rely on the identification of impaired process, by which the expression of candidate transgenes could complement the defective functions.

As described in the previous section, a large number of genes have target sites for HSA21 miRNAs. The validation of the targets is revealing that certain genes (such as *MeCP2*) are underexpressed in DS. Normalizing MeCP2 expression by gene therapy could potentially rescue associated phenotypes and ameliorate specific DS defects. According to bioinformatic predictions, many other genes may be interesting candidates.

In the past years, several studies supported the hypothesis that the alteration of the excitatory–inhibitory neurotransmitters balance in DS murine models might be involved in the cognitive deficits of trisomic models. This imbalance might be caused by an alteration of neurotransmitters release, the activation of their receptors, and/or downstream signaling pathways.

Deregulated genes outside the HSA21 chromosome might also be involved in the neurotransmitter release alterations in DS, and therefore, their correction could complement this neurotransmitter imbalance. In this line, pharmacological evidence-based treatments targeting different neurotransmitter systems have been published along the past decade. These studies reported partial corrections of specific defects observed in mouse models of DS and identified target genes as candidates for gene therapy approaches. In a recent work of Bianchi and collaborators, the authors reported that early administration of fluoxetine in Ts65Dn mice rescued the proliferation defects in different brain areas of Ts65Dn mice. The treatment restored the expression of serotonin 5-HT1A receptors and brain-derived neurotrophic factor (BDNF), together with the correction of behavioral impairment in specific tasks (Bianchi et al., 2010).

Another pharmacological strategy consisted in targeting the glutamatergic neurotransmission system. Different works showed that acute injections of memantine (NMDA receptor antagonist) into Ts65Dn mice promoted an improvement in the behavioral performance in cognitive tests and restored the levels of vesicular glutamate transporter-1 expression and BDNF, suggesting possible mechanisms of behavioral modification (Costa et al., 2008; Lockrow et al., 2011; Rueda et al., 2010). These data support the modulatory role of the glutamatergic system in DS murine models, which is in accordance with the proposed model of an imbalance between excitatory and inhibitory neurotransmitter systems in the brain of trisomic mice. This model is also supported by the work of Fernández and collaborators, which demonstrated that treating Ts65Dn mice with $GABA_A$ antagonists improved the cognition abilities of trisomic mice (Fernandez et al., 2007).

All these identified deregulated genes could potentially be expressed or silenced by gene therapy strategies, thus enlarging the therapeutic arena of molecular targets that could be tested in preclinical studies by gene therapy approaches.

Potentially, the application of gene therapy could provide with some advantages versus conventional pharmacology. Gene therapy could be selectively applied to a specific area, and to specific cell types what could minimize side effects. Additionally, the effect could be sustained for a long period of time after a single injection and would not require of repeated administrations. Proper control of the therapeutic transgene would allow its expression for a length of time and at a level as to have maximal therapeutic impact. Nevertheless, it would be necessary that the delivery vector does not introduce vector-related pathologies or strong host immune responses.

Delivery systems. Safety and toxicity issues

In addition to the target gene, the delivery system is also a key determinant on the efficacy of gene therapy. Currently, two major areas based on viral and nonviral approaches are under development. A broad number of gene therapy vectors to target defective functions in DS might be considered. In this chapter, we emphasize on those vectors more suitable for central nervous system (CNS) delivery, given that CNS defects are among the more disabling conditions associated with DS.

Viral systems

Viral vectors are the more effective means of gene transfer to modify specific cell types and can be manipulated to express therapeutic genes. The choice of a viral vector type depends on multiple factors, such as the target cell type(s), suited efficiency of transgene expression, toxicity, stability, safety, and easy production.

Adeno-associated virus

The AAV vector system has been a favored gene delivery tool, with demonstrated therapeutic effect in a variety of preclinical disease models of CNS dysfunction (McCown, 2011) and in patients with Parkinson's and Batten's diseases (Kaplitt et al., 2007; Worgall et al., 2008). The ability of AAV to transduce both neuronal and nonneuronal cells in brain, together with the associated long-term stable gene expression and very little toxicity, has made them a very attractive tool for neurobiological studies. Recombinant AAV vectors based on serotype 2 expressing an shRNA against Dyrk1A have been studied as a proof of principle to correct motor defects in DS. This gene therapy was found to be devoid of toxicity and succeeded in normalizing Dyrk1A protein levels in TgDyrk1A mice. Behavioral studies of adult TgDyrk1A mice treated showed a reversal of corticostriatal-dependent phenotypes, as revealed by the attenuation of their hyperactive behavior, the restoration of motor coordination defects, and an improvement in sensorimotor gating (Ortiz-Abalia et al., 2008). In this approach, as in the majority of AAV-CNS gene therapy, viral vectors were delivered by intracranial injections, which has limited potential for treating DS individuals. Intravenous AAV injections into neonatal disease mouse models had shown the potential to correct a genetic defect affecting CNS (Heldermon et al., 2010). This could be allowed by the biomechanical properties of blood–brain barrier (BBB) that may still be permeable to AAVs in neonatal mice. However, in humans, BBB is nonpermeable at birth. The development of specific AAV chimeras, based on capsid serotypes able to cross the BBB, offers the potential to further refine targeting strategies. AAV9 serotype has been shown to enable transgene expression upon intravascular delivery in neonatal and adult brain cells

(Duque et al., 2009; Foust et al., 2009). Importantly, efficient correction of lysosomal storage pathology upon systemic delivery of therapeutic AAV9 vectors in adult MPSIIIB mice has been reported (Fu et al., 2011). These data open the possibility of using AAV9 vectors, in a noninvasive manner, for many other diseases involving CNS; in particular, it would be interesting to study its potential therapeutic efficacy in DS mouse models.

Lentivirus

Lentiviral vectors have been extensively used and characterized for the delivery of transgenes into target tissues or organs. Along with the development of highly reliable biosafe vectors, lentivectors are now entering to clinical trials for the treatment of genetic conditions, such as X-linked adrenoleukodistrophy (Cartier et al., 2009) or Parkinson's disease (PD). Regarding the latter, a phase I/II clinical trial (reported at the "American Society for Gene and Cell Therapy, 2011 edition") indicated that PD patients having received a lentivector expressing genes involved in dopamine biosynthesis, exhibited an improvement of motor performance. Indeed, patients exhibited an increase of the motor performance score 6 months after lentiviral infusion, with an average of 41% (with a maximum of 61% in one patient), indicative of the efficacy of the therapy.

Although no data on DS models exist, the characteristics of the vector are attractive to achieve the fine control of gene expression required in DS. The expression of shRNAs with lentiviral vectors is useful to induce stable RNAi. Few copies of the vector are incorporated per cell, facilitating the control of gene expression by optimal cassette designs nearly to physiological conditions. In addition to their ability to transduce nondividing cells, lentivectors are very suitable tools to express transgenes in neuronal cells. In addition to the sustained expression of the transgene, a very low toxicity has been reported. Lentiviruses can also be used *ex vivo* by an initial modification step of the cells *in vitro* and a subsequent step of infusion of transduced cells. Regarding *in vivo* application, because intravenous deliver mostly targets to liver (Bielicki et al., 2010), the delivery requires the direct injection at the desired organ or tissue.

Nonviral approaches

Nonviral systems, based on cationic lipids, polymers, dendrimers, or different types of nanoparticles, are attractive gene delivery vectors because they are nonimmunogenic, easy to prepare, and nonexpensive. Despite active research is directed to develop more efficient systems to deliver transgenes or siRNA molecules, the main drawback of nonviral approaches remains their low transfection efficiency. Bearing in mind that some of the target phenotypes in DS are brain associated, a major impediment for the vectors delivery is the crossing of BBB (Posadas et al., 2010). Kumar and collaborators identified a short 29-amino-acids peptide derived from rabies virus glycoprotein that enabled transvascular delivery of siRNA into the brain. Moreover, repeated administration did not induce inflammatory response, providing a safe and noninvasive approach for the delivery of siRNAs (Kumar et al., 2007). The development of specific nanoparticles harboring the capacity to efficiently cross BBB will facilitate therapeutics of neurological dysfunctions (Perez-Martinez et al., 2011).

Perspectives

The advances in the characterization of the multiple pathways perturbed in DS contributed to improve our understanding of DS molecular etiology. Consequently, potential therapies targeting specifically these identified systems shed light on the possibility to treat alterations of DS murine models and identified candidate genes as modulators of the beneficial effects.

Gene therapy strategies in preclinical studies might be, at some point, translated into patients. Up to now, the majority of human clinical trials based on gene therapy (65%) have been performed in the context of cancer. Not surprisingly, the risk/benefit considerations in cancer make it the logical target for novel biological therapies. However, good and promising progresses in gene therapy clinical trials for specific forms of retinal and neuronal degeneration as well as for inherited immunodeficiencies have been recently reported (Bainbridge et al., 2008; Hacein-Bey-Abina et al., 2010). Although gene therapy for DS is not yet at the corner, DS-associated phenotypes (such as Alzheimer Disease in non-DS individuals) are entering clinical trials. NGF gene therapy with AAV is currently in phase II clinical trials, although it is too early to know its impact on the course of the disease (ClinicalTrials.gov Identifier: NCT00087789 and NCT00876863). It is then possible that, in the future, DS individuals will enter into gene therapy protocols. Meanwhile, we still need to improve our understanding of the pathophysiology of DS phenotypes and to evaluate the impact of potential gene therapies using preclinical mouse models.

Acknowledgments

Research in the authors' laboratories was supported by the Spanish Ministry of Science and Innovation, BIO2008-04,692-C03-02, Fondation Jérôme, Lejeune, Instituto de Salud Carlos III (PS09/00,733, MS10/00,548) and received partial support from the Generalitat de Catalunya SGR091527. C. F.'s group is also partially funded by CIBERER and by IIS10/00014 from the Instituto de Salud Carlos III and cofinanced by Fondo Europeo de Desarrollo Regional, FEDER.

References

Altafaj, X., Dierssen, M., Baamonde, C., Marti, E., Visa, J., Guimera, J., et al. (2001). Neurodevelopmental delay, motor abnormalities and cognitive deficits in transgenic mice overexpressing Dyrk1A (minibrain), a murine model of Down's syndrome. *Human Molecular Genetics*, *10*, 1915–1923.

Antonarakis, S. E., Lyle, R., Dermitzakis, E. T., Reymond, A., & Deutsch, S. (2004). Chromosome 21 and down syndrome: From genomics to pathophysiology. *Nature Reviews. Genetics*, *5*, 725–738.

Arque, G., de Lagran, M. M., Arbones, M. L., & Dierssen, M. (2009). Age-associated motor and visuo-spatial learning phenotype in Dyrk1A heterozygous mutant mice. *Neurobiology of Disease*, *36*, 312–319.

Bainbridge, J. W., Smith, A. J., Barker, S. S., Robbie, S., Henderson, R., Balaggan, K., et al. (2008). Effect of gene therapy on visual function in Leber's congenital amaurosis. *The New England Journal of Medicine*, *358*, 2231–2239.

Bianchi, P., Ciani, E., Guidi, S., Trazzi, S., Felice, D., Grossi, G., et al. (2010). Early pharmacotherapy restores neurogenesis and cognitive performance in the Ts65Dn mouse model for Down syndrome. *The Journal of Neuroscience*, *30*, 8769–8779.

Bielicki, J., McIntyre, C., & Anson, D. S. (2010). Comparison of ventricular and intravenous lentiviral-mediated gene therapy for murine MPS VII. *Molecular Genetics and Metabolism*, *101*, 370–382.

Brennecke, J., Stark, A., Russell, R. B., & Cohen, S. M. (2005). Principles of microRNA-target recognition. *PLoS Biology*, *3*, e85.

Brown, B. D., Gentner, B., Cantore, A., Colleoni, S., Amendola, M., Zingale, A., et al. (2007). Endogenous microRNA can be broadly exploited to regulate transgene expression according to tissue, lineage and differentiation state. *Nature Biotechnology*, *25*, 1457–1467.

Brown, B. D., & Naldini, L. (2009). Exploiting and antagonizing microRNA regulation for therapeutic and experimental applications. *Nature Reviews. Genetics*, *10*, 578–585.

Cartier, N., Hacein-Bey-Abina, S., Bartholomae, C. C., Veres, G., Schmidt, M., Kutschera, I., et al. (2009). Hematopoietic stem cell gene therapy with a lentiviral vector in X-linked adrenoleukodystrophy. *Science*, *326*, 818–823.

Costa, A. C., Scott-McKean, J. J., & Stasko, M. R. (2008). Acute injections of the NMDA receptor antagonist memantine rescue performance deficits of the Ts65Dn mouse model of Down syndrome on a fear conditioning test. *Neuropsychopharmacology*, *33*, 1624–1632.

Dorsett, Y., & Tuschl, T. (2004). siRNAs: Applications in functional genomics and potential as therapeutics. *Nature Reviews. Drug Discovery*, *3*, 318–329.

Duque, S., Joussemet, B., Riviere, C., Marais, T., Dubreil, L., Douar, A. M., et al. (2009). Intravenous administration of self-complementary AAV9 enables transgene delivery to adult motor neurons. *Molecular Therapy*, *17*, 1187–1196.

Elton, T. S., Sansom, S. E., & Martin, M. M. (2010). Trisomy-21 gene dosage over-expression of miRNAs results in the haploinsufficiency of specific target proteins. *RNA Biology*, *7*, 540–547.

Fedorov, Y., Anderson, E. M., Birmingham, A., Reynolds, A., Karpilow, J., Robinson, K., et al. (2006). Off-target effects by siRNA can induce toxic phenotype. *RNA*, *12*, 1188–1196.

Fernandez, F., Morishita, W., Zuniga, E., Nguyen, J., Blank, M., Malenka, R. C., et al. (2007). Pharmacotherapy for cognitive impairment in a mouse model of Down syndrome. *Nature Neuroscience*, *10*, 411–413.

Fillat, C., Dierssen, M., de Lagran, M. M., & Altafaj, X. (2010). Insights from mouse models to understand neurodegeneration in Down syndrome. *CNS & Neurological Disorders Drug Targets*, *9*, 429–438.

Foust, K. D., Nurre, E., Montgomery, C. L., Hernandez, A., Chan, C. M., & Kaspar, B. K. (2009). Intravascular AAV9 preferentially targets neonatal neurons and adult astrocytes. *Nature Biotechnology*, *27*, 59–65.

Fu, H., Dirosario, J., Killedar, S., Zaraspe, K., & McCarty, D. M. (2011). Correction of neurological disease of mucopolysaccharidosis IIIB in adult mice by rAAV9 trans-blood-brain barrier gene delivery. *Molecular Therapy*, *19*, 1025–1033.

Hacein-Bey-Abina, S., Hauer, J., Lim, A., Picard, C., Wang, G. P., Berry, C. C., et al. (2010). Efficacy of gene therapy for X-linked severe combined immunodeficiency. *The New England Journal of Medicine*, *363*, 355–364.

Heldermon, C. D., Ohlemiller, K. K., Herzog, E. D., Vogler, C., Qin, E., Wozniak, D. F., et al. (2010). Therapeutic efficacy of bone marrow transplant, intracranial AAV-mediated gene therapy, or both in the mouse model of MPS IIIB. *Molecular Therapy*, *18*, 873–880.

Jackson, A. L., & Linsley, P. S. (2010). Recognizing and avoiding siRNA off-target effects for target identification and therapeutic application. *Nature Reviews. Drug Discovery*, *9*, 57–67.

Kaplitt, M. G., Feigin, A., Tang, C., Fitzsimons, H. L., Mattis, P., Lawlor, P. A., et al. (2007). Safety and tolerability of gene therapy with an adeno-associated virus (AAV) borne GAD gene for Parkinson's disease: An open label, phase I trial. *Lancet*, *369*, 2097–2105.

Kittler, R., Surendranath, V., Heninger, A. K., Slabicki, M., Theis, M., Putz, G., et al. (2007). Genome-wide resources of endoribonuclease-prepared short interfering RNAs for specific loss-of-function studies. *Nature Methods*, *4*, 337–344.

Kuhn, D. E., Nuovo, G. J., Martin, M. M., Malana, G. E., Pleister, A. P., Jiang, J., et al. (2008). Human chromosome 21-derived miRNAs are overexpressed in down syndrome brains and hearts. *Biochemical and Biophysical Research Communications*, *370*, 473–477.

Kuhn, D. E., Nuovo, G. J., Terry, A. V., Jr., Martin, M. M., Malana, G. E., Sansom, S. E., et al. (2010). Chromosome 21-derived microRNAs provide an etiological basis for aberrant protein expression in human Down syndrome brains. *The Journal of Biological Chemistry*, *285*, 1529–1543.

Kumar, P., Wu, H., McBride, J. L., Jung, K. E., Kim, M. H., Davidson, B. L., et al. (2007). Transvascular delivery of small interfering RNA to the central nervous system. *Nature*, *448*, 39–43.

Le Pecheur, M., Bourdon, E., Paly, E., Farout, L., Friguet, B., & London, J. (2005). Oxidized SOD1 alters proteasome activities in vitro and in the cortex of SOD1 overexpressing mice. *FEBS Letters*, *579*, 3613–3618.

Lockrow, J., Boger, H., Bimonte-Nelson, H., & Granholm, A. C. (2011). Effects of long-term memantine on memory and neuropathology in Ts65Dn mice, a model for Down syndrome. *Behavioural Brain Research*, *221*, 610–622.

McCown, T. J. (2011). Adeno-Associated Virus (AAV) vectors in the CNS. *Current Gene Therapy*, *11*, 181–188.

Ortiz-Abalia, J., Sahun, I., Altafaj, X., Andreu, N., Estivill, X., Dierssen, M., et al. (2008). Targeting Dyrk1A with AAVshRNA attenuates motor alterations in TgDyrk1A, a mouse model of Down syndrome. *The American Journal of Human Genetics*, *83*, 479–488.

Perez-Martinez, F. C., Guerra, J., Posadas, I., & Cena, V. (2011). Barriers to non-viral vector-mediated gene delivery in the nervous system. *Pharmaceutical Research*, *28*, 1843–1858.

Porta, S., Serra, S. A., Huch, M., Valverde, M. A., Llorens, F., Estivill, X., et al. (2007). RCAN1 (DSCR1) increases neuronal susceptibility to oxidative stress: A potential pathogenic process in neurodegeneration. *Human Molecular Genetics*, *16*, 1039–1050.

Posadas, I., Guerra, F. J., & Cena, V. (2010). Nonviral vectors for the delivery of small interfering RNAs to the CNS. *Nanomedicine (London, England)*, *5*, 1219–1236.

Pucharcos, C., Estivill, X., & de la Luna, S. (2000). Intersectin 2, a new multimodular protein involved in clathrin-mediated endocytosis. *FEBS Letters*, *478*, 43–51.

Pucharcos, C., Fuentes, J. J., Casas, C., de la Luna, S., Alcantara, S., Arbones, M. L., et al. (1999). Alu-splice cloning of human Intersectin (ITSN), a putative multivalent binding protein expressed in proliferating and differentiating neurons and overexpressed in Down syndrome. *European Journal of Human Genetics*, *7*, 704–712.

Rueda, N., Llorens-Martin, M., Florez, J., Valdizan, E., Banerjee, P., Trejo, J. L., et al. (2010). Memantine normalizes several phenotypic features in the Ts65Dn mouse model of Down syndrome. *Journal of Alzheimer's Disease*, *21*, 277–290.

Seregaza, Z., Roubertoux, P. L., Jamon, M., & Soumireu-Mourat, B. (2006). Mouse models of cognitive disorders in trisomy 21: A review. *Behavior Genetics*, *36*, 387–404.

Sethupathy, P., Borel, C., Gagnebin, M., Grant, G. R., Deutsch, S., Elton, T. S., et al. (2007). Human microRNA-155 on chromosome 21 differentially interacts with its polymorphic target in the AGTR1 3′ untranslated region: A mechanism for functional single-nucleotide polymorphisms related to phenotypes. *American Journal of Human Genetics*, *81*, 405–413.

Tsujimura, K., Abematsu, M., Kohyama, J., Namihira, M., & Nakashima, K. (2009). Neuronal differentiation of neural precursor cells is promoted by the methyl-CpG-binding protein MeCP2. *Experimental Neurology*, *219*, 104–111.

Voronov, S. V., Frere, S. G., Giovedi, S., Pollina, E. A., Borel, C., Zhang, H., et al. (2008). Synaptojanin 1-linked phosphoinositide dyshomeostasis and cognitive deficits in mouse models of Down's syndrome. *Proceedings of the National Academy of Sciences of the United States of America*, *105*, 9415–9420.

Worgall, S., Sondhi, D., Hackett, N. R., Kosofsky, B., Kekatpure, M. V., Neyzi, N., et al. (2008). Treatment of late infantile neuronal ceroid lipofuscinosis by CNS administration of a serotype 2 adeno-associated virus expressing CLN2 cDNA. *Human Gene Therapy*, *19*, 463–474.

Yamada, H., Yanagisawa, K., Tokumaru, S., Taguchi, A., Nimura, Y., Osada, H., et al. (2008). Detailed characterization of a homozygously deleted region corresponding to a candidate tumor suppressor locus at 21q11-21 in human lung cancer. *Genes, Chromosomes & Cancer*, *47*, 810–818.

Yu, Y., Chu, P. Y., Bowser, D. N., Keating, D. J., Dubach, D., Harper, I., et al. (2008). Mice deficient for the chromosome 21 ortholog Itsn1 exhibit vesicle-trafficking abnormalities. *Human Molecular Genetics*, *17*, 3281–3290.

Subject Index

Note: Page numbers followed by "*f*" indicate figures, and "*t*" indicate tables.

Volume 149: Cortical Function: A View from the Thalamus, by V.A. Casagrande, R.W. Guillery and S.M. Sherman (Eds.) – 2005 ISBN 0-444-51679-4.
Volume 150: The Boundaries of Consciousness: Neurobiology and Neuropathology, by Steven Laureys (Ed.) – 2005, ISBN 0-444-51851-7.
Volume 151: Neuroanatomy of the Oculomotor System, by J.A. Büttner-Ennever (Ed.) – 2006, ISBN 0-444-51696-4.
Volume 152: Autonomic Dysfunction after Spinal Cord Injury, by L.C. Weaver and C. Polosa (Eds.) – 2006, ISBN 0-444-51925-4.
Volume 153: Hypothalamic Integration of Energy Metabolism, by A. Kalsbeek, E. Fliers, M.A. Hofman, D.F. Swaab, E.J.W. Van Someren and R.M. Buijs (Eds.) – 2006, ISBN 978-0-444-52261-0.
Volume 154: Visual Perception, Part 1, Fundamentals of Vision: Low and Mid-Level Processes in Perception, by S. Martinez-Conde, S.L. Macknik, L.M. Martinez, J.M. Alonso and P.U. Tse (Eds.) – 2006, ISBN 978-0-444-52966-4.
Volume 155: Visual Perception, Part 2, Fundamentals of Awareness, Multi-Sensory Integration and High-Order Perception, by S. Martinez-Conde, S.L. Macknik, L.M. Martinez, J.M. Alonso and P.U. Tse (Eds.) – 2006, ISBN 978-0-444-51927-6.
Volume 156: Understanding Emotions, by S. Anders, G. Ende, M. Junghofer, J. Kissler and D. Wildgruber (Eds.) – 2006, ISBN 978-0-444-52182-8.
Volume 157: Reprogramming of the Brain, by A.R. Møller (Ed.) – 2006, ISBN 978-0-444-51602-2.
Volume 158: Functional Genomics and Proteomics in the Clinical Neurosciences, by S.E. Hemby and S. Bahn (Eds.) – 2006, ISBN 978-0-444-51853-8.
Volume 159: Event-Related Dynamics of Brain Oscillations, by C. Neuper and W. Klimesch (Eds.) – 2006, ISBN 978-0-444-52183-5.
Volume 160: GABA and the Basal Ganglia: From Molecules to Systems, by J.M. Tepper, E.D. Abercrombie and J.P. Bolam (Eds.) – 2007, ISBN 978-0-444-52184-2.
Volume 161: Neurotrauma: New Insights into Pathology and Treatment, by J.T. Weber and A.I.R. Maas (Eds.) – 2007, ISBN 978-0-444-53017-2.
Volume 162: Neurobiology of Hyperthermia, by H.S. Sharma (Ed.) – 2007, ISBN 978-0-444-51926-9.
Volume 163: The Dentate Gyrus: A Comprehensive Guide to Structure, Function, and Clinical Implications, by H.E. Scharfman (Ed.) – 2007, ISBN 978-0-444-53015-8.
Volume 164: From Action to Cognition, by C. von Hofsten and K. Rosander (Eds.) – 2007, ISBN 978-0-444-53016-5.
Volume 165: Computational Neuroscience: Theoretical Insights into Brain Function, by P. Cisek, T. Drew and J.F. Kalaska (Eds.) – 2007, ISBN 978-0-444-52823-0.
Volume 166: Tinnitus: Pathophysiology and Treatment, by B. Langguth, G. Hajak, T. Kleinjung, A. Cacace and A.R. Møller (Eds.) – 2007, ISBN 978-0-444-53167-4.
Volume 167: Stress Hormones and Post Traumatic Stress Disorder: Basic Studies and Clinical Perspectives, by E.R. de Kloet, M.S. Oitzl and E. Vermetten (Eds.) – 2008, ISBN 978-0-444-53140-7.
Volume 168: Models of Brain and Mind: Physical, Computational and Psychological Approaches, by R. Banerjee and B.K. Chakrabarti (Eds.) – 2008, ISBN 978-0-444-53050-9.
Volume 169: Essence of Memory, by W.S. Sossin, J.-C. Lacaille, V.F. Castellucci and S. Belleville (Eds.) – 2008, ISBN 978-0-444-53164-3.
Volume 170: Advances in Vasopressin and Oxytocin – From Genes to Behaviour to Disease, by I.D. Neumann and R. Landgraf (Eds.) – 2008, ISBN 978-0-444-53201-5.
Volume 171: Using Eye Movements as an Experimental Probe of Brain Function—A Symposium in Honor of Jean Büttner-Ennever, by Christopher Kennard and R. John Leigh (Eds.) – 2008, ISBN 978-0-444-53163-6.
Volume 172: Serotonin–Dopamine Interaction: Experimental Evidence and Therapeutic Relevance, by Giuseppe Di Giovanni, Vincenzo Di Matteo and Ennio Esposito (Eds.) – 2008, ISBN 978-0-444-53235-0.
Volume 173: Glaucoma: An Open Window to Neurodegeneration and Neuroprotection, by Carlo Nucci, Neville N. Osborne, Giacinto Bagetta and Luciano Cerulli (Eds.) – 2008, ISBN 978-0-444-53256-5.
Volume 174: Mind and Motion: The Bidirectional Link Between Thought and Action, by Markus Raab, Joseph G. Johnson and Hauke R. Heekeren (Eds.) – 2009, 978-0-444-53356-2.
Volume 175: Neurotherapy: Progress in Restorative Neuroscience and Neurology — Proceedings of the 25th International Summer School of Brain Research, held at the Royal Netherlands Academy of Arts and Sciences, Amsterdam, The Netherlands, August 25–28, 2008, by J. Verhaagen, E.M. Hol, I. Huitinga, J. Wijnholds, A.A. Bergen, G.J. Boer and D.F. Swaab (Eds.) –2009, ISBN 978-0-12-374511-8.
Volume 176: Attention, by Narayanan Srinivasan (Ed.) – 2009, ISBN 978-0-444-53426-2.
Volume 177: Coma Science: Clinical and Ethical Implications, by Steven Laureys, Nicholas D. Schiff and Adrian M. Owen (Eds.) – 2009, 978-0-444-53432-3.
Volume 178: Cultural Neuroscience: Cultural Influences On Brain Function, by Joan Y. Chiao (Ed.) – 2009, 978-0-444-53361-6.
Volume 179: Genetic models of schizophrenia, by Akira Sawa (Ed.) – 2009, 978-0-444-53430-9.
Volume 180: Nanoneuroscience and Nanoneuropharmacology, by Hari Shanker Sharma (Ed.) – 2009, 978-0-444-53431-6.

Volume 181: Neuroendocrinology: The Normal Neuroendocrine System, by Luciano Martini, George P. Chrousos, Fernand Labrie, Karel Pacak and Donald W. Pfaff (Eds.) – 2010, 978-0-444-53617-4.
Volume 182: Neuroendocrinology: Pathological Situations and Diseases, by Luciano Martini, George P. Chrousos, Fernand Labrie, Karel Pacak and Donald W. Pfaff (Eds.) – 2010, 978-0-444-53616-7.
Volume 183: Recent Advances in Parkinson's Disease: Basic Research, by Anders Björklund and M. Angela Cenci (Eds.) – 2010, 978-0-444-53614-3.
Volume 184: Recent Advances in Parkinson's Disease: Translational and Clinical Research, by Anders Björklund and M. Angela Cenci (Eds.) – 2010, 978-0-444-53750-8.
Volume 185: Human Sleep and Cognition Part I: Basic Research, by Gerard A. Kerkhof and Hans P.A. Van Dongen (Eds.) – 2010, 978-0-444-53702-7.
Volume 186: Sex Differences in the Human Brain, their Underpinnings and Implications, by Ivanka Savic (Ed.) – 2010, 978-0-444-53630-3.
Volume 187: Breathe, Walk and Chew: The Neural Challenge: Part I, by Jean-Pierre Gossard, Réjean Dubuc and Arlette Kolta (Eds.) – 2010, 978-0-444-53613-6.
Volume 188: Breathe, Walk and Chew; The Neural Challenge: Part II, by Jean-Pierre Gossard, Réjean Dubuc and Arlette Kolta (Eds.) – 2011, 978-0-444-53825-3.
Volume 189: Gene Expression to Neurobiology and Behaviour: Human Brain Development and Developmental Disorders, by Oliver Braddick, Janette Atkinson and Giorgio M. Innocenti (Eds.) – 2011, 978-0-444-53884-0.
Volume 190: Human Sleep and Cognition Part II: Clinical and Applied Research, by Hans P.A. Van Dongen and Gerard A. Kerkhof (Eds.) – 2011, 978-0-444-53817-8.
Volume 191: Enhancing Performance for Action and perception: Multisensory Integration, Neuroplasticity and Neuroprosthetics: Part I, by Andrea M. Green, C. Elaine Chapman, John F. Kalaska and Franco Lepore (Eds.) – 2011, 978-0-444-53752-2.
Volume 192: Enhancing Performance for Action and Perception: Multisensory Integration, Neuroplasticity and Neuroprosthetics: Part II, by Andrea M. Green, C. Elaine Chapman, John F. Kalaska and Franco Lepore (Eds.) – 2011, 978-0-444-53355-5.
Volume 193: Slow Brain Oscillations of Sleep, Resting State and Vigilance, by Eus J.W. Van Someren, Ysbrand D. Van Der Werf, Pieter R. Roelfsema, Huibert D. Mansvelder and Fernando H. Lopes da Silva (Eds.) – 2011, 978-0-444-53839-0.
Volume 194: Brain Machine Interfaces: Implications For Science, Clinical Practice And Society, by Jens Schouenborg, Martin Garwicz and Nils Danielsen (Eds.) – 2011, 978-0-444-53815-4.
Volume 195: Evolution of the Primate Brain: From Neuron to Behavior, by Michel A. Hofman and Dean Falk (Eds.) – 2012, 978-0-444-53860-4.
Volume 196: Optogenetics: Tools for Controlling and Monitoring Neuronal Activity, by Thomas Knöpfel and Edward S. Boyden (Eds.) – 2012, 978-0-444-59426-6.

Kolibri 360 GmbH
Ferdinand-Jühlke-Straße 7
99095 Erfurt, Deutschland
produktsicherheit@kolibri360.de